Reading Sounds

Reading Sounds

*Closed-Captioned Media
and Popular Culture*

SEAN ZDENEK

The University of Chicago Press   Chicago and London

SEAN ZDENEK is associate professor of technical communication
and rhetoric at Texas Tech University.

The University of Chicago Press, Chicago 60637
The University of Chicago Press, Ltd., London
© 2015 by The University of Chicago
All rights reserved. Published 2015.
Printed in the United States of America

24 23 22 21 20 19 18 17 16 15   1 2 3 4 5

ISBN-13: 978-0-226-31264-4 (cloth)
ISBN-13: 978-0-226-31278-1 (paper)
ISBN-13: 978-0-226-31281-1 (e-book)
DOI: 10.7208/chicago/9780226312811.001.0001

Library of Congress Cataloging-in-Publication Data

Zdenek, Sean, author.

   Reading sounds : closed-captioned media and popular culture /
Sean Zdenek.

       pages ; cm

   Includes bibliographical references and index.

   ISBN 978-0-226-31264-4 (cloth : alk. paper)—ISBN 978-0-226-31278-1
(pbk. : alk. paper)—ISBN 978-0-226-31281-1 (ebook)   1. Closed captioning.
2. Visual communication.   I. Title.

   P93.5.Z37 2015

   302.23—dc23

                                        2015014458

♾ This paper meets the requirements of ANSI/NISO Z39.48-1992
(Permanence of Paper).

FOR PIERCE

Contents

Preface

Growing up in Southern California in the 1970s and '80s, I watched a lot of television but never with closed captioning. (I was eleven years old when closed-captioned TV was introduced in March 1980.) I didn't have any deaf or hard-of-hearing friends. No deaf neighbors or relatives that I knew of. I grasped what closed captioning was at a rudimentary level but didn't have any real experiences with it. I grew up a hearing kid in a hearing family.

Everything changed in 1997, with the birth of our second son. When he was about eight months old, audiological tests confirmed what we had suspected: Pierce was born with profound hearing loss in both ears. Hearing aids and other accommodations, including closed captioning, quickly followed. We started watching everything with closed captions, even though he was, at first, too young to read them. Because it's a hassle to toggle TV captions on and off without a handy "cc" button on the remote, we left the captions on all the time. We watched DVDs with closed captions too, and at some point—I can't recall the exact moment when this first happened, but it feels like forever ago—his mom and I began watching DVDs with captioning even when the kids were out of the room. While these days I can still suffer through an uncaptioned movie at the theater if I have to, I much prefer to watch everything with closed captioning. I have come to rely on captions not only to catch what I've missed but also to make sense of what I'm hearing. There's something about *reading a movie*—i.e., experiencing it through the rhetorical transcription of its soundtrack—that provides a level of access and satisfaction

0.1   Reading words and reading sounds.

In this frame from *The Grand Budapest Hotel*, Zero (Tony Revolori) and his fiancée Agatha (Saoirse Ronan) stand in a two-shot on a lighted carousel during a winter's night. The young woman sits on a carousel horse while the young man faces her with his hands resting on the horse's head. She reads the inscription he wrote for her in a book of romantic poetry he just gave her. A closed caption, (READING), is stamped inelegantly on her forehead. Printed in open subtitles with a script typeface at the bottom of the screen are the words she is reading: "For my dearest, darling." Fox Searchlight Pictures, 2014. Blu-Ray. http://ReadingSounds.net/preface/#figure1.

I can't quite seem to reach without captions. I don't have to work as hard to follow what's going on. I don't have to worry so much about the "vocal epidemic" of actors who are increasingly prone to "mumble" their way through their lines (Simkins 2013). With captions I don't miss characters' names. Or, put another way, captions tell me which sounds are proper nouns. Captions can be a helpful lifeline in movies with strange or unusual names for characters, places, and things. In the *Harry Potter* movies, for example, viewers who haven't read the books (ahem) are plunged into a strange world of new nouns (see figure 0.2).

Captions also allow me to focus on important nonspeech sounds because the captioner has rescued them from the teeming soundscape and made them visible in writing, which, as we will see in chapter 5, can be a mixed blessing. I'm not alone in this feeling. People who have difficulty processing sensory or speech information, or who process it differently, have reportedly found similar benefits from closed captioning, because it allows them to bypass the cognitive interference and tap into the meaning of the sound through a different channel. For example, Judith Garman (2011) has argued that closed captioning can help people on the autistic spectrum who have difficulty discerning significant sounds, because captioning "gives a greater depth of understanding and context by providing a second input stream."

This example and others are intended to remind us that hearing viewers can benefit from closed captioning too. Popular examples include the child learning to read, the adult or child learning a second language, the individual with a cognitive disability who may benefit from having

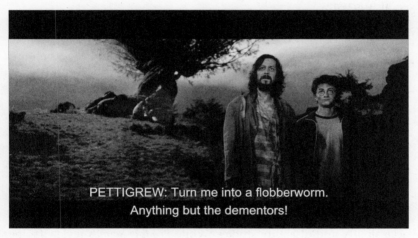

PETTIGREW: Turn me into a flobberworm.
Anything but the dementors!

0.2   Captions clarify.
Unusual words and neologisms are made more accessible to caption viewers in this frame from *Harry Potter and the Prisoner of Azkaban*. Harry Potter (Daniel Radcliffe) stands next to Sirius Black (Gary Oldman) as they both face the camera. The scene is a dark countryside, with a large tree and foreboding sky in the background. The tree is the Whomping Willow, and four people kneel at the base of it, though it is hard to make them out in the darkness of the scene and at the distance from which Harry and Sirius are standing. The frame's caption is: "PETTIGREW: Turn me into a flobberworm. Anything but the dementors!" This faint line of dialogue is uttered by Peter Pettigrew (Timothy Spall), one of the characters kneeling at the tree. (In the DVD version, there's no speaker identifier attached to this line.) When the focus shifts from Harry and Sirius to the characters at the tree, we begin to understand. This low line of dialogue in the distance, with its unusual neologisms, is made loud and clear when captioned. Warner Bros, 2004. Blu-Ray. http://ReadingSounds.net /preface/#figure2.

access to multiple streams (audio and text), the college student review-
ing and searching recorded lectures prior to an exam, the night owl who
doesn't want to wake a sleeping partner or child, the treadmill jogger
at the neighborhood gym facing a bank of muted TVs, and anyone trying to watch TV in a noisy environment (airport, restaurant, nightclub,
etc.). When "changes in one's abilities based on environment, device, or
other temporary conditions" create *situational disabilities* for able-bodied
people (Chisholm and May 2009, 12), closed captioning can step in to
provide access for a wide range of viewers, regardless of hearing ability. Literacy studies of hearing children have suggested a correlation between reading captions and more effective foreign language learning
(Winke, Gass, and Sydorenko 2010), increased word recognition, vocabulary learning, inference generation (Linebarger, Piotrowski, and Greenwood 2010), and even gains in motivation with "students who have been
difficult to reach with traditional methods and materials" (Koskinen
et. al. 1993, 42).

But I keep coming back to my son. I've watched intently over the years
as he has made daily use of captioned media. Captions provide essential
access for him. Through him, I am reminded constantly of the millions
of people in the United States and around the world who require quality
captioning. My research is motivated by a desire to advocate on behalf
of people like my son. I don't claim to know what it's like to be deaf
or hard of hearing, and I have no experience working as a professional
captioner. I've simply paid attention over the years to the closed captions themselves, to what other advocates and experts have said about
them, and to my own experiences as a hearing parent of a deaf child.
I've also interviewed closed captioners, surveyed regular viewers of captioning, kept up with the scholarly discussions about captioning in disability and deaf studies, and taught a yearly graduate seminar on web
accessibility and disability studies. While *Reading Sounds* reviews some
of the research on how deaf and hard-of-hearing viewers experience captions, and while it reflects on the relationships between accessibility and
literacy, it is deeply rooted in my own daily experiences listening to
movies and reading captions simultaneously. Out of these experiences,
I return again and again to the productive tensions between sound and
writing, tensions that are not yet well understood but become palpable
in the closed-captioned text. This book is concerned with the technical
specifications of analog and digital captioning (see Robson 2004) only
insofar as they impose space and time constraints on the captioner and
the reader. This book is neither a technical manual nor a how-to guide
on using captioning software. Rather, *Reading Sounds* is a meditation on

the possibilities and challenges of transforming sound into accessible writing for time-based reading.

It took me about a decade of watching closed-captioned programming every day to begin to realize that something pretty interesting was going on, that captioning was potentially much more complex than we've ever considered. Definitions of closed captioning too often stress the technology of "displaying" text on the screen over the complex practice of selecting sounds and rhetorically inventing words for them. In most definitions, the practice itself is simplified, reduced to a mechanical process of unreflective transcription. No one has really treated captioning as a significant variable in multimodal analysis, on par with image, sound, and video. No one has considered the possibility that captions might be as potent and meaningful as other kinds of texts we study in the humanities. In short, we don't yet have a good understanding of the rhetorical work captions do to construct meaning and negotiate the constraints of space and time.

Through captions, the names of characters, locations, and actions appear before our eyes. Unusual or foreign-sounding names, which may be difficult for hearing and hard-of-hearing viewers to make out through listening alone, are clarified in writing. The same goes for thick accents, which are converted (or reduced) to standard English (chapter 8). In the case of song titles and lyrics, caption viewers may have access to information that noncaption (hearing) viewers do not have. Just because a hearing viewer can hear a sound doesn't mean she knows what it is, even when presented with that sound in context. In the case of "mondegreens" (misheard lyrics), she may even think she knows and yet still be wrong. The captioner knows, but only because the captioner has consulted published lyrics (which may also be incorrect). What the singer is saying may be inaccessible to the captioner in the absence of written lyrics because the sung words may sound nothing like the published lyrics. As one captioner explained to me, this situation becomes even more complex when competing versions of the same lyrics vie for attention (e.g., multiple versions of the same lyrics posted online). Consider the competing interpretations of Stewie's line in the opening theme of *Family Guy*: Is it "laugh and cry" or "effin' cry"? As figure 0.3 shows, there's evidence in the official closed captions for both interpretations, despite creator Seth MacFarlane's protestations that the former is correct (Aberdeen Captioning 2011).

In various ways, then, captions have the potential to convey new knowledge to viewers by imbuing sounds with new or revised meanings, countering the popular misconception that captions simply repeat

0.3 **"Laugh and cry" or "effin' cry"?**
"Effin' cry" is admittedly rare but can be found in the closed captions for some early epi-
sodes of Fox's *Family Guy*. In these two frames from two different episodes of the opening
theme song, identical except for the captions, Lois (voiced by Alex Borstein) holds baby
Stewie (voiced by Seth MacFarlane). Both are wearing identical yellow tuxedos with yellow
top hats. A white stairway with blue risers fills the background. Left frame's caption: "♪
Laugh and cry." Right frame's caption: "♪ EFFIN' CRY ♪" Source (left): season 7, episode 10,
"Fox-y Lady," 2009, DVD. Source (right): season 1, episode 7, "Brian: Portrait of a Dog,"
1999, cable TV. (To create a high quality image suitable for print publication, the left DVD
frame was duplicated and substituted for the low quality image in the TV original.) http://
ReadingSounds.net/preface/#figure3.

information that's already present on the sound layer. By inverting the
usual relationship between primary text (movie) and secondary accommodation (captions), scholars of caption studies—and sound and disability scholars more generally—can generate insights about the role
of captions in meaning making. Rather than leaving out accessibility
(and people with disabilities) from our discussions of multimodality, or
starting from the assumption that captions only offer pale or mirrored
reflections of the soundscape, *Reading Sounds* inserts closed captioning
and its affordances into the heart of the multimodal landscape for the
first time.

For other kinds of nonspeech sounds such as sound effects, knowing
what a sound *is*—who or what produces the sound—will not necessarily
provide enough information for it to be captioned effectively (chapter 3).
For example, consider a sound produced by a certain kind of turbine
engine. Even knowing what specific engine produces the sound is not
enough; we need to know how that sound is situated in a specific context, because the same sound could conceivably support a number of divergent visual contexts. I like to joke that *captioners don't caption sounds*.
But behind this seemingly nonsensical claim is a truth that reminds us
that meaning develops out of the interplay of sounds, moving images,

and evolving contexts and narratives. The sound alone may not provide enough information for it to be captioned effectively.

Captioning is a subjective and interpretative practice, one that, at least ideally, strives to bring the producer's vision before our eyes. But captioners work under a different set of constraints than the producers—spatial, temporal, economic, rhetorical, technological, and institutional. Captioners are typically independent contractors who, not unlike some technical communicators, are hired or brought on after the main work has already been completed. They often work under extremely tight deadlines and manage slim profit margins. Their contact with the content producers is usually limited or nonexistent. Captioners ostensibly serve deaf and hard-of-hearing viewers, and yet prerecorded captioning is typically done without any input or feedback from these users. For this reason, J. P. Udo and D. I. Fels (2010, 211) suggest that "the primary user is actually much more covert: the broadcaster and media producer, who use these services as a means of placating governmental requirements." Yet satisfying the producers doesn't usually require negotiating with them over questions of caption quality. To address the disconnect between those who create the content and those who caption it, Udo and Fels (2010) recommend making captioners integral members of the creative team and giving producers more creative control over the design of captions. Though idealistic, these recommendations are reminiscent of similar proposals to integrate technical communicators into decision making teams rather than bringing them on board at the end of the project to write up user guides.

Captioners produce interpretations, drawing on a range of materials, including scripts and new media detective work (i.e., Google searches). Stylistic differences between one captioner and another, or one captioning company and another, are often subtle, revealing themselves in small changes over time in how a recurring sound on a television series is captioned (chapter 3), or in large differences across media formats for the same movie (DVD, Netflix, and broadcast TV versions). For example, I was initially drawn to *BloodRayne 2: Deliverance* (2007) when I first saw it on cable TV for no other reason than its abundant and often creative nonspeech descriptions: [children's screams continue grating], [solemn whistling], [sensuously panting]. But when I ordered it on DVD from Netflix, it was clear that another captioner, and possibly another company, had been responsible for the DVD captions. Far fewer nonspeech captions were included on the DVD version of the same movie. This situation is actually fairly common, as I soon discovered, and points to the deeply subjective nature of the captioning process: multiple, official caption files will be in circulation for any TV show or movie that has been subject to redistribution (see figure 0.4 for another example). Official caption files for the same content will vary noticeably, even sometimes radically, from each other. We have never attended to these differences before, but it is through them that we can begin to understand the influences of the captioner's agency on the practices of captioning.

There is no perfect or objective reading of a film or TV show, just as there is no objective meaning for any sound (Schafer 1977, 137). Context matters. At their best, captioners are ideal readers—rhetorical proxy agents, I like to say—who listen closely, size up the situation, and determine the best way to convey its meaning. A rhetorical view of captioning applies to speech sounds as well as nonspeech sounds. Whether speech sounds are captioned verbatim or edited for speed and content (Szarkowska et al. 2011; Ward et al. 2007) matters little, rhetorically speaking. Verbatim captioning is still interpretative, because it involves making decisions about how to represent standard and nonstandard speech (i.e., regional dialects, manners of speaking). Even in those moments of seeming objectivity and simplicity, captioning is rhetorical through and through.

What's so hard about writing down what people are saying? The sounds are right there, dripping with meaning, right? Not exactly. *Reading Sounds* aims to deepen and complicate a process that has too often been dismissed as straightforward, simple, and objective. The same interpretative flexibility and multiplicity that inform the act of making sense of any text (novels, plays, images, TV shows, speeches, music, etc.) also inform the act of converting sound into writing. Captioning is a subjective and highly contextual act. This book is motivated by a central tension between, on the one hand, theoretical approaches to sound and aurality that stress the very limits (and even the impossibility) of representing sound in the face of its transcendent, intangible, heterogeneous, uncanny, and immersive qualities (Dyer 2012; Dyson 2009) and, on the other hand, approaches to closed captioning that present the process of rhetorical invention as simple and straightforward. *Reading Sounds* explores the complexities of making sound accessible, using these complexities to forge a new, deeper understanding of quality captioning and the relationships between sound and word.

Every example in this book is accompanied by a media clip or image on the book's website. To begin exploring the examples, including the examples described in the figures above, go to http://ReadingSounds.net. Direct links are included throughout the book. Because I am continually

Leeloo Minai Lekarariba-Laminai-Tchai
Ekbat De Sebat.

[Speaking
Unknown Language]

0.4 **Is there more than one way to caption a character's name?**

You don't have to search far to find multiple official caption files for the same movie. The DVD for Gaumont's *The Fifth Element* (1997) contains two caption tracks: a bitmap track of speech-only subtitles (top frame) and a text track of closed captions (bottom frame). In both frames from the movie, which are identical except for the captions, Leeloo (Milla Jovovich) aims a pistol-like weapon at Korben Dallas (Bruce Willis). The camera is positioned over Korben's right shoulder. These two tracks were most likely created by different captioning companies at different times for different formats (DVD, VCR). That one track contains speech only and the other contains full closed captions (speech and nonspeech) doesn't explain why one track fails to caption a main character's full name. Names of main characters always need to be captioned verbatim. That's not an unknown language but her name. Top frame's caption: "Leeloo Minai Lekarariba-Laminai-Tchai Ekbat De Sebat." Bottom frame's caption: [Speaking Unknown Language]. (To create a high quality image suitable for print publication, the Blu-Ray version was used to create the figure, even though the Blu-Ray captions are not the same as the DVD captions.) http://Reading Sounds.net/preface/#figure4.

finding examples from movies and TV shows that support, deepen, and/ or challenge my ideas, the website also includes additional examples not discussed in the book. The book and website contain hundreds of captioned examples from popular TV shows and movies, but a word of caution: A few of the examples contain spoilers, while others contain potentially offensive language and adult themes, such as curse words and graphic violence (but no nudity). All of the examples come from shows intended for a mature audience (no children's shows are analyzed, although cartoons such as *Family Guy* and *South Park* make more than one appearance). Every example serves a purpose and, more importantly, reflects the world of pop culture, which, for better or worse, is sometimes crude, discriminatory, violent, and highly sexualized. To understand how captions make meaning in pop culture, we need to explore the full range of themes and genres in programming for adults, from *The Artist* to *Zombie Apocalypse*.

ONE

A Rhetorical View of Captioning

Four New Principles of Closed Captioning

Closed captioning has been around since 1980—it's not "*new* media" by any means—but you wouldn't know it from the passionate captioning advocacy campaigns, new web accessibility laws, revised international standards, ongoing lawsuits, new and imperfect web-based captioning solutions, corporate feet dragging, and millions of uncaptioned web videos. Situated at the intersection of a number of competing discourses and perspectives, closed captioning offers a key location for exploring the rhetoric of disability in the age of digital media. *Reading Sounds* offers the first extended study of closed captioning from a humanistic perspective. Instead of treating closed captioning as a legal requirement, a technical problem, or a matter of simple transcription, this book considers how captioning can be a potent source of meaning in rhetorical analysis.

Reading Sounds positions closed captioning as a significant variable in multimodal analysis, questions narrow definitions that reduce captioning to the mere "display" of text on the screen, broadens current treatments of *quality* captioning, and explores captioning as a complex rhetorical and interpretative practice. This book argues that captioners not only select which sounds are significant, and hence which sounds are worthy of being captioned, but also rhetorically invent words for sounds. Drawing on a number of examples from a range of popular movies and television shows,

1.1   **Captioners offer interpretations within the constraints of time and space.**
A frame from *21 Jump Street* showing police officers Schmidt (Jonah Hill) and Jenko
(Channing Tatum), dressed in black uniforms with matching black shorts and helmets,
riding their police bicycles side by side on the park grass. The bike cops are heading
straight for the viewer. A small red light is visible on each bicycle's handlebars. The cops
are peddling to confront a small biker gang smoking pot on the other side of the park.
Pounding rock music (uncaptioned) accompanies the pursuit but cuts out momentarily
to call attention to the faint bicycle sirens, which sound like children's toys. The sirens are
captioned as (SIRENS WHOOPING SOFTLY), which is supposed to capture the ridiculousness of
the scene. Packed into this single caption, then, is the reminder that these are not real
cops because real cops would be burning rubber in a patrol car and blaring their sirens.
Columbia Pictures, 2012. Blu-Ray. http://ReadingSounds.net/chapter1/#figure1.

Reading Sounds develops a rhetorical sensitivity to the interactions among
sounds, captions, contexts, constraints, writers, and readers.

This view is founded on a number of key but rarely acknowledged
and little-understood principles of closed captioning. Taken together,
these principles set us on a path towards a new, more complex theory
of captioning for deaf and hard-of-hearing viewers. These principles also
offer an implicit rationale for the development of theoretically informed
caption studies, a research program that is deeply invested in questions of
meaning at the interface of sound, writing, and accessibility.

1. Every sound cannot be closed captioned.

Captioning is not mere transcription or the dutiful recording of every
sound. There's not enough space or reading time to try to provide captions for every sound, particularly when sounds are layered on top of
each other in the typical big-budget flick. Multiple soundtracks create

a wall of sound: foreground speech, background speech, sound effects, music with lyrics, and other ambient sounds overlap and in some cases compete with each other. Sound is simultaneous; print is linear. It's not possible to convert the entire soundscape of a major film or TV production into a highly condensed print form. It can also be distracting and confusing to readers when the caption track is filled with references to sounds that are incidental to the main narrative. Caption readers may mistake an ambient, stock, or "keynote" sound (Schafer 1977, 9) for a significant plot sound when that sound is repeatedly captioned. A professional captioner shared the following example with me: Consider a dog barking in an establishing shot of a suburban home. When the dog's bark is repeatedly captioned, one may begin to wonder if there's something wrong with that dog. Is that sound relevant to this scene? (See figure 1.2.) Very few discussions of captioning acknowledge or even seem to recognize that captioning, done well, must be a selective inscription of the soundscape, even when the goal is so-called "verbatim captioning."

2. Captioners must decide which sounds are significant.

If every sound cannot be captioned, then someone has to figure out which sounds should be. Speech sounds usually take precedence over nonspeech sounds, but it's not that simple. What about speech sounds in the background that border on indistinct but are discernable through careful and repeated listening by a well-trained captioner? Should these sounds be captioned (1) verbatim, (2) with a short description such as (indistinct chatter), or (3) not at all? Answering this question by appealing to volume levels (under the assumption that louder sounds are more important) may downplay the important role that quieter sounds sometimes play in a narrative (see figure 1.3). What is needed is an awareness of how sounds are situated in specific contexts. Context trumps volume level. Only through a complete understanding of the entire program can the captioner effectively interpret and reconstruct it. Just as earlier scenes in a movie anticipate later ones, so too should earlier captions anticipate later ones. In the case of a television series, the captioner may need to be familiar with previous episodes (including, when applicable, the work of other captioners on those episodes) in order to identify which sounds have historical significance. The concept of significance (or "relevant" sounds [see Sydik 2007, 181]) shifts our attention away from captioning as copying and toward captioning as the creative selection and interpretation of sounds.

1.2 All dog sounds are not created equal.
The top row contains two frames from an episode of *Grimm* (2011, season 1, episode 1, NBC). In the top left frame, Nick (David Giuntoli) is shown in profile walking at night on a suburban street. A home in the background is lit by porch light. Large trees provide an ominous backdrop. The caption, [dog barking], is more than a stock sound to provide suburban ambience. A few seconds later in this scene, the same dog seems to be suffering, drawing the attention of Nick, who turns to face the camera in the top right frame. The accompanying caption is [dog yelps, whines, goes silent]. The bottom row contains two frames from *Extract* (2009, Ternion Pictures), both of which are taken during a dinner table scene at night. In the bottom left frame, Joel (Jason Bateman) and Suzie (Kristen Wiig) are eating at their dining table with the [DOG BARKING IN DISTANCE]. In the bottom right frame, Suzie stares blankly after Joel walks away from the table upset. The accompanying caption: [CRICKETS CHIRPING]. The dog barking in *Extract* is part of a stock soundscape that includes crickets chirping, whereas the dog sounds are an integral element of the horror storyline in the *Grimm* episode. The animal and insect captions in *Extract* end up intruding into the serious dinner discussion. TV source: *Extract* rebroadcast on *Comedy Central* and *Grimm* rebroadcast on the *Syfy* channel. http://ReadingSounds.net/chapter1/#figure2.

3. Captioners must rhetorically invent and negotiate the meaning of the text.

The caption track isn't a simple reflection of the production script. The script is not poured wholesale into the caption file. Rather, the movie is transformed into a new text through the process of captioning it. In fact, as we will see in chapter 4, when the captioner relies too heavily on the script (for example, mistaking ambient sounds for distinct speech sounds), the results can be disastrous. In other cases, words must be rhetorically invented, which is typical for nonspeech sounds. I don't mean that the captioner must invent neologisms—I issue a warning about neologistic onomatopoeia in chapter 8. Rather, the captioner must choose the best word(s) to convey the meaning of a sound in the context of a scene and under the constraints of space and time. The best way to understand this process, as this book argues throughout, is in terms of a rhetorical negotiation of meaning that is dependent on context, purpose, genre, and audience.

4. Captions are interpretations.

Captioning is not an objective science. The meaning is not waiting there to be written down. While the practice of captioning will present a number of simple scenarios for the captioner, the subjectivity of

(GRACE CONTINUES CHATTERING)

1.3   **Where does distinct speech shade off into indistinct chatter?**
In this frame from *Avatar*, Jake Sully (Sam Worthington), inhabiting his Na'vi avatar body, is shown in a mid-shot looking slightly off camera to the viewer's right. Jake has just wandered away from the scientists Grace (Sigourney Weaver) and Norm (Joel Moore), who are busy taking plant root samples. Captions create a clear line between distinct speech and indistinct background chatter, even though, sonically speaking, the dividing line is not always quite so obvious. Chattering is also a popular option for describing indistinct crowd noise, but captioners need to be mindful of the term's gendered implications. The conversations of women have at times been described dismissively as chattering. Caption: (GRACE CONTINUES CHATTERING). Twentieth Century Fox, 2009. Blu-Ray. http://ReadingSounds .net/chapter1/#figure3.

the captioner and the ideological pressures that shape the production of closed captions will always be close to the surface of the captioned text. The practice of captioning movies and TV shows is typically performed independently, as contract work by captioning companies for major production studios, with little oversight, interest, or input from the content producers beyond the need to ensure legal compliance (Udo and Fels 2010, 209). In the case of nonspeech sounds, these independent contractors possess near-total control over the selection of significant sounds and the creation of captions for them. The resulting caption track is not an objective reflection of the text but what Abé Mark Nornes (2007, 15) calls, in the context of foreign language subtitling, a "new text." This view of captioning as rhetorical invention or textual performance, with the captioner serving as a rhetorical proxy agent, is likely to seem at odds with the goal of "equal access for all." But access to captioned content will never, strictly speaking, be the same as access to the sonic landscape. Rather, the captioned text will always be inflected by the captioners' interpretative powers and the different affordances of sound and writing.

These four principles will need a book to explain and defend. They are new and challenge the conventional wisdom about closed captioning. They have the potential to transform how we think about captioning, accessibility for deaf and hard-of-hearing viewers, and the relationships between sound and writing in the digital age. Researchers in rhetorical studies and disability studies have yet to provide a sustained analysis of closed captioning. (For exceptions, see Lueck 2011, Lueck 2013, and my own previous research: Zdenek 2011a, Zdenek 2011b, and Zdenek 2014.) We haven't paused to pay attention to captioning as rhetoric, even as we've held up captioning as one of the centerpieces of an accessible web. By rhetoric, I don't simply mean language pressed into the service of persuasion but, more broadly, signs and symbols that construct worlds of meaning for us to inhabit. Closed captions are not windowpanes on a sonic reality but mediate that reality in the course of providing access to it (cf. Miller 1979, 611). The conventional view of closed captioning tends to simplify questions of *quality* and focus on questions of *quantity*. For example, the Twenty-First Century Communications and Video Accessibility Act of 2010 (CVAA) requires that only certain types of TV-like content on the Internet be closed captioned, leading advocates to ask: How do we compel producers of independent web series, which aren't covered under the new law, to caption their programs? Quality tends to be defined narrowly in terms of *completeness* (Is the entire show captioned?) and *accuracy* (Is every speech sound captioned correctly? Are any captions garbled as a result of poor autotranscription?). Just as quality in foreign language subtitling too often gets reduced to mistranslation or "misprision"—what Nornes (2007, 16) calls "red meat" for critics of subtitling—quality in closed captioning too often gets reduced to questions of accuracy (e.g., "caption fails"). This book offers a new approach to quality in captioning by considering how captions create new meanings, manipulate space and time, call attention to productive tensions between sound and writing, and reflect captioners' subjectivities and interpretative skills. In short, this book offers a humanistic rationale for closed captioning—the first of its kind—by countering the popular perception that captioning is straightforward, objective, or simple. If captioning can be shown to be a complex rhetorical practice, then universal design advocates will have even more ammunition to argue that closed captioning should be an integral aspect of the production cycle, not an add-on or afterthought (see Udo and Fels 2010).

Despite the age of captioning technology, we still do not have a comprehensive approach to caption quality that goes beyond important but basic issues of typography, placement, accuracy, timing, and presentation rate. Current practice, at least on television, is too often burdened by a legacy of styling captions in all capital letters with centered alignment, among other lingering and pressing problems. Caption quality has been evaluated in terms of visual design—how legibility and readability interact with screen placement, timing, and caption style (e.g., scroll-up style vs. pop-on style). What we do not have yet is a way of thinking about captioning as a rhetorical and interpretative practice that warrants further analysis and criticism from scholars in the humanities and social sciences. In short, while we have captioning style guidelines for quality, we have not explored quality *rhetorically*. A rhetorical perspective recasts quality in terms of how writers and readers make meaning: What do captioners need to know about a text or plot in order to provide access to it? Which sounds are essential to the plot? Which sounds do not need to be captioned? How should genre, audience, context, and purpose shape the captioning act? What are the differences between making meaning through reading and making meaning through listening? Given the inherent differences between, and different affordances of, writing and sound, how can captioners ensure that deaf and hard-of-hearing viewers are sufficiently accommodated? The concepts that structure these questions—effectiveness, meaning, purpose, context, genre, audience—are of abiding interest to rhetoricians.

My argument, developed over the following chapters, is that a rhetorical view of captioning calls attention to seven transformations of meaning:

1. *Captions contextualize.* Captioning is about meaning, not sound per se. Captions don't describe sounds so much as convey the purpose and meaning of sounds in specific contexts. The meaning of a sound in a particular context may transcend its origins. The precise sonic qualities of a squeaky water tap may be less significant than the act of turning the tap off: (TURNS TAP OFF). In such cases, the action trumps the sound. Additional examples include [TURNS OFF RADIO], [unbuckles seat belt], [BLADE PULLS FREE], [Snaps Oscar's Neck], and [HITS CYMBAL]. Onomatopoeia has a role to play in captioning, but it must be used with care and when the visual context clearly informs the meaning of the captions. **Media**: http://ReadingSounds.net /chapter1/#contextualize.
2. *Captions clarify.* Captions tell us which sounds are important, what people are saying, and what nonspeech sounds mean. As a hearing viewer, I continually find myself relying on captions to learn characters' names and apprehend unusual words such as "flobberworms." (So that's what Peter Pettigrew just said in the background of the Harry Potter movie!) Reading provides superior access over listening, particularly when a noisy environment may work against the listener's ability to clearly make out what people are saying. The same goes for music lyrics that are transcribed on the screen for easy reading, as lyrics are well known for being misinterpreted by hearing fans. **Media**: http://ReadingSounds.net/chapter1/#clarify.
3. *Captions formalize.* Captions tend to be presented in standard written English, with information about manner of speaking relegated to identifiers such as (drunken slurring). Nothing else about the speech will mark it as inflected or accented (e.g., drunk) except for a lone identifier at the beginning of the first speech caption. While standard English provides the fastest access to information, it comes at the expense of conveying the embodied aspects of speech. Embodiment is carried almost entirely by manner of speaking identifiers or simple phonetic transformations (e.g., *gonna, can't*). While it is easy to find examples of substandard or phonetic spellings in speech captions, even these examples are informed by a desire to make the captions as fast to read as possible. Phonetic transcriptions are rhetorical insofar as they balance accuracy with accessibility. In this way, we might say that captions *rationalize* the teeming soundscape. Sounds that resist easy classification or simple description, such as mood music, are tamed or ignored altogether. **Media**: http://ReadingSounds.net/chapter1/#formalize.
4. *Captions equalize.* Every sound tends to play at the same "volume" on the caption track. While there are ways of modulating the volume of captioned sounds and differentiating background from foreground sounds in the captions, these ways are limited and space consuming. As a result, every sound tends to occupy the same

sonic plane, making every sound equally "loud." **Media**: http://ReadingSounds.net/chapter1/#equalize.

5. *Captions linearize*. Sounds that are heard simultaneously cannot be read simultaneously. Captions linearize by presenting the soundscape in a form that can be read one sound/caption at a time. Although it is unusual, multiple nonspeech parentheticals can be presented on the screen at the same time. Multiple sounds can also occupy the same caption—see, for example, *District 9*'s (2009) [ALIEN GROWLS AND PEOPLE SHOUTING INDISTINCTLY] and [RAPID GUNFIRE AND MEN SHOUTING IN DISTANCE]. Multiple, simultaneous sounds can also be reduced to single captions such as [overlapping chatter] and [overlapping shouts] from *Silver Linings Playbook* (2012). But simultaneous sounds must still be read one at a time. The caption reader thus experiences the film soundscape as a series of individual captions. **Media**: http://ReadingSounds.net/chapter1/#linearize.

6. *Captions time-shift*. Viewers do not necessarily read at the same rate as characters speak. Speech captions don't always start precisely on the first beat of the utterance being captioned. The same is true for nonspeech captions, which may precede or follow the sounds being captioned. I devote chapter 5 to exploring some of the ways in which captions give advance notice to readers. Even something as seemingly innocuous as a dash at the end of a caption can alert caption readers to a forthcoming interruption in speech. Names in nonspeech captions can also give away plot details. For example, when [GINA SCREAMS] in *Unknown* (2011), caption readers can guess that Gina is more than an insignificant taxi driver. Readers not only learn the taxi driver's name before listeners do but also venture a guess that Gina will return later in the narrative. I coin the term "captioned irony"—adapting the concept of dramatic irony—to describe cases in which caption readers know more or sooner than listeners who are watching with the captions turned off. **Media**: http://ReadingSounds.net/chapter1/#time-shift.

7. *Captions distill*. The soundscape is often pared down to its essential elements in the caption track. Only the most significant sounds are represented. Exceptions abound, as when ambient PA announcements are overcaptioned as verbatim speech. But for the most part, ambient sounds tend to be reduced to single captions or not captioned at all. Music is distilled to a simple description and/or captioned music lyrics. Captions reconstruct the narrative as a series of elemental sounds. This process also transforms sustained sounds—instrumental music, environmental noise, ambient sounds—into discrete, one-off captions. Consider a tense scene in *Terminator 3* (2003) in which the evil terminator (Kristanna Loken) has broken into a veterinarian clinic looking to kill the vet, Kate Brewster (Claire Danes). As Kate confronts John Connor (Nick Stahl), whom she has trapped in a dog cage in one of the exam rooms, the commotion in other areas of the clinic is reduced to a series of elemental sounds: [GLASS BREAKING], [DOGS BARKING], [DOGS BARKING], [WOMAN SCREAMS], [GUNSHOTS], [GASPING]. In this example, the captions construct

a narrative out of key sounds: the terminator breaks a window to gain entry to the clinic, the dogs react, a customer screams before being shot, and Kate gasps when she sees the customer's body fall. These are the essential moments of the scene, each of which is mapped onto a corresponding caption. **Media**: http:// ReadingSounds.net/chapter1/#distill.

In the context of accessible media, these seven interlocking transformations provide a way of accounting for the differences between sound and writing, listening and reading. They also complicate our notions of universal design, which are sometimes based on general pronouncements about the benefits of captioning for all without accounting for the accompanying changes in meaning and experience that captions support.

Digital Rhetorics and Disability Awareness

Despite the growth of disability studies over the last fifteen years, scholarship in the humanities continues to assume, for the most part, that the world is made up of only hearing, seeing, walking, mouse-using, able-bodied technology users and students. In this world, everyone is fleet-footed, nimble-fingered, and tech-savvy. They are not disabled. They do not require assistive technologies or feel left out because of inaccessible interfaces. They use computers, mobile devices, and the Internet "out of the box." They feel at home on the web. In this world, "access" tends to refer only to inequalities based on class and race—i.e., the haves and have-nots of the digital age (Moran 1999)—and not to disability and ability.

This imagined world of normalcy is segregated from the assistive technologies used by people with disabilities, thus reinforcing "the binary between normal and assistive technologies" (Palmeri 2006, 58). Consider Jonathan Alexander's (2008, 2) depiction of a vibrant, exciting scene populated entirely by able-bodied young people expertly and nimbly remixing and repurposing multimedia content. In one breathtaking passage, the able-bodied digital natives use a range of technologies and skills to take control of their digital environments:

Pictures, sound clips, and video clips captured with cell phones are nearly instantaneously uploaded to blogs; IM chats are scooped up for dissemination on listservs and web sites; podcasts offer a medley of sound, sight, and text; and computer games immerse players in rich multimodal experiences that many gamers manipulate for their own ends and purposes.

Assistive technologies are not a part of Alexander's imagined world because no one in this world is disabled. The implied subjects of the passive constructions ("are . . . uploaded," "are scooped up") are able-bodied dynamos. Indeed, disabled bodies are not simply segregated from these normal bodies; disability doesn't seem to exist at all.

Within this context, it's easy to understand how an appeal to "accessibility" can paradoxically exclude people with disabilities. In "You-Tutorial: A Framework for Assessing Instructional Online Video," Matt Morain and Jason Swarts (2012) develop a rubric for evaluating instructional videos uploaded to YouTube. Their rubric is defined along three dimensions: physical design, cognitive design, and affective design. "Accessibility" is one criterion for evaluating a video's physical design, but because every user is assumed to be able-bodied, the rubric is limited to how well able-bodied users can "access" (see, hear, perceive) digital content:

Accessibility issues concerned how well the video helped viewers focus on the topic of instruction (e.g., the gradient tool). In other words, what efforts were made through screencasting technique, voice-over, or postproduction editing to direct a viewer's attention to the site of instructional action? (9)

Morain and Swarts do not actively exclude disabled users; disability is off their radar altogether. Every "viewer" is assumed to be sighted and hearing. This move is all too common: The nondisabled user, writer, or student is inscribed as the default subject, the unmarked norm, through which claims are tested and judgments rendered. Even a seemingly disability-friendly term like "accessibility" can't rescue Morain and Swarts's evaluative rubric and turn it towards a more diverse set of users—those who can't see the "screencasting technique" because they use an assistive technology such as a screen reader, or those who need closed captioning to access the "voice-over" technique. The presumption of normalcy is so deeply ingrained in Morain and Swarts's article that nothing appears to be able to challenge it, not even an appeal to "accessibility." This reduction of "accessibility" to able-bodied users is not simply the result of confusing *access* for *accessibility* (see Porter 2009, 216). Rather, accessibility is defined in such a way that no one is disabled to begin with.

The distinction between mainstream and margin remains sharp despite recent efforts to integrate disability into studies of digital rhetoric. Disability is often segregated from the "normal science" we do. Disability is a special topic. It reappears in our journals at regular intervals—perhaps not unlike the regularly recurring "tic" of disability that, according to Lennard Davis (2010a, 15), serves a "patrolling function" in Conrad's *Heart of Darkness*. For example, disability has been a regular, if rare, "tic" in *Technical Communication Quarterly* since 2000 (see Wilson 2000; Salvo 2001; O'Hara 2004; Kain 2005; Palmeri 2006; Walters 2010). Even when accessibility and disability are recognized as important subjects, they may be rhetorically shelved for another time, another researcher. For example, when accessibility makes an appearance in Heidi McKee's otherwise excellent essay on "Sound Matters" (2006, 335), it is pushed aside, squeezed out from the world of normal bodies, relegated to a footnote that acknowledges accessibility as an "important issue" but presumably not important enough to break through into the dominant discourse of the essay:

An important issue I do not discuss is accessibility. None of the works I analyze follows the principles of Universal Design and the guidelines of such organizations as W3C (e.g., providing subtitles for all sounds, text-reader alternatives for all photographic images).

The footnote does not explain (1) why "none of the works" analyzed in McKee's essay "follow the principles of Universal Design and the guidelines of such organizations as W3C," (2) why subtitling and not closed captioning is the term of choice, or (3) how it's even possible to "provid[e] subtitles for all sounds," since subtitling (or, more accurately, captioning) always involves a complex process of selection, deflection, negotiation, and invention. But these questions fall outside the scope of the article's inscribed world of hearing bodies, even as the footnote seems to justify the absence of accessibility by calling it "important." Here we find another instance of what, in a different context, Gunther Kress (2010, 59) calls a "backhanded theoretical compliment": "a recognition of the phenomenon in the same moment as its instant dismissal . . . . I notice you and you're not significant enough for me to bother."

An able-bodied norm can make invisible and justify the exclusion of any differences related to disability. Concepts such as "ableism" (Linton 1998) and "aversive disablism" (Deal 2007) call attention to forms of discrimination, sometimes subtle, and the ways in which normalcy is constructed at the expense of people with disabilities. In technical communication and rhetoric and composition studies over the last decade, a growing number of books and articles have participated in the turn towards disability studies and aging studies (e.g., Arduser 2011; Bayer and Pappas 2006; Booher 2011; Brueggemann 1999, 2009; Chisnell, Redish, and Lee 2006; Duffy and Yergeau 2011; Dunn and Dunn De Mers 2002; Hewett and Ball 2002; Kain 2005; Lewiecki-Wilson and Brueggemann 2007; Meloncon 2012; O'Hara 2004; Oswal 2013; Palmeri 2006; Price 2011; Salvo 2001; Theofanos and Redish 2003, 2005; Van Der Geest 2006; Van Horen et al. 2001; Walters 2010; Wilson 2000; Wilson and Lewiecki-Wilson 2001; Yergeau 2009; Zdenek 2009, 2011b, 2014). Disability studies aims to make visible the assumptions that support ableist attitudes and produce and patrol the "normal" body. Disability scholars are concerned with how normalcy is culturally and institutionally constructed and maintained; how differences between the normal and the abnormal body are created, policed, and rhetorically mediated; and how normalcy is mandated and imposed on different (so-called "deviant") bodies through medical interventions that seek to erase rather than accommodate disability (see Wilson 2010, 59). When we leave out disabled people and disabled perspectives from our scholarship, we reinscribe the assumption that only nondisabled people matter, that disability is marginal, or, worse, doesn't exist.

A representative and more accurate account of how our students and technology users interact with multimedia texts, then, must include people with disabilities. A sampling of statistics about hearing and deafness in the United States suggests that a significant percentage of the country needs or may benefit from closed captioning:

- In the United States, approximately thirty-six million adults—about 11 percent of the population—"report some degree of hearing loss" (NIDCD 2010).
- The number of closed caption users in the United States is estimated at fifty million (CaptionsOn 2008)—about one in six Americans.
- The number of US students with disabilities going to college "more than tripled" between 1978 and 1996 (OCR 1999).
- "According to the Deafness Research Foundation, hearing loss is the No. 1 diagnosis for U.S. soldiers in Afghanistan and more than 65 percent of Afghan war veterans are suffering from hearing damage" (Hemstreet 2010).
- The number of Americans sixty-five years of age and older—a population group more likely to benefit from accommodations such as closed captioning—is projected to rise from 13 percent in 2010 to 20 percent by 2050 (US Census Bureau 2008).

- "One third of all senior citizens have hearing problems" (CaptionsOn 2008). When we focus only on so-called digital natives and millennials, we risk ignoring the needs of this fast-growing group of older Americans.

To take these numbers seriously requires us to reorient our research studies towards universal design and away from an able-bodied, youth-oriented norm. We do a disservice to all of our students and users when we assume that captions and other accommodations can only benefit people who are disabled (when accessibility is considered at all).

In disability studies, disability is treated as a potentially transformative construct, not a mere add-on within identity politics or another thing for scholars to worry about. A disability studies perspective can "transform [our] pedagogical practice" (Palmeri 2006, 52). It can serve as a master lens for critiques of science by "connect[ing]" these critiques (Wilson 2000, 150, 159). Because disability "is the most human of experiences, touching every family and—if we live long enough—touching us all" (Garland-Thomson 2010, 356), it "pervades all aspects of culture" (355). Disability scholars invert the binary between impairment and normalcy to argue that, after postmodernism, "[i]mpairment is the rule, and normalcy is the fantasy" (Davis 2010b, 314). Disability studies is not limited to disabled bodies and disability rhetorics but explores norms and normalcy: how norms are constructed, how they are maintained and resisted, and what assumptions we make about the bodies, minds, and abilities of our students and technology users within the "ability/disability system" (Garland-Thomson 2010, 355). In short, disability studies is also normalcy studies (see Davis 2010a).

What would sound studies look, sound, or feel like if it took disability and web accessibility seriously? What if accessibility was not treated as an aside or an objective process of "providing subtitles for all sounds" but had the potential to be transformative? What if we inverted the relationship between the primary text of popular culture and the secondary accommodation—what if we put captioning first? What if we didn't simply argue for the importance of closed captioning but treated it as a significant variable in our multimodal analyses and productions? This book considers the unacknowledged richness of closed captioning as a laboratory for studying:

- *How sound and writing interact.* The differences between speech and writing, listening and reading, and sound and text can be productively explored in the spaces and moments between sounds and their captions. This book explores a number of issues at the intersection of sound and writing: ironic tensions, multiple meanings, temporal shifts (e.g., reading ahead into the future), the deeply rhetorical nature of so-called neutral descriptions, the constraints of space and time, and the negotiations every captioner must make when faced with complex, co-occuring sounds and limited space. In short, how do writers who are constrained by the linearity of print negotiate an auditory space in which "multiple registers can co-exist simultaneously" (Bull and Back 2003, 15)?

· *How to transcribe rhetorically.* By attending closely to which sounds in a movie or TV show are captioned and how they are captioned, we move away from the simplistic view that captioning is straightforward, all-encompassing, or objective. We become aware of different approaches to captioning as reflections of different subjectivities and organizational values. We become aware of how cultural literacy allows captioners to identify allusions to sounds of the past. Narrow views of captioning as objective transcription become troubled by an increasing awareness of how contexts, constraints, and captioners' life experiences shape caption design (and shape, more broadly, any act of description). What the linguist Ronald Macaulay (1991, 281) says about the "inherent limits of any transcription" is true for captioning as well: "A ~~transcript~~ *caption track* should be appropriate for the specific purpose for which it is to be used" and "The aim of any ~~transcription~~ *caption track* is to make the reader's task as simple as possible" (287, cross-outs and italicized additions mine). Both guidelines involve making rhetorical choices in the process of sizing up and accounting for the sonic environment.

· *How to deepen and complicate our understanding of audience and access.* By including people with disabilities in our thinking about sound and multimodality, we deepen our understanding of audience. Rather than paying lip service to audience, or failing to consider the needs of users with disabilities, we open ourselves up to more diverse and inclusive notions of audience. For example, Jody Shipka's (2006) theory of multimodal soundness shows no awareness of disability, defines "access" merely in terms of students' access to resources (361, 371), and at times risks burying the concept of audience under the weight of letting students do what they want with their "time, talent, desire, and access to resources" (371). Such an approach could be productively reimagined by asking these students to consider the needs of more diverse audiences.

· *How to listen deeply.* Asking our students and each other to caption the opening scene of a favorite movie, and then to compare their captions against the official DVD captions, can be a productive activity in deep listening. In the classroom, this activity helps students become more acutely aware of the challenges involved in discriminating among multiple layers of sound, determining which sounds are significant, and describing nonspeech sounds in short, accessible linguistic bursts. (Asking students to write alternative text for images can generate a similar awareness of the rhetorical challenges of making images accessible to web users who are blind.) Deaf and hard-of-hearing students who use live captioning services in the classroom can also participate in and benefit from this activity when asked to interpret the visual field of a film clip, compare how sounds in that clip are captioned both by other students and their own interpreter, and then evaluate the range of linguistic possibilities. "[A]ttuning our ears to listen again to the multiple layers of meaning potentially embedded in the same sound" is associated with agile listening (Bull and Back 2003, 3). In a small way, the amateur practice of closed captioning can help "to clean the sludge out of [our] ears and regain the talent for clairaudience—clean hearing" (Schafer 1977, 11).

Caption studies has the potential to make a number of contributions to the study of multimodal composition. Captioners and captioning researchers study the meaning and significance of sounds, the relationships between writing and sound, the visual and typographic display of sound, the remediation of sound into writing, the challenges of making sound accessible, ambient sounds and music, sonic intertextuality and juxtaposition, sonic allusions and cultural literacy, nonstandard dialects and language varieties, and the design of accessible pedagogical soundscapes. These topics complement and can potentially inform sound studies in composition, technical communication, and related fields. But more importantly, caption studies can call attention to our underlying beliefs and assumptions about technology users and our students. If we start from the assumption that our pedagogies and multimodal compositions need to be accessible, if we assume that not all of our students are able-bodied digital natives, we can develop richer, more informed, more robust, and more accessible pedagogies, tools, technologies, and texts. We limit our theories when we assume that all of our students are hearing, or when we recognize the "important issue" of accessibility but simply choose not to discuss it. A robust account of sound in multimodal composition and technical communication must be attuned to the ways in which sound is made accessible.

A Brief History of Closed Captioning

The history of closed captioning goes all the way back to the silent film era of the early twentieth century. The use of protosubtitles or intertitles—"printed cards that were photographed and integrated with the film itself" (Downey 2008, 19)—allowed audiences to access and enjoy silent movies regardless of hearing ability. As George Downey (2008, 20) puts it in his excellent history of closed captioning (a text I lean on heavily in the next few paragraphs), the silent era of cinema was a "golden age" for deaf audiences.

The introduction of "talkies" in 1927 disenfranchised deaf movie-goers but also planted the seeds for the development of a worldwide subtitling industry that would eventually benefit deaf film audiences. Early foreign-language subtitlers established a development process that persists today, along with a number of guidelines for reading speed, line length, matching new subs to shot changes, and timing subs to appear just after the start of dialogue to allow viewers to visually identify speakers before they start speaking (Downey 2008, 29–30). During this time, deaf film audiences had access to foreign films subtitled in English but not enough films to meet demands (37). The creation of a program to make books available to blind readers on 33-rpm recording discs—the Talking Books project in the 1930s—inspired the development of a similar program devoted to captioning films for deaf audiences (39). Deaf organizations—oralist and manualist—came together to support film captioning and, later, federal support for it (42). In the early days of the Captioned Films for the Deaf (CFD) program, films were purchased that had already been transcribed in English by professional subtitlers. (English transcription was a necessary step in the process of translating a film into a foreign language.) CFD could thus hire deaf instructors from Gallaudet University to work as film captioners (43), a remarkable moment in the early history of captioning in which some of the very first captioners of the deaf were themselves deaf. The CFD captioners also worked out guidelines to edit captions for reading comprehension: "the dialogue in a captioned film for the D/HOH audience had to be drastically modified for word difficulty and reading speed—a foot of film per word, at a fourth-grade reading level, was the target" (43). Editing captions for comprehension, though not popular with deaf audiences in the United States today, would persist into the early days of TV captioning (Earley 1978) and through to the present, particularly in Europe where edited/summary captioning is still common (Schilperoord, de Groot, and van Son 2005). CFD was federally subsidized in 1959. Additional federal funding in the 1960s turned the program increasingly away from entertainment and towards educational media (Downey 2008, 44).

Captioning advocates turned their attention to television in the late 1960s and 1970s. Experiments in open-captioned television in the 1970s led, first, to captioned reruns of *The French Chef* with Julia Child starting in August 1972. Open captions can't be turned off (i.e., closed) but are displayed on the screen for all to see. The Public Broadcasting System affiliate WGBH was contracted by the US Department of Education to spearhead the project (Peltz Strauss 2006, 207). On the heels of the piloted success of open captions, WGBH, through its new division The Caption Center, contracted with ABC to begin offering open-captioned reruns of *The ABC Evening News* in December 1973. The team at WGBH had four and a half hours each night to prepare the captions, starting at 6:30 pm, when the live ABC broadcast ended, until 11:00, when the program was rebroadcast on PBS with open captions (Earley 1978). The captioned version spread quickly, from three Eastern cities to ten after the first week. By August 1974, the program went national to 56 PBS stations (Earley 1978, 3) and after eight years to "more than 190 public stations" (Peltz Strauss 2006, 208).

Closed captioning, which debuted in March 1980, addressed the problem of how to make TV accessible to deaf audiences without making it "unpalatable to the hearing" (Downey 2008, 55). Closed captioning could be turned on and off with the use of a separate decoder attached to a TV. Television captions are delivered on Line 21 of the vertical blanking interval of the analog television signal and were thus invisible until decoded. The Line 21 standard is also known as EIA-608. ABC and NBC initially agreed to five hours of closed-captioned programming a week, and PBS to twelve and a half hours (Peltz Strauss 2006, 211). CBS initially refused to participate in the deal because the network was banking on the success of an alternative Teletext model that was already being used in Europe. (In 1984, CBS gave in and adopted the Line 21 standard, in addition to Teletext, following protests organized by the National Association of the Deaf.) The decoders were expensive, starting at $249 in 1980 and dropping to a still pricey $160 by 1986. While the decoders sold quickly at first, by 1988 "[f]ewer than 200,000 had been purchased during the entire eight-year period that these devices had been on the market" (Peltz Strauss 2006, 220). By the end of the 1980s, 200 hours of programming each week ("less than one-third of all programming shown on the three major broadcast networks" [220]) were being offered by the networks with closed captions, all on a voluntary basis.

Two problems—slow decoder sales and no legal mandate to compel broadcasters to provide captions—were addressed in the 1990s. First, the Decoder Circuitry Act of 1990, signed into law not long after the signing of the Americans with Disabilities Act, required all televisions with screens thirteen inches or larger to be equipped with internal decoders for displaying closed captioning—thus eliminating the need for separate decoders and presumably increasing the market for captioned programming. Indeed, the same arguments used today in the name of increasing the number of captioned web videos were first used in the early 1990s to advertise the new decoder chip requirement. Accordingly, captions were claimed to help children to read, assist anyone with learning English as a second language, allow sports lovers to watch the game even in noisy environments, and give families the pleasure of reading together (Peltz Strauss 2006, 237–38). The Decoder Circuitry Act was a "victory for D/HOH interests [that] was only accomplished through the rhetorical widening of the captioning audience to encompass hearing viewers" (Downey 2008, 19).

Rhetorical widening is a crucial component of captioning awareness campaigns today. It is similar to "interest convergence," which, according to Jay Dolmage (2005), drawing on critical race theory, is the "idea that conditions change for minorities only when the changes can be seen (and promoted) as positive for the majority group as well." Closed captioning advocates, in the name of universal design, argue today that captions benefit everyone, regardless of age or ability. Studies exploring the effects of captioning on reading comprehension have been particularly popular since the 1980s. One of the cosponsors of the Decoder Circuitry legislation, Senator Harkin, appealed to reading studies in 1989 to garner support for the newly introduced bill, "not[ing] the ability of captioning to increase reading comprehension, language retention, and word recognition" (Peltz Strauss 2006, 230). Studies of reading comprehension have accumulated over the decades, providing some of the most persuasive evidence to support the rhetorical widening of the captioning audience (see Bean and Wilson 1989; Cambra, Silvestre, and Leal 2008/9; Goldman and Goldman 1988; Koskinen et al. 1993; Linebarger 2001; Linebarger, Piotrowski, and Greenwood 2010; Markham 1989). The challenge for universal design, as Dolmage (2005) puts it, is to "avoid this problem" of defining captioning as positive and desirable only when it applies to the majority population of hearing viewers.

The second problem—that only a percentage of programming was being voluntarily captioned in the late 1980s—was addressed by the 1996 Telecommunications Act, which required all new, nonexempt English language programming to be closed captioned by 2006. By 2008, 75 percent of a "programming distributor's pre-rule nonexempt video programming" needed to be closed captioned (FCC 2010). Captioning rules for Spanish-language programming were also established at this time. Exemptions were made for TV commercials: Anything under five minutes in duration was not required to be captioned. As a result, fewer than 50 percent of TV commercials are voluntarily captioned today, except during the Super Bowl, which has achieved 100 percent captioning on national Super Bowl commercials in the last couple years (NAD 2012). Other exemptions were made for late-night programming, some locally produced programming, programming on new or small-revenue networks, and in languages other than Spanish or English, among others (Peltz Strauss 2006, 263). As with the Americans with Disabilities Act, the "undue burden" exemption was also available by petition to the FCC. Despite these exemptions and the lack of a consistent body of quality captioning standards, the Telecomm Act was a major victory for captioning advocates as "America became the first country in the world to require all new television programs, with few exceptions, to be closed captioned" (Peltz Strauss 2006, 268).

Within a few years of the passage of the Telecomm Act, the FCC released decoder standards for digital TVs (DTVs) which "incorporated sections of industry standard EIA-708-B" (FCC 2000). This standard, also known as CEA-708, allowed DTV owners to choose from a range of options when styling how the captions appear on their TVs:

- Three font sizes: small, standard, and large
- Eight fonts: Default (undefined), Monospaced Serif (similar to Courier), Proportional Serif (similar to Times), Monospaced Sans Serif (similar to Helvetica), Proportional Sans Serif (similar to Arial and Swiss), Casual (similar to Dom and Impress), Cursive (similar to Coronet and Marigold), and Small Capitals (similar to Engravers Gothic) (see CEA R4.3 Work Group 1 2003/4)
- Eight background and eight foreground colors: "white, black, red, green, blue, yellow, magenta and cyan" (Goldberg 2007, 23)
- Four character edges: "none, right drop shadow, raised, depressed, or uniform (outline)" (Clark 2006b)
- Four foreground and four background opacity options: translucent, transparent, solid, and flashing
- Support for semantic coding tags. Caption providers can tag the function of any string of caption text. Joe Clark (2006b) lists the possible tags, which include dialogue (default setting), source or speaker ID, electronically reproduced voice, second-language dialogue, voiceover, dubbing, subtitling, voice quality, song lyrics, sound effects, music description, and expletive. It's doubtful, however, whether professional captioners are making use of function tagging. As Clark (2006b) puts it, "This is a tremendously useful feature that will, presumably, never be used."
- Support for up to six "services" or streams, to be displayed one at a time. For example, one service could be devoted to verbatim closed captioning, and the other services . . . well, the possibilities are only limited by our imaginations: an edited/near-verbatim track, an easy-reading stream for young viewers or viewers learning English, a speech-only (subtitle) stream, foreign language subtitles, a director's

commentary track, a parody or "resistive" stream, a Klingon language stream, and so forth. (In the concluding chapter, I return to parody and the potential for DTVs to provide a space for the kinds of online parodies that rely on open captions, such as "literal music videos" and *Downfall* parodies [theamishaugur 2008]). There's untapped potential here, especially in educational contexts, for the development of easy-reading tracks aimed at elementary school children. In 2007, Ward et al. (21) described how "the children's animated television series *Arthur* began to be broadcast with two streams of captions for viewers who are deaf or hard of hearing: Near-verbatim captions and edited captions." But at a time when captioning advocates are still fighting to bring *any* captions to uncaptioned web videos, the idea of a second or third caption stream for the same show, while enticing, seems idealistic.

The user can select the caption provider's settings or save their own preferences in the TV. As Larry Goldberg (2007, 23) describes it: "Decoders must include an option that permits a viewer to choose a setting that will display captions as intended by the caption provider (a default). Decoders must also include an option that allows a viewer's chosen settings to remain until the viewer chooses to alter these settings, including during periods when the television is turned off."

The current fight over access to video content is taking place online. Until recently, Internet videos were not required to be closed captioned. The Communications and Video Accessibility Act (CVAA), signed into law by President Obama in 2010 and rolled out by the FCC starting in 2012, changed everything. Prior to the CVAA, programs that were captioned on TV were not likely to be captioned on the web when redistributed through official channels (e.g., Hulu, iTunes, the websites of TV networks, etc.). These same TV shows and movies are now required to be closed captioned when redistributed over the Internet. The CVAA also authorizes the FCC to introduce quality standards, which it had declined to do in the wake of the Telecomm Act of 1996. The FCC (2012) currently defines quality captioning for TV shows and movies delivered over the Internet through a "consideration of such factors as completeness, placement, accuracy, and timing." Web captioning for content covered under the CVAA is required to be "at least the same quality as the television captions provided for that programming." The CVAA represents another major victory for captioning advocates and those who rely on captioning. As Internet video grows in popularity and as more people watch movies and TV shows on the web, the web needs to be as accessible as television. As the former director of the Law and Advocacy center for the National Association of the Deaf put it, "We do not want to be left behind as television moves to the Internet" (quoted in Stelter 2010).

Many types of video programming are not covered by the CVAA, from user-generated content, which includes millions of YouTube and Facebook videos, to independently produced web series, regardless of their quality or number of viewers. Whether a video has ten views or ten million doesn't matter to the law. Captioning awareness campaigns continue to target types of video content that fall outside the law's blanket of coverage (e.g., see Jamie Berke's web series advocacy [Mbariket 2011]). Advocates have allies among distributors of user-generated video content. For example, Google has developed an accessible YouTube interface for toggling captions on and off, a relatively simple process for video owners to add closed captions to their videos, and an automated (albeit much ridiculed) speech-to-text system for adding captions to any video. At the same time, however, advocates have also faced resistance from other major content distributors. For example, Vimeo, a video distributor similar to YouTube, took four years to act on repeated requests in their discussion forums to build support for closed captioning into their video player (Vimeo 2012; Summers 2014). Without support for captioning at the interface level (e.g., integrated *cc* button and controls), the only option available is open captions burned into the video.

Only a few years ago (circa 2009), Netflix might have been placed in the same boat as Vimeo. At the time, Netflix offered no support for closed captioning on their streaming videos, and even claimed, in a move some interpreted as a blatant disregard for the needs of deaf and hard-of-hearing viewers (e.g., NAD 2009), that captioning was still a year away (Netflix 2009). A year later, Netflix rolled out initial support for captions, but only for 100 programs total (the majority of which were episodes of *Lost*) and with no way to search for them since captions had not been tagged as a variable in Netflix's search engine (Netflix 2010). Then, slowly, support for captions improved, from 30 percent of Netflix's streaming content at the beginning of 2011 to a projected 80 percent by the end of that year (Netflix 2011). Through a settlement reached in a lawsuit filed by the National Association of the Deaf, Netflix agreed to caption 100 percent of its library by 2014 (Mullin 2012). According to the agreement, new content added to the Netflix library will also be captioned "within 30 days by 2014" and "within 14 days by 2015" (Mullin 2012). To reach this goal, Netflix formed a partnership with Amara in 2012, an online tool that allows anyone to caption online videos. Volunteers were solicited to caption Netflix videos using Amara, but the results of their efforts have been, at times, disappointing. In describing how Netflix has "alienated and insulted its deaf subscribers," Jon Christian (2014) provides a few examples of poor Netflix captions (although it's unclear whether these are the result of crowdsourcing):

"ALL STATIONS PREPARE FOR A HAPPY BIRTHDAY"—Commander Riker in *Star Trek: The Next Generation*

"Report to the new Quartermaster for ur documentation."—M in *Skyfall* (2012)

Christian lists other problems as well, including censored words in the captions that aren't censored in the original speech, miscaptioned sound effects, inconsistently identified speakers, skipped words and sentences, poorly timed captions, and poorly placed captions covering subtitles. Perhaps for these reasons, the crowdsourcing experiment didn't last long. Amara is no longer soliciting volunteers on behalf of Netflix, and according to a Netflix spokesperson, captions today "are typically the same subs or captions that would be used on traditional broadcast TV" (quoted in Christian 2014). While it's not known whether the problems discussed by Christian have been corrected, what seems clear is that any large-scale captioning solution will require editorial oversight. As Christian (2014) puts it, "the most prominent errors could be fixed on a single read-through by a competent copyeditor."

Other laws in the United States, particularly the Americans with Disabilities Act (ADA), *may* also require closed captioning for the Web's private sector. For example, the judge presiding over the landmark *National Federation of the Blind v. Target Corporation* case ruled in 2006 that the ADA, which was signed into law before the advent of the Web, applies to private businesses regardless of whether goods and services are offered in brick-and-mortar stores or online: "Judge Marilyn Patel rejected Target's position that their site couldn't be sued under the ADA because the services of *Target.com* were separate from Target's brick-and-mortar stores" (Chisholm and May 2009, 16). But because Target settled the case in 2008 "without admitting any wrongdoing," "the question of the ADA's applicability to the Web [is] somewhat unresolved" (16). Regardless, the Department of Justice has declared that the ADA does indeed apply to the Internet. According to Thomas E. Perez (2010), Assistant Attorney General in the DOJ's civil rights division, "It is and has been the position of the Department of Justice since the late 1990s that Title III of the ADA applies to websites. We intend to issue regulations under our Title III authority in this regard to help companies comply with their obligations to provide equal access."

The web's public sector is regulated by Section 508 of the Rehabilitation

Act of 1973, as amended in 1998, which requires federal agencies that "develop, procure, maintain, or use electronic and information technology" to make their products and services, including their websites, accessible (Section 508). §1194.22b of Section 508 mandates the use of synchronized alternatives (e.g., open or closed captions) for video content: "Equivalent alternatives for any multimedia presentation shall be synchronized with the presentation" (Section 508). In the private sector, businesses that have contracts or hope to have contracts with the federal government must ensure that the products they deliver to the government comply with Section 508. State universities that receive federal money, even indirectly (e.g., through federal student loan programs), are also responsible for adhering to Section 508. State agencies, including state universities, may have additional or different obligations. In Texas, for example, where I have lived for fifteen years, the Texas Administrative Code regulates, among other things, website accessibility for state agencies, including state institutions of higher education. The regulations for higher education websites in Texas follow Section 508 almost to the letter, with one major exception: §1194.22b of Section 508, which mandates synchronized alternatives such as captions for multimedia content, is compulsory in Texas Administrative Code only *after* an institution of higher education receives a request from a web visitor for "alternative form(s) of accommodation." For example, an informational video on the public website of a state university in Texas only needs to be accessible to deaf and hard-of-hearing users *after* a request for accommodation is made by a visitor to the site. But inside the classroom, all materials need to be accessible. According to my university's Student Disability Services office, "all class-related materials (specifically, PowerPoint voice-overs, online courses, audio recordings) must be closed captioned, subtitled or transcribed. Videos, movies, and internet clips must be captioned or subtitled as a transcription is not sufficient. This applies to both on-campus and online courses" (Texas Tech University Student Disability Services 2014). This mandate for video captioning is authorized by the ADA and Section 504 of the Rehabilitation Act. According to my university's operating policies, "Both ADA and Section 504 are civil rights statutes that prohibit discrimination on the basis of disability, obligate colleges and universities to make certain adjustments and accommodations, and offer to persons with disabilities the opportunity to participate fully in all institutional programs and activities" (Texas Tech University 2013).

Despite the legal advances, new digital technologies, and growing awareness of disability issues, closed-captioned TV hasn't changed much since its debut in 1980. Watching closed-captioned television can sometimes feel like going back in time. On any given night in living rooms across America, captioned TV for nonlive programming may look like an ancient relic from a predigital era. As I will discuss in detail in chapter 2, captions for many prerecorded TV shows continue to be displayed in all uppercase letters, even though the original reason for all-caps styling (i.e., low-resolution fonts) has not survived advances in technology (Clark 2008). Scroll-up styling is becoming more and more popular for prerecorded content, even though it is less readable than the more expensive pop-on style of captioning. In other words, cost cutting is driving down quality. In the absence of any sustained opposition—I have read few complaints about all-caps styling on the captioning mailing lists—such practices will continue and may even be increasing today.

On the web, closed captioning and foreign-language subtitling are playing increasingly vital roles in video search and retrieval. Captions and subtitles are stored as plain text files and thus can be indexed on sites like YouTube to provide a better search experience for users (Ballek 2010; Sizemore 2010). Imagine leveraging this power for students who are searching recorded lecture videos to prepare for the final exam. Or giving web visitors the power to search inside the full archive of videos stored on a company's servers. Robust search capabilities in the age of web video and user-generated content require an infrastructure built on captioning technology, because search engines are not able to determine the specific contents of web video from audiovisual content alone. In terms of Search Engine Optimization (SEO), closed captioning can drive more targeted traffic to business websites by providing more information to Google about the contents of videos stored on those sites. Captioning technology is also producing richer experiences on the Web through interactive transcripts, which allow users to click anywhere in a video transcript and be transported to that moment in the video where the clicked words are spoken. Interactive transcripts, like enhanced TV episodes that rely on captioned commentary, suggest an expanded role for captions that could one day make them part of the standard topography of the typical web video. The same technology can also support the translation of video into multiple foreign language subtitle files for a global marketplace (e.g., see the crowdsourcing model of translation at Dotsub.com). Indeed, it may be more accurate to say that accessibility is a by-product, not a driver, of subtitling efforts on TED.com and other sites.

Brief histories have a tendency to downplay the human struggles, protests, conflicts, and advocacy campaigns that produced the world we

know today. Brief accounts exert a strong pull towards a set of decontextualized facts, events, and laws. A narrative of progress is implied,
one that seems to move us inexorably from experiments in open captioned television to laws requiring 100 percent captioning on all new,
nonexempt programming, or from an untamed, uncaptioned web to a
regulated digital environment in which captioned programming (albeit
still only certain types) is more plentiful than ever. But we have to be
careful about implying a "progressiveness that is inexorable," as Greg
Dickinson, Brian L. Ott, and Eric Aoki (2013, 346) put it in their rhetorical analysis of the Plains Indian Museum, in which the "inexorability of
history's progress . . . absolves Euro-Americans of guilt over the violence
done to Plains Indians." Disability, like technology (see Bijker 1997),
is socially constructed. Fundamental concepts of normalcy and difference (Davis 2010a), ability and disability (Linton 1998), and access and
accessibility (Titchkosky 2011) are continually being renegotiated even
as their meanings seem, at times, obvious and natural. The claim to
objectivity, to naturalness, to an absence of rhetoric is itself a rhetorical
achievement, as rhetoricians of science have argued (e.g., Gross 1996).
Lennard Davis's (2010a) groundbreaking work on the social construction of normalcy suggests how the boundaries between disabled and
normal bodies are not inescapable but arise (and are policed) in specific
historical, economic, and political contexts. The notion of "rhetorical
widening," discussed briefly above, begins to gesture towards that more
complex history, but for the most part I've muted and distilled the enormous complexity of the history of captioning developments and laws
in order to make room for another kind of complexity in the chapters
that follow. (For a fascinating account of the economic, legislative, and
social debates over telecommunications access, see Karen Peltz Strauss's
[2006] *A New Civil Right: Telecommunications Equality for Deaf and Hard of
Hearing Americans*. For a history of closed captioning, see Gregory John
Downey's [2008] *Closed Captioning*.) In what follows, I embrace the uncertainty, creativity, and transformative potential of closed captioning
in order to dislodge some of the assumptions rooted in an overly simplistic transmission model of meaning-making.

A Note on Method

Reading Sounds draws on a number of mainstream Hollywood movies
and popular TV shows. It offers an in-depth study of prerecorded, nonlive captioning across a number of contexts and genres, supplemented

by interviews with professional captioners and survey responses from regular viewers of closed-captioned content (chapter 2). English-language closed captioning is the book's focus, with a special emphasis on non-speech sounds. Foreign-language subtitling practices are discussed only when they come into meaningful contact with the caption track (see chapter 8). This book is motivated by my perspective as a hearing parent of a deaf child and my abiding interest in the relationships between sound and writing. My hope is that this book will nudge captioning towards the scholarly mainstream. While *Reading Sounds* is not the result of a collaboration with deaf and hard-of-hearing viewers, it values their experiences and preferences. As suggested by the disability rights slogan "Nothing About Us without Us" (see Charlton 2000), captioning advocacy campaigns need to be authorized and led by those with the greatest stake in the outcome: viewers who are deaf and hard of hearing. While I am not deaf or hard of hearing, I offer this book as a small contribution to captioning advocacy from the perspective of a scholar, parent, and demanding caption viewer who cares deeply about accessibility.

Rhetorical criticism, the approach I use in this book, tends to proceed inductively. The critic builds and tests theories by example but offers only probable conclusions—arguments—based on the critic's interpretations and evidence from the text. The critic's interpretative skills and insights play a central role in rhetorical criticism, because, as Edwin Black (1978, x) proclaimed, "critical method is too personally expressive to be systematized." Black argued:

criticism, on the whole, is near the indeterminate, contingent, personal end of the methodological scale. In consequence of this placement, it is neither possible nor desirable for criticism to be fixed into a system, for critical techniques to be objectified, for critics to be interchangeable for purposes of replication, or for rhetorical criticism to serve as the handmaiden of quasi-scientific theory (xi).

Preexisting theory can become "fateful" in rhetorical criticism (Black 1980, 331) when "the critic is disposed to find exactly what he or she expected to find. The epistemological constraints imposed by a theoretical orientation inhibit critics from seeing new things, from making new discoveries. Such criticism tends much more to be a confirmation than an inquiry. It is, in the strictest sense, a *prejudice*" (333). Black referred to theory-driven criticism as "detached" and "sterile" (333). In contrast, he argued that criticism should aim to provide "*singular* access to its subject" (1978, xii). Such access is only possible when the critic does not mechanically apply preexisting theories but "undertake[s] to see the object on its own terms—to see it with the utmost sympathy and compassionate understanding" (Black 1980, 334).

This perspective continues to inform rhetorical criticism even as scholars build increasingly sophisticated theoretical accounts of how texts make meaning and influence audiences. I am motivated by the perspective that the best criticism will provide *"singular* access to its subject," not "predictable or conventional" arguments based on what we expect to find (Black 1978, xii). Because a number of the ideas in this book were first tried out on my blog—captioned irony, series awareness, cultural literacy, logocentrism, and others—the book is grounded in specific examples. A couple of years before I started thinking about writing a book on captioning, examples started to accumulate in my blog posts and private notebooks, which eventually pointed to larger themes, and finally evolved into chapter ideas for the book. As will soon become clear, this book is driven by literally hundreds of examples. I couldn't presuppose a large number of preexisting theories about captioning because, while the research literature from captioning scholars is fairly robust in the areas of reading comprehension, vocabulary and foreign-language learning, visual design, and presentation rate (verbatim vs. edited captioning), the literature on how captions make meaning and represent sounds is scant. No study has yet to offer a humanistic rationale for captioning, one grounded in a penchant for reading texts with "sympathy and compassionate understanding." Rhetorical studies can inform disability studies by offering illuminating analyses of captioned texts to complement and challenge discussions of universal and retrofitted design (see Dolmage 2005). A rhetorical approach contextualizes captioning by showing us how captions work. It also "authorizes" the critic's interpretation. As Bonnie Dow (2013, 147) puts it in her argument for criticism in the artistic mode, "it is the critic, not the text or the audience or the method, who authorizes the interpretation. That interpretation may be focused on a single speech, a scientific treatise, a public monument, vernacular rhetorics, or performative traditions, but its aim is not truth, or representation, but illumination." To Dow's list of interpreted texts I would add, of course, closed captions. The critic's reading "makes [the object] *interesting* in a way that it was not before." Illuminating closed captioning, offering a sympathetic reading, providing singular access, telling us something we didn't know before, "mak[ing] us think about our world in new ways" (148)—these are all aims of criticism I identify with.

At some point, every rhetorical critic's interpretations, even when they seem to be unencumbered by theoretical presuppositions, begin to act as inevitabilities. There's a tension between bottom-up (or "emic") criticism and top-down (or "etic") criticism. An engagement with a rhetorical artifact that strives to be sympathetic can become detached as the critic's interpretations become controlling. "[W]hile still engaged in the interpretative act," according to Michael Leff (1980, 345),

the critic constructs a meaning for the object, an hypothesis or model that explains what it is. And the emic critic is now caught up in the same process that circumscribes the thought of his or her etic colleagues. Just as the etic critic, once convinced that his or her theory accounts for the rhetorical dimensions in a work, assimilates the text into the theoretical categories, so also the emic critic, once convinced of the inevitability of his or her conception of the work, forces the particular features of the work into conformity with the conceptual representation of its meaning. Since interpretation is inherently circular, this problem is unavoidable.

This "problem" can be productive, allowing the emic critic to "circle closer to the ground" (Leff 1980, 345) even as early conceptions of the text direct the critic's later attentions. In my approach to closed captioning, tentative interpretations, based on specific examples I encountered and analyzed on my blog, gained more influence under the weight of a growing collection of examples. These interpretations became hypotheses that closed off or directed my attention away from other potential avenues of analysis. My own interests, particularly in cultural literacy (chapter 7) and the creative description of nonspeech sounds, deflected my gaze away from other potential topics that another researcher might have pursued more earnestly.

To provide some counterbalance to an inductive approach, I began compiling a corpus of DVD caption files that could be searched and organized in new ways. The subtitle-sharing websites, populated with content uploaded by fans, proved insufficient to this task. These sites are typically used to share not closed captions but foreign-language subtitles for viewers who are presumed to be hearing and thus don't require special access to nonspeech sounds such as thunder and grunting. As I also discovered, the files marked "hearing impaired" on the subtitle-sharing sites couldn't always be trusted to match the official closed captions on the DVD versions. I tried repeatedly to find official DVD captions on the subtitle-sharing sites. Finally, and most importantly, it is not possible to search *inside* the public subtitle-sharing databases. Users can only search these sites by movie or program title. As a result of these limitations, I decided to extract DVD captions manually. Extraction is a labor-intensive process that involves multiple steps. First, a DVD decrypter is used to extract the Video Object (VOB) files to the user's hard drive.

Then, the extracted VOB files need to be run through a subtitle ripper program. The ripper program asks the user to identify each bitmap letter, storing that information for the next time the same bitmap image shape appears in the files. The program slowly learns to identify each letter, working faster and faster until the full bitmap caption file has been extracted as a plain text SubRip (.srt) caption file, complete with an ordered list of captions and start and end times for each caption. The "character matrix" for each movie can be saved and reused with other DVDs that use the same typeface and type size. In the absence of a previously saved or usable character matrix, the same process of identifying each character by hand will have to be repeated anew for each DVD, requiring the user to patiently respond to numerous requests from the program to identify letters, numbers, and punctuation. At the end of this process, careful editing with a plain text editor may be needed to ensure the ripper didn't make any mistakes, such as confusing a lower case l for an uppercase I (an error so common that there's a menu button in the subtitle ripper program to fix it). Finally, I used a DVD converter to extract a copy of the DVD movie with closed captions burned in, in the event that I needed to pull video clips from the movie later on. If there were timestamp discrepancies between the extracted captions and the original DVD captions, I ran the extracted caption file through a small software program my older son wrote in Perl which resets each timestamp by an amount of time I designated. For example, the last command I sent to this program reset the first caption in *Star Wars: Episode 1—The Phantom Menace* (1999) to start at 00:01:59,900 and adjusted each of the remaining caption times accordingly: "Adjust-Timestamps.pl 'Star Wars—The Phantom Menace.txt' 00:01:59,900." To use this program, I don't need to know how inaccurate every timestamp is, only the correct start time of the first caption, and the program does the rest. I uploaded each extracted caption file to a web interface, also written by my older son. I used this interface to search inside my growing collection of closed-caption files and also to reorder each caption file according to type of caption (speech or nonspeech).

Because I am interested in nonspeech sounds and nonspeech information (NSI) such as speaker IDs, I needed a way to separate NSI from the speech captions in each file and then further separate speaker IDs from nonspeech captions. The custom software program searches for punctuation marks such as colons and parentheses to identify which captions include nonspeech descriptions and/or speaker IDs. While the program isn't always perfect—some editing in a spreadsheet program may be required to remove speech captions that contain colons from the list of nonspeech

Table 1.1 Nonspeech captions from a single movie displayed in table format.

Caption number	No.	Start time	End time	Text
1	1	00:01:19,184	00:01:20,845	(THUNDER RUMBLING)
2	2	00:01:32,130	00:01:33,620	(MEN CLAMORING)
3	3	00:01:41,206	00:01:43,197	(YELLING)
4	4	00:01:46,411	00:01:48,504	(SCREAMING)
5	5	00:01:59,524	00:02:00,684	(GROANING)
59	6	00:04:12,824	00:04:15,054	My last barber hanged himself. (CHUCKLES)
62	7	00:04:20,832	00:04:22,129	(CHUCKLES)
70	8	00:04:36,414	00:04:39,383	Yeah. We heard you speak . . . (STAMMERING) Goddamn.
86	9	00:05:03,207	00:05:04,401	(STAMMERING)
146	10	00:08:24,175	00:08:25,506	(LINCOLN SIGHS)
169	11	00:09:43,421	00:09:44,888	-(DOOR CREAKING) -Oh!
175	12	00:10:06,677	00:10:07,644	(DOOR OPENS)
176	13	00:11:23,888	00:11:25,116	(KISSES)
183	14	00:11:50,614	00:11:52,275	(PLAYING A MARCH)
184	15	00:12:02,092	00:12:03,286	(MUSIC STOPS)
190	16	00:12:42,533	00:12:43,932	(AUDIENCE LAUGHING)
191	17	00:12:44,034	00:12:45,126	(AUDIENCE CLAPPING)
192	18	00:12:45,269	00:12:47,567	ALL: (SINGING) We are coming, Father Abraham
204	19	00:13:15,499	00:13:16,488	(LAUGHS) "Only twenty?"
237	20	00:14:33,444	00:14:35,378	-It's too important. -(KNOCKING ON DOOR)

Note: A table showing the first twenty search results for the nonspeech DVD captions in *Lincoln*. Touchstone Pictures, 2012. http://ReadingSounds.net/chapter1/#table1.

captions, for example—it nevertheless does an admirable job of creating two HTML tables, one containing all of the nonspeech descriptions in a movie and another containing all of the speaker IDs (see table 1.1). Each HTML table also includes hyperlinked caption numbers for viewing each example in the full context of the caption file associated with it. The database can be searched, with the results delivered in table format, also with hyperlinked caption numbers (see table 1.2). I've increasingly relied on these tools to follow up on hunches, find additional or related examples, compare totals across multiple files, and so on. In my comparison of four sci-fi action movies in chapter 2, for example, these tools were indispensable. My arguments about boilerplate captions, such as [INDISTINCT CHATTER] in chapter 6, started as hunches that were pursued with the help of this program's search function. Because the program was designed to output data in HTML tables, search results or even entire caption files can be pasted directly into Microsoft Excel for coding, filtering, and sorting. Sorting all of the nonspeech captions from a movie into alphabetical

Table 1.2 Search results for "indistinct chatter" displayed in table format

Caption number	No.	Source	Start time	End time	Text
1167	1	*21 Jump Street* (2012)	00:50:06,684	00:50:08,151	(INDISTINCT CHATTER)
511	2	*Aliens vs. Predator: Requiem* (2007)	00:38:34,700	00:38:38,329	(indistinct chatter in distance)
777	3	*Aliens vs. Predator: Requiem* (2007)	00:56:57,402	00:57:00,530	(indistinct chatter in distance)
1271	4	*Argo* (2012)	01:39:56,295	01:39:58,160	[INDISTINCT CHATTER OVER RADIO]
388	5	*Beasts of the Southern Wild* (2012)	00:43:19,677	00:43:21,645	[INDISTINCT CHATTER]
97	6	*CSI: NY,* "Unspoken" (2012)	00:13:37,346	00:13:38,973	(muffled, indistinct chatter)
140	7	*CSI: NY,* "Unspoken" (2012)	00:20:10,005	00:20:11,996	(indistinct chatter, phones ringing)
1127	8	*Cloud Atlas* (2012)	01:16:19,820	01:16:21,549	[INDISTINCT CHATTER OVER DEVICE]
1153	9	*Django Unchained* (2012)	01:28:11,577	01:28:12,771	[indistinct chatter]
966	10	*Inglourious Basterds* (2009)	01:55:47,387	01:55:48,877	(INDISTINCT CHATTERING)
57	11	*Killing Them Softly* (2012)	00:04:11,121	00:04:12,452	[indistinct chatter]
148	12	*Killing Them Softly* (2012)	00:08:27,043	00:08:28,032	[indistinct chatter]
178	13	*Killing Them Softly* (2012)	00:09:50,793	00:09:52,021	[indistinct chatter]
284	14	*Killing Them Softly* (2012)	00:15:51,120	00:15:53,953	- [indistinct chatter, laughter] - Go, go, go, go, go.
1217	15	*Killing Them Softly* (2012)	01:21:18,536	01:21:20,697	[indistinct chatter]
1331	16	*Killing Them Softly* (2012)	01:33:05,709	01:33:08,678	[indistinct chatter]
1335	17	*Killing Them Softly* (2012)	01:34:14,645	01:34:18,638	[indistinct chatter]
1345	18	*Killing Them Softly* (2012)	01:36:14,431	01:36:17,423	[indistinct chatter, birds chirping]
982	19	*Les Miserables* (2012)	01:10:52,228	01:10:54,093	(INDISTINCT CHATTERING)

Note: A table showing the first nineteen search results for "indistinct chatter" using all the movies in the corpus. http://ReadingSounds.net/chapter1/#table2.

order in Excel, for example, is guaranteed to generate insights or at least raise compelling questions for the researcher. If it weren't for copyright concerns, I would make this software tool available to the public. Caption studies should make it a priority, perhaps as part of a crowdsourcing project, to develop databases of official closed captioning files for big data studies that aren't possible currently. Finding patterns and trends across thousands of caption files is the necessary complement to inductive arguments that proceed by example.

TWO

Reading and Writing Captions

Closed captioning is usually defined as a mechanical, objective process of unreflective transcription. When the subjective, creative work of meaning making is diminished or elided, captioning becomes little more than a "technique used to display text" (WiseGeek 2014). Indeed, "display" appears as a main verb in a number of definitions of closed captioning (Leibs 2014; FCC 2014; Media Access Group 2002a; Wikipedia 2014). This emphasis on display is most likely an artifact of the original analog specification that allowed captions to be embedded in the TV signal (EIA-608, or more commonly referred to as Line 21 captions). In other words, what was and continues to be noteworthy about the technology of closed captioning is that it is indeed a technology of displaying and hiding text on the screen. But when captioning is treated primarily or only as a technology of display, the processes of writing and reading captions are made invisible.

A number of definitions of closed captioning also naïvely imply that *all* sounds in a program are captioned. For example, the Federal Communications Commission (2014) defines captioning as providing access "by displaying the audio portion of a television program as text on the television screen." Wikipedia (2014) similarly defines closed captioning as a "transcription of the audio portion of a program." WhatIs.com (2014) explains that "closed captions are a text version of the spoken part of a television, movie, or computer presentation." Putting aside for now the fact

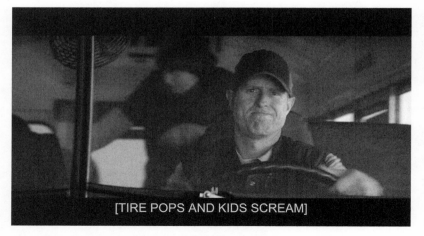

[TIRE POPS AND KIDS SCREAM]

2.1   **Single captions may be packed with implied meaning.**
       A frame from *Man of Steel* shows a school bus driver with pursed lips wearing a baseball
       cap gripping the steering wheel in concentration. A boy stands in the aisle a few rows
       back, head turned away and down. The camera aims straight through the front wind-
       shield. The caption is a two-sounder, discussed in more detail in this chapter: [TIRE POPS
       AND KIDS SCREAM]. When *and* is used as a connector instead of *then*, the logical relationship
       between the two sounds may be obscured. The kids scream because the tire on the bus
       pops. Readers must supply the missing inference. Warner Bros, 2013. Blu-Ray. http://
       ReadingSounds.net/chapter2/#figure1.

that captions should account for much more than just "the spoken
part," these definitions imply that captioning accounts for *every* sound:
[all of] "the speech,"[all of] "the spoken part," "the [entire] audio portion." But it doesn't take more than a cursory listen to any TV show or
movie to recognize that captioning cannot and does not account for
all sounds. It will take longer to recognize that all sounds should *not* be
captioned and that definitions simplify the complex process of manag-
ing sonic uncertainty.

The definition of closed captioning I adopt in this book is decidedly
open ended: *closed captioning provides access to audiovisual content for deaf
and hard-of-hearing viewers*. Because silences sometimes need to be captioned (see chapter 6), definitions should avoid reducing captioning to
sound only and stress instead how captioning is about meaning, not
sound per se. The term "transcription" should also be used with care, as
it may imply an overly mechanical process that any machine or typist
can perform with little training (see Neves 2008, 135). Captioners need
more than a basic familiarity with writing conventions and grammar.

Finally, as a viewer based in the United States, I distinguish between sub-titling and closed captioning, reserving the former for on-screen transla-tions of the spoken language into the reader's written language and the latter for the full complement of sounds, both speech and nonspeech, that need to be made accessible to deaf and hard-of-hearing viewers. Subtitles are burned onto the surface of the video. They can't be turned off, unlike closed captions, which can either be opened for viewing or closed and hidden. In the United States, subtitling usually implies a hear-ing audience that doesn't understand the target language and is in need of on-screen translations. Closed captioning usually implies a deaf or hard-of-hearing audience. A movie may present subtitles and closed cap-tions to the reader simultaneously in two separate tracks (see chapter 8): English subtitles placed at the bottom of the screen for translations of the foreign speech and closed captions placed at the top of the screen for the name of the language being spoken. In *Inglourious Basterds* (2009), for example, the caption (SPEAKING FRENCH) accompanies the English sub-title "What starts tomorrow?" In Europe, the preferred term for closed captioning is Subtitling for the Deaf and Hard-of-Hearing (SDH). While Josélia Neves (2008, 130) suggests that closed captioning and SDH are "quite the same thing," I maintain a distinction between captioning and subtitling because they are distinct tracks in US DVD captioning that place special demands on readers over either captions or subtitles alone.

When it comes to exploring meaning and difference, subtitling and translation theorists have much to offer caption studies. A number of concepts from the research on translating cinematic speech can be trans-lated for caption studies, including domestication and foreignization, "culture bumps," and "abusive subtitling" (e.g. see Chiaro, Heiss, and Bucaria 2008; Díaz Cintas 2009; Díaz Cintas and Anderman 2009; and Nornes 2007). For example, in Italy, a "dubbing country" that "broad-casts more than 350 weekly hours of dubbed programmes," most of which originate from the United States (Antonini 2008, 135), research-ers have explored the "turbulence" and "culture bumps" that are cre-ated at the intersection of the original language (e.g., English) and the dubbed/spoken translation (e.g., Italian). Rachele Antonini (2007, 153) defines "dubbese" as a flattening of language varieties in the course of translation:

It is quite common in Italy to watch a TV episode of *The Practice* or *JAG* (but the same situation applies to all genres of imported fictional products) and hear the characters (e.g., a member of a street gang, his/her lawyer and the judge) use the same sociolect, the same register, and the same vocabulary. This happens because they are speaking *dubbese* (in Italian "doppiaggese"), the language variety that Italian audiences have grown accustomed to hearing on the big and small screens.

Likewise, Diana Bianchi (2008, 184) explores how dubbese reduces linguistic variation and standardizes speech in a case study of *Buffy the Vampire Slayer*, an American TV show, adapted for Italian TV audiences. In particular, Bianchi shows how "Buffyspeak," the variety of slang associated with the show's youth culture and teenage resistance to authority, is "tamed" through a "normalizing drive." The informal slang of Buffy and her friends is formalized when dubbed into Italian. Differences are blurred "between the way in which teenagers and adults speak" (188). The teenagers adopt the "formulaic language which characterizes both adults and vampires" (189). Gender is also normalized as well. The "adaptation radically changes the image of active 'female sexuality' projected in the original lines to one that is closer to traditional sexual images" (191). Bianchi (2008, 192) suggests that the normalizing effects of adapting Buffyspeak can't be explained only by reference to "technical and linguistic constraints." Linguistic standardization and gender normalization are driven by an ideological "resistance towards representing [young people] on [Italian] TV in ways that are too radical" (194). While captioning and dubbing are different processes with different goals, they both involve, as we will see, a flattening of the source material, a tendency to reduce and disembody varieties of language. In the case of captioning, spoken discourse is transformed into standard written English.

"Culture bumps" are felt by readers when "culture-bound elements hinder communication of the meaning to readers in another language culture" (Leppihalme 1997, viii). In her book *Culture Bumps*, Ritva Leppihalme (1997, viii) studies allusions as "one type of culture bound elements in a text" that "depend largely on familiarity to convey meaning." In other words, the meaning of an allusion isn't inscribed in the text but "goes beyond the mere words used." An example is a journalist's allusion to the Cheshire Cat in an article on health care reform (viii–ix), in which the journalist expects the audience to fill in the unwritten meaning of the allusion with their knowledge of the character from Lewis Carroll's *Alice's Adventures in Wonderland*. Translation researchers focus on intercultural bumps—turbulence at the intersection of two languages—but bumps can also be produced intraculturally in closed captioning, when, for example, a captioner doesn't recognize that a five-note melody in *Paul* (2011), a movie that pays homage to 1970s and 80s sci-fi classics, is in fact the well-known five-note motif from *Close Encounters of the Third Kind* (1977). By captioning this musical allusion as (ELECTRONIC MUSIC PLAY-ING), the captioner misses the unwritten reference, the deeper meaning that stitches the notes together (see chapter 7). An intracultural bump is not created in the tension between different languages or cultures but instead between different levels or expectations of cultural literacy in the same (American) culture.

Research on translation and subtitling calls attention to the ways in which, as Abé Mark Nornes (1999, 17) puts it, "all subtitles are corrupt." The practice is corrupt when subtitlers

accept a vision of translation that violently appropriates the source text, and in the pro-cess of converting speech into writing within the time and space limits of the subtitle they conform the original to the rules, regulations, idioms, and frame of reference of the target language and its culture. It is a practice of translation that smoothes over its textual violence and domesticates all otherness while it pretends to bring the audience to an experience of the foreign (Nornes 1999, 18).

Nornes (1999, 18) contrasts the corrupt subtitler who "disavow[s] the violence of the subtitle" with the "abusive translator" who "revel[s]" in that violence. An abusive practice is one "that accounts for the unavoid-able limits in time and space of the subtitle, a practice that does not feign completeness, that does not hide its presence through restrictive rules" (28). Subtitles always do violence to the source text:

Even the subtitles for the most nondescript, realist film tamper with language us-age and freely ignore or change much of the source text; however, corrupt subtitlers suppress the fact of this violence necessitated by the apparatus, while the abusive translator enjoys foregrounding it, heightening its impact and testing its limits and possibilities (Nornes 1999, 29).

This distinction between corrupt and abusive practices is foreign in cap-tion studies. Or rather, discussions of captioning tend to promote a "cor-rupt practice" (without using this term), starting with how definitions and style guides treat captioning as simple and unreflective transcrip-tion (see below). This view undergirds the dream of fully automated cap-tioning, which is based on the assumption that captioning is objective, reducible to clear rules, and only concerned with speech (the "spoken part"). Yet captioning also produces textual violence, transforming the very text it claims to be transparently representing. Caption studies have yet to explore the nature of these changes. They are initiated by the captioner, who is situated between the producers and the needs of caption users. The captioner channels the text but also interprets it, domesticating, standardizing, distilling, and leaving "culture bumps" in her wake. Readers produce interpretations of their own, which captioning scholars have explored through studies of reading comprehension, vocabulary learning, and literacy development (Burnham et al. 2008; Cambra, Silvestre, and Leal 2008/9; Koskinen et al. 1993; Linebarger 2001; Linebarger, Piotrowski, and Greenwood 2010; Ward et al. 2007). Captioning is also shaped by technical and linguistic constraints, just as subtitling is. What Nornes (1999, 28) calls "the unavoidable limits in time and space of the subtitle" also applies to captioning. These "time-and-space constraints" produce what Kristiina Taivalkoski-Shilov (2008, 253) calls a "rather telegraphic form of writing." Captioning is likewise telegraphic. When we consider the potential intersections between translation/subtitling and captioning, we nudge caption studies towards a deeper appreciation of the transformative, even radically interpretative, power of captioning to shape the meaning and experience of the text.

Anatomy of a Sound Description

Closed captioners must account for every significant sound. At a basic level, we can distinguish between speech sounds and nonspeech sounds. The former are captioned as basic text on the screen, akin to subtitles and marked up with standard punctuation, while the latter are placed inside brackets or parentheses to indicate that the words are not spoken but descriptive. Parentheses are the standard notation for nonspeech sounds, but brackets are sometimes used because captioning companies do not agree on typographic standards. In DVD captioning, speech is presented in sentence case, while nonspeech sounds are usually presented in all capital letters. The category of nonspeech also includes information, when relevant, about who is speaking, what language they are speaking, and the manner in which they are speaking, which is why the term "nonspeech information," or NSI, has been offered as an umbrella term for everything that falls outside of the category of captioned speech (Harkins et al. 1995). Table 2.1 divides NSI into a number of types, based on my analysis of DVD captioning in the United States: speaker identifiers, language identifiers, sound effects, paralanguage, manner of speaking identifiers, music, and channel identifiers. In addition to the major types of NSI, a number of conventional tools (also included in definitions of NSI) are used to help readers make meaning:

Table 2.1 Major types of nonspeech information (NSI).

Type of NSI	Description	Examples
Speaker identifiers	Identifies the name of the speaker, usually formatted in DVD captioning using all capital letters followed by a colon and the accompanying speech. This is the classic form, but speakers may also be identified in any other nonspeech caption. Speaker IDs are necessary when it is not clear who is speaking, a speaker is offscreen, etc.	BOND: Have you got him? NARRATOR: This is the island of New Penzance. CROWD: Envy! Envy! [Pat] I don't have an iPod.
Language identifiers	Identifies the language spoken. May be accompanied by a separate, hard-coded subtitle track. In such cases, the language identifier is placed at the top of the screen so as not to interfere with the subtitles at the bottom.	[IN FOREIGN LANGUAGE] [SPEAKING IN ALIEN LANGUAGE] (speaking french) [SPEAKS IN NYANJA]
Sound effects	Includes a wide range of nonspeech sounds. I make a distinction between sound effects and nonspeech sounds that emanate from a speaker's vocal chords (paralanguage).	(RAINDROPS PATTERING) [Plane Passing Overhead] [EGGS SQUEALING THEN POPPING]
Paralanguage	Includes sounds made by speakers that can't or shouldn't be transcribed as distinct speech.	[CROWD SCREAMS] (GRUNTS IN ALARM) [ANGELIQUE LAUGHS] (PANTING) [PAVEL YELLING]
Manner of speaking identifiers	Describes a speaker's significant way of pronouncing words. For the sake of definition, manner identifiers are preceded or accompanied by the speech they qualify.	(WHISPERS) Don't go! (sobbing deeply): She's dead! She's dead . . . (drunken slurring): It's a little late, isn't it? [French accent] With pleasure.
Music	Includes song titles, music notes, music lyrics, and descriptions of music.	(SOUS LE CIEL DE PARIS PLAYING ON GRAMOPHONE) [vivacious, sparkling melody continues] [♪♪♪] ♪Searchin' for light in the darkness♪
Channel identifiers	Identifies the medium of communication such as TV, PA, radio, etc. May be combined with speaker IDs or other NSI.	WOMAN [OVER PA]: Your attention, please. PIENAAR [OVER RADIO]: He's not gonna talk. WOMAN [ON TV]: This is remarkable.

Note: http://ReadingSounds.net/chapter2/#table1.

colons (popular with speaker IDs but also sometimes used with manner IDs), italics (for offscreen speech), parentheses or brackets to signal nonspeech, upper- and lowercase letters (especially all uppercase for speaker IDs), quotation marks, musical notes (usually eighth notes), screen placement to signal which person is speaking, and preceding dashes in lieu of screen placement to distinguish multiple speakers in the same caption (see Harkins et al. 1995, 1). Other tools, while not used in DVD captioning in the United States, are common in live captioning and some television captioning, such as double chevrons (>>) for speaker identification. In the United States, color options are typically limited to user-defined foreground/background combinations, although the HDTV standard does support the option to color individual characters in their bounding boxes (see Clark 2006b). In Europe, multicolored subtitles and animations are common (Rashid et al. 2008, 506). **Media:** http://ReadingSounds.net/chapter2/#NSI-types.

This list of common NSI types is not meant to oversimplify a complex practice in which combinations of different types are routine and necessary. Channel identifiers may be nested inside speaker identifiers in order to name both the speaker and the technology through which she is speaking: "WOMAN [OVER PA]: Security breach. . . ." Multiple sounds from different categories may occupy the same caption: [RAPID GUNFIRE AND MEN SHOUTING IN DISTANCE] and [SPEAKING IN ALIEN LANGUAGE AS WIKUS GRUNTS], both from *District 9* (2009). Sounds may also be qualified by volume level, direction, amount of emotion, speaker, and so on: (WAILING LOUDLY), (RHYTHMIC MUSIC PLAYING IN THE DISTANCE), (CHUCKLING APPRECIATIVELY), (NEYTIRI SPEAKING SOOTHINGLY IN NA'VI). In this last example from *Avatar* (2009), the speaker (Neytiri) and the foreign language (Na'vi) are identified, and the speech is qualified (soothingly), all in a single complex, but not unusual, caption. This caption stands alone, which is to say that it describes foreign speech—a few words spoken by Neytiri to a horselike creature called a Direhorse—while her speech is not accompanied by English subtitles. **Media:** http://ReadingSounds.net/chapter2/#complex-NSI.

Captioners also need to distinguish between discrete and sustained sounds. Discrete sounds are one-off, such as a single cough. Sustained sounds are continuous, such as a fit of coughing. The convention for signaling a discrete sound is the third-person present tense form of the verb: (LAUGHS), (GRUNTS), (COUGHS). The convention for signaling a sustained, continuous, or repeating sound is the present participle form of the verb: (LAUGHING), (GRUNTING), (COUGHING). In *Man of Steel* (2013), for example, [LOIS GRUNTING] is immediately followed in the caption track by [CLARK GRUNTS]. The captioner follows convention: Lois's multiple vocalizations (including panting) are contrasted with Clark's single grunt. Readers learn these conventions, processing them quickly despite their complexity. **Media:** http://ReadingSounds.net/chapter2/#discrete-sustained.

Nonspeech descriptions can be defined as single sound or multisound. Single-sound captions are standard in DVD captioning: One sound, either discrete or continuous, is identified in each nonspeech caption. See the sound effects and paralanguage examples in table 2.1: (RAINDROPS PATTERING), (GRUNTS IN ALARM), and so on. While raindrops technically comprise multiple sounds, they are captioned as a single continuous sound (i.e., the sound of pattering). In contrast, multisound captions describe two or more different sounds, usually occurring simultaneously (connected with "and" or "as"), or in sequence (connected with "then"): [WOMAN SCREAMS AND ALARM WAILS], [TIRE POPS AND KIDS SCREAM], [LEADER SHOUTS IN FARSI THEN GUNS CLICKING], [SPEAKING IN ALIEN LANGUAGE AS WI-KUS GRUNTS]. Two-sound captions are unusual. Of the 1,026 closed captions in *Oblivion* (2013) on DVD, for example, none are multisound captions among the 143 NSI captions. Of the 1,469 closed captions in *Man of Steel*, five are multisound captions among the 411 NSI captions: [TIRE POPS AND KIDS SCREAM], [TIRES SCREECHING AND HORN HONKING], [CAR ALARM WAILING AND HORNS HONKING LOUDLY], [PEOPLE CLAMORING AND SCREAMING], [SU-PERMAN GRUNTS AND NECK SNAPS]. Nearly every multisound caption is a two-sounder, as these examples suggest. The three-sound caption is rare in both DVD and TV captioning, most likely because of the cognitive effort it places on readers. One exception is a nonspeech caption from an episode of *American Dad* ("Pulling Double Booty," 2008): (panicked shouts, screams, alarm ringing). This three-sounder is intended to capture the chaos and cacophony when Haley riots inside a shopping mall after her boyfriend breaks up with her. Another three-sounder is from *Aliens vs. Predator: Requiem* (2007): (sirens, horns honking, gunshots in distance). A final example was mentioned in chapter 1, from the TV show *Grimm*: [dog yelps, whines, goes silent]. **Media:** http://ReadingSounds.net/chapter2/#single-multiple.

At the heart of almost every sound description is a verb form: banging, shouts, growls, ringing, speaks, shouting, wails, honks, and so forth. This verb may stand alone in the caption or be qualified in various ways. For example, a verb can be qualified to the left with nouns: *alien* growls, *horn* honking, *alarm* ringing. A verb can be qualified to the right with adverbs and adverbials that answer questions about manner, location, channel, language, and quality: honking *loudly*, shouts *in Farsi*, chuckling *appreciatively*, rhythmic music playing *in the distance*, Neytiri speaking *soothingly in Na'vi*, rap music playing *over speakers*. A single sound description may thus hit on multiple NSI categories. For example, [MAN SINGING IN FOREIGN LANGUAGE ON RADIO], a caption from *Inception* (2010), would be triple coded for music ("man singing"), language ("in foreign language"), and channel of communication ("on radio"). When absent, a verb may nevertheless still be implied, which is the case with some basic language

ID captions: [IN ZULU], [IN ALIEN LANGUAGE]. In these examples, "speaking" is an implied verb that doesn't need to be made explicit (because speech that follows language IDs is already assumed to be spoken). Just as we don't need to be told that a nonspeech caption describes "sounds," we also don't need to be told that someone is "speaking" in a foreign language. Unless the speaker is doing something other than speaking, such as shouting, the "speaking" can often be left off for the sake of efficiency. Contrast (SHOUTING IN GERMAN) or [PEOPLE SHOUTING IN FOREIGN LANGUAGE NEARBY], which rely on a verb to convey how the foreign words are pronounced. Nouns may sometimes be used in place of verbs to build a foundation for NSI captions, as in "panicked shouts" above. Indeed, sound effects are more likely than other types of NSI to be structured around nouns: [GUNSHOT], [RAPID GUNFIRE], [DIAL TONE], [silence], [APPLAUSE]. But overall, verbs dominate, even in sound effects captioning: [DEFIBRILLATOR CHARGING], [GUN COCKS], [WEAPON POWERING UP], [THUNDER CRASHING]. **Media:** http://Reading-Sounds.net/chapter2/#verb-structure.

The NSI categories in table 2.1 account for every nonspeech caption found in DVD captioning today. (Prerecorded TV captioning is no different, with the exception of a heavier reliance on both scroll-up styling and uppercase lettering for speech captions. DVD captioning, by contrast, relies solely on pop-on style captioning and sentence case speech captions. Because my corpus of searchable closed captioning files is entirely composed of feature-length DVD films, I limit my discussion in this section to DVD captioning.) Just how prevalent is each type? If we double or triple count when necessary to account for single captions that span two or more categories (e.g., a channel identifier nested inside a speaker identifier), or for two of the same category that are placed within a single caption (e.g., two speaker IDs in one caption), we can calculate some rough percentages for four sci-fi action movies: *District 9* (2009), *Inception* (2010), *Man of Steel* (2013), and *Oblivion* (2013) (see table 2.2). I've divided speaker identification into two categories: (1) the classic speaker ID set in all capital letters followed by a colon, and (2) all other names that are used in sound descriptions to identify speakers. This second category includes proper names as well as generic placeholders such as "all," "both," "people," and so forth. Coding NSI captions is complicated by the fact that meaning is dependent on a context that is removed when captions are dumped into an Excel spreadsheet. The sound of groaning, for example, usually denotes paralanguage—a sound emanating from human vocal chords but not reducible to distinct speech. But animals groan too, sometimes. Consider [H'RAKA GROANS], which I coded as a "sound effect" and not as "paralanguage" because H'Raka is the dragon that Jor-El (Russell Crowe) rides at the beginning of *Man of Steel*. Dragon sounds, like other animal sounds such as a dog barking, were coded as sound effects. Similarly, growling is typically associated with animals. But humans growl too, sometimes. Consider [ZOD GROWL-ING], also from *Man of Steel*. I coded this sound as paralanguage because while Zod, the evil warlord from Krypton, is not technically human, he appears as a human to Superman, and he's played by an actor (Michael Shannon) who doesn't wear any significant makeup or prosthetics in the film. In contrast, I coded [ALIEN GROWLS] in *District 9* as a sound effect on the assumption that an alien growling is similar to a dog growling. Note that the origin of the sound matters little. The same sound can be interpreted (coded and captioned) differently depending on the visual context, a phenomenon in film sound that Michel Chion (2012) calls synchresis. The coding process was also most sensitive to captions that were accompanied by notation particular to nonspeech captions. The software program I used, discussed in chapter 1, looked for punctuation associated with NSI captions, including colons, all-caps styling, paren-theses, brackets, and music notes. Usually, captioned lyrics are accom-panied by musical notes, but there is one exception in *Man of Steel*, a captioned lyric that wasn't picked up by my software program because no symbols identified it as nonspeech. I happened upon this lyric manu-ally. In short, this process of placing captions into categories, while it aimed for some level of objectivity, can neither be performed outside of the contexts in which the captions occur nor eliminate the interpretative work required to map captions onto categories. Nevertheless, the calcu-lations in table 2.2 remain valuable, I would argue, because they give us a rough breakdown of the different functions of NSI, which have never been offered by captioning advocates. **Media:** http://ReadingSounds .net/chapter2/#contextual-coding.

The statistical differences among these four movies are striking. Some differences can be explained by the movies themselves. For ex-ample, *District 9*, set in Johannesburg, relies heavily on foreign speech in multiple languages, including Afrikaans, Zulu, and the alien's own unique language. Hence the need for a large number of foreign language identifiers (roughly 10 percent of the NSI captions in *District 9* include a language identifier). When it isn't clear in polyglot movies like *District 9* that people are speaking English, even English speech sometimes needs to be identified for caption readers of English language DVDs. Four of the sixty-one language identifiers in *District 9* refer to English, including:

Table 2.2 A comparison of nonspeech information across four sci-fi action movies in DVD format

Type of NSI	District 9 (2009)	Inception (2010)	Man of Steel (2013)	Oblivion (2013)
	1,680 closed captions 680 NSI captions NSI: 40.5 percent of CC 1:52:16 runtime	1,741 closed captions 314 NSI captions NSI: 18 percent of CC 2:28:07 runtime	1,469 closed captions 399 NSI captions NSI: 27.2 percent of CC 2:23:02 runtime	1,026 closed captions 141 NSI captions NSI: 13.7 percent of CC 2:04:42 runtime
Classic speaker identifiers	516 (75.7 percent)	160 (51 percent)	222 (55.6 percent)	56 (39.7 percent)
Other references to names	51 (7.5 percent)	31 (9.9 percent)	49 (12.3 percent)	6 (4.3 percent)
Foreign language identifiers	61 (9 percent)	2 (0.6 percent)	4 (1 percent)	0
Sound effects	37 (5.4 percent)	80 (25.5 percent)	73 (18.3 percent)	27 (18.4 percent)
Paralanguage	90 (13.2 percent)	71 (22.6 percent)	108 (27.1 percent)	43 (30.5 percent)
Manner of speaking identifiers	2 (0.3 percent)	1 (0.3 percent)	1 (0.2 percent)	11 (7.8 percent)
Music	32 (4.7 percent)	11 (3.5 percent)	1 (0.2 percent)	5 (3.5 percent)
Channel identifiers	51 (7.5 percent)	3 (1 percent)	17 (4.3 percent)	3 (2.1 percent)

Note: Each percentage refers to the proportion of that film's NSI captions that correspond to each category. The number is the raw number of NSI captions in that category. Percentages will not add up to 100 percent because single captions can be tagged in multiple categories. http://ReadingSounds.net/chapter2/#table2.

"[IN ENGLISH] Does it look like a hamburger shop, eh?" *Oblivion* includes no foreign speech (and hence no language identifiers), while *Man of Steel* includes only four language identifiers (captions 707–710). All four identifiers are identical, generic, and occur consecutively over thirteen seconds: [ZOD SPEAKING IN FOREIGN LANGUAGE ON TV]. During this sequence, General Zod addresses different groups around the world in their own languages. The specific languages are not identified, which, as I will argue in more detail later (see chapter 8), flattens the soundscape, reducing multilingual complexity to a generic placeholder.

Other differences in these four movies can also help to account for the wide disparity in NSI numbers. For example, *District 9* is set in a densely packed urban city, including a sprawling alien refugee district, whereas *Oblivion* is set in a decimated postapocalyptic landscape where few humans have survived an alien invasion. No wonder *District 9* requires a much larger number of speaker IDs (a whopping 76 percent of NSI to *Oblivion's* 40 percent). *District 9* also contains a large number of mediated or channeled speech captions to account for characters' reliance on a range of technologies to deliver speech and music: speakers, monitors, PA systems, TVs, radios, microphones, phones, headsets, and an answering machine. Forty-five of the fifty channel identifiers in *District 9* are nested inside speaker IDs, as in "MAN [OVER RADIO]: All civilian boats, stand clear." Of the remaining five, four are identical—[RAP MUSIC PLAYING OVER SPEAKERS]—while the fifth is a speech caption qualified with the name of the mediating technology: "[ON MONITOR] Just untie that guy. Untie that guy, sir."

But appealing to the content of the movies themselves will not fully account for the statistical differences among them. Every action movie relies heavily on instrumental music to create mood and maintain suspense. But only *District 9* makes use of musical notes. In fact, twenty-six of the thirty-two captions coded for music in *District 9* are lone music note captions: [♪ ♪ ♪]. Before we consider whether musical notes by themselves are effective in conveying meaning, we should ask why only one of the four movies used them. Indeed, the *Man of Steel* caption track gives readers the impression that there is no instrumental or nondiegetic music in the movie. The only music caption is: "WOMAN [SINGING]: Bound by wild desire," which describes live music playing inside a bar. The caption that follows this one is a music lyric, but it is not accompanied by any musical notes, potentially causing confusion. *Inception* includes eleven music captions, ten of which are associated with the crucial piece of diegetic music that alerts the characters to prepare for the "kick" that will drop them out of the dream space and back into reality: ["NON, JE NE REGRETTE RIEN" PLAYING]. For another notable difference, contrast the low percentage of sound effects captions in *District 9* (5.4 percent) with the percentages in the other three movies (20.7 percent sound effects on average). In raw numbers, *District 9* only includes thirty-seven sound effects captions to eighty and seventy-three for *Inception* and *Man of Steel*, respectively. These differences can't be easily explained by appealing to the inherent differences among the films themselves. *District 9* is not more or less likely to rely on sound effects than other action movies.

Beyond these differences are a number of broad similarities that cut across all four examples. First, speaker IDs are the most popular type of NSI. In three of the four movies, classic speaker IDs occur twice as often as any other type of NSI. That shouldn't surprise us. According to Judith Harkins et al. (1995: 12), "Speaker identification is one of the most important categories of non-speech information, because problems in identifying the speaker are frequent and can cause confusion." Sound effects and paralanguage are the next most popular categories in all four movies, beating out the other categories by a usually wide margin. The consistency between *Man of Steel* and *Oblivion* in the category of sound effects is noteworthy, as these two movies are separated by only 0.1 percent (at 18.3 percent and 18.4 percent respectively). Finally, we might be surprised by the very low number of music and manner captions in the four movies, given that popular movies rely so heavily on background music to provide ambience and modulate mood. In addition, every person has a unique way of pronouncing words, making manner of speaking always potentially relevant and yet rarely indicated in the captions. But according to Raisa Rashid et al. (2008, 505), "Closed captioning currently provides only the verbatim equivalent of the dialogue and ignores most nonverbal information such as music, sound effects, and intonation. Much of this missing sound information is used to establish mood, create ambiance, and complete the television and film viewing experience." While Rashid et al.'s claim is reinforced by the small number of music and intonation (manner) captions in the four movies in table 2.2, their claim doesn't explain the significant number of sound effects captions in these movies, particularly in *Inception* and *Man of Steel*. Regardless, now we have some statistics to support or challenge claims about the prevalence or lack of certain kinds of nonspeech captions. More work along these lines is needed.

When we extract the NSI captions from a closed captioning file and code them by hand in a spreadsheet program using the major categories of NSI, we can visualize patterns that might be otherwise difficult to see. When codes are conditionally formatted in an Excel spreadsheet, patterns (or apparent patterns) reveal themselves as streams and rainbows of color. For example, I set up Microsoft Excel to highlight sound effects in yellow, music captions in green, paralanguage in red, and so on. When individual captions were double- or triple coded (e.g., sound effect and paralanguage together in [GUNFIRE THEN EAMES GRUNTS]), I added additional coding cells above or below the caption in the coding column. Usually, sound effects and paralanguage captions tend to alternate randomly in coded files of NSI captions. In *Inception* (2009), however, randomness gives way to seeming order in a couple key places. *Inception* is a sci-fi heist movie about highly skilled corporate thieves who are hired to plant information in the subconscious mind of their target. Through "shared dreaming" and multiple dream levels (dream within a dream), the team sets out to plant an idea so deep in the target's mind that he believes it was his own idea. The target "gives himself the idea," because as Eames (Tom Hardy) puts it, "That's the only way it will stick. It has to seem self-generated." In every dream sequence in *Inception*, a song—Édith Piaf's "Non, Je Ne Regrette Rien"—is played to warn the participants of an upcoming "kick" that will wake them up or, in the case of a dream within a dream, elevate them to the next higher dream level. During the second dream sequence in which Cobb (Leo DiCaprio) begins to train Ariadne (Ellen Page) on the rules of the dream space, two music captions—["NON, JE NE REGRETTE RIEN" PLAYING] and [MUSIC STOPS]— are preceded by seven sound effects captions that, while they are interrupted by speech captions, are not interrupted by any other type of NSI (with the exception of speaker IDs). Following the music captions are four additional, uninterrupted sound effects. The pattern—seven sound effects captions, two music captions, and four additional sound effects— stands out against a background of coded randomness:

[PHONE RINGS]

[DIAL TONE]

[KNOCKING ON DOOR]

[HELICOPTER STARTING]

[BELLS TOLLING]

[CUP RATTLING]

[RUMBLING]

["NON, JE NE REGRETTE RIEN" PLAYING]

[MUSIC STOPS]

[MACHINE HISSING]

[RUMBLING THEN METAL GROANING]

[METAL CREAKING]

The build up of sound effects around the music captions allows us to see the action in the movie in a new way, through the lens of the closed captions. Granted, this particular dream sequence is only associated with the final two of the first seven descriptions that make up the first sound effects block: [CUP RATTLING] and [RUMBLING]. Nevertheless, when we extract NSI from speech captions and code them, we become more sensitive to potential patterns that are easy to miss otherwise. The same patterning persists during the next captioned reference to the "kick" music, this time with a band of five sound effects captions preceding one music caption for Piaf's song. A third notable band of sound effects captions—six in a row—occurs during the transition from the second-level dream to the third-level dream, with some cutaways to the van taking on gunfire in the first-level dream. The dream space is violent, volatile, and frenetic—no wonder sound effects dominate during the scenes when characters are collaborating inside a shared dream. But an analysis of NSI captions can assist us in revealing that volatility, demarcating the dream space, and reinforcing a possible hierarchy in which characters are at the mercy of external forces when they enter a subject's dream. The movie explains the violence inside the dream world in terms of the subconscious of the subject reacting to the presence of intruders by trying to kill them. ("The subject realized he was dreaming and his subconscious tore us to pieces.") These external forces are expressed as a preponderance of sound effects captions. **Media:** http://Reading-Sounds.net/chapter2/#inception.

Coding NSI captions can also help us identify and analyze recurring verbs in sound descriptions. Repetition of the same or similar captions, especially in a short duration, may signal the captioner's overreliance on certain verbs at the expense of a more careful engagement with the soundscape. For example, in *Man of Steel*, General Zod's initial attack on Metropolis is accounted for, in part, by four consecutive nonspeech captions (numbers 1185–88) that describe the crowd's curious and then panicked reaction to the terraforming "world engine": [ALL SCREAMING], [ALL SCREAMING], [ALL CLAMORING], [ALL SCREAMING]. These four consecutive captions span seventeen seconds, during a time when the terraforming ships begin wreaking havoc on the planet, turning the Earth's gravity, mass, and atmosphere into that of Krypton's. Two massive beams of light and energy repeatedly send destructive pulses into the surface of the earth. The scene is visually stunning, complete with sounds of annihilation and an ominous electrical buzzing coming from the terraforming energy pulse. While captioners need to ensure that readers have time away from reading to take in the awe and spectacle of this scene, they should also be careful not to reduce such an impressive sonic cacophony to a copy-and-paste caption, a boilerplate or default placeholder that risks doing disservice to the movie's sonic complexity. Yes, people are screaming and running. But so much else is going on visually and sonically that it seems unfair to include only the sound of people running away from the pulse in a panic. Do we need to be told every time we see people fleeing that they are screaming? Or can we be alerted to their screams the first time only, freeing up some space and time on the caption track for a description of what are arguably more significant sonic experiences for listeners during this scene: the sounds of awesome destruction and the unsettling sounds of the massive energy pulse?

When we step back from this scene of captioned repetition, we find that "scream/ing" is a fairly popular descriptor in the nonspeech captions of *Man of Steel*, occurring twenty times over 108 paralanguage captions. In other words, over 18 percent of the paralanguage NSI captions refer to "scream/ing," including five consecutive doubles—for example, [SCREAMING], [SCREAMING]—and one consecutive triple: [SCREAMING], [PEOPLE SCREAMING], [ALL SCREAMING]. These consecutive NSI captions may be interrupted by speech captions, but they are not interrupted by any other sound descriptions (not including speaker IDs). Why should it matter how many times "screaming" is included in the NSI captions if screaming is an accurate representation of the sounds people are making? It matters because it may signal that some sounds have been privileged by the captioner at the expense of others, as in the terraforming scene. More broadly, it matters because when we pay attention to patterns, repetitions, and the popularity of certain descriptors in NSI captioning, we can better evaluate them. Regular caption readers are accustomed to seeing the same paralinguistic descriptors over and over: grunting, laughing, sighing, screaming, (indistinct) chattering, and so on. They may have even experienced a particularly vivid scene, such as a climactic rescue scene, reduced to a mundane series of boilerplate descriptors. At the end of *Man of Steel*, for example, as two minor characters are trying to rescue a woman trapped in the concrete rubble created by the terraforming world engine, three consecutive, identical nonspeech captions describe their efforts: [LOMBARD & PERRY GRUNTING], [LOMBARD & PERRY GRUNTING], [LOMBARD & PERRY GRUNTING]. This repetition, like the repetition of [ZOD SPEAKING IN FOREIGN LANGUAGE ON TV], threatens to deflate the scene's intensity, because it fails to make room, once again, for the destructive sonic force of the energy beam quickly approaching Lombard, Perry, and the trapped woman. **Media:** http://ReadingSounds.net/chapter2/#man-of-steel.

How prevalent are common paralinguistic descriptors such as grunt/ing? Which verbs are most popular in paralinguistic NSI? Table 2.3 lists the most common verbs for each of the four movies discussed in this section. The data show a preponderance of "grunt" and "scream" paralanguage captions across all four movies. Both verbs were the top choices in *District 9* and *Man of Steel* and the top and third choices in *Inception* and *Oblivion*. Grunt and scream comprise, on average, 41 percent of the paralinguistic NSI in these four movies. Put another way, close to half of all paralanguage nonspeech captions in these movies will include some reference to grunting or screaming. That's a remarkable percentage, one that, I would argue, is easy to miss when viewers are caught up in the regular flow of a captioned movie experience. The preponderance of grunts and screams becomes palatable when we extract and code NSI. These two verbs serve as catch-all terms in the four movies, standing in for a range of emotional outburst sounds (pain, yelling, exertion, groaning, panting). The close agreement among the four

Table 2.3 The most popular verbs used to describe paralinguistic sounds in four sci-fi action movies.

District 9 (2009)	Inception (2010)	Man of Steel (2013)	Oblivion (2013)
Total paralinguistic NSI captions: 90	Total paralinguistic NSI captions: 71	Total paralinguistic NSI captions: 108	Total paralinguistic NSI captions: 43
Grunt: 24	Grunt: 20	Grunt: 28	Grunt: 11
Scream: 15	Yell: 9	Scream: 20	Chuckle: 8
Laugh: 12	Scream: 6	Clamor: 9	Scream: 6
Cough: 8	Cough: 7	Groan: 5	Gasp: 5
Shout: 6	Shout: 7	Laugh: 5	Pant: 4
Speak: 6	Gasp: 3	Murmur: 5	
Cry: 4	Groan: 3	Speak: 5	
Sob: 4	Laugh: 3	Chatter: 4*	
Chatter: 3	Chatter: 2	Chuckle: 4	
Gasp: 3	Giggle: 2	Gasp: 4	
Yell: 3	Scoff: 2	Pant: 4	
	Sigh: 2	Cough: 3	
		Sigh: 3	
		Coo: 2	

Note: *3 of the 4 uses of "chatter" in Man of Steel are nouns, not verbs.
All verb forms are counted in the results. For example, [grunt], [grunts], and [grunting] were included in the same category ("Grunt"). Only verbs used more than once in each movie are listed. If a verb was also used for sound effects captioning (e.g., "metal groaning"), it was not included in the totals. http://ReadingSounds.net/chapter2/#table3.

movies is offset by some significant differences, starting with the contrast between *Oblivion* and the other movies in the number and variety of paralanguage NSI. Granted, *Oblivion* features only a handful of characters because an alien invasion has annihilated nearly the entire population. Still, the lack of variety is noteworthy and warrants further study from captioning advocates. How much variety should we expect? Do default placeholders such as [GRUNTING] take precious caption space away from other significant sounds, including sound effects? Do common descriptors ever start to feel stale and overused? Do captioners sometimes lean more heavily on certain stock terms as a result of pressure to get the job done? More research is needed using a much larger corpus of caption files. I've already hinted at my own position, namely that variety in NSI speaks to the captioner's creativity and is more likely to reflect the captioner's closer engagement with the soundscape than NSI captions that rely more heavily on a few standbys or boilerplate captions. At the same time, it is important to remember that the raw data only tell part of the story. One way to qualify or modulate action words at the heart of an NSI caption is with trailing adverbials. Qualifiers can help embody and clarify meaning and thus distinguish captions that share the same verb. For example, *Oblivion* includes three nonspeech captions that are qualified with adverbs: (EXHALES SHARPLY), (CHUCKLES SOFTLY), and, from the sound effects list, (WHIRRING LOUDLY). None of the (GRUNTS) are qualified, however, while the (SCREAMING) captions are only qualified with identifiers, such as (ALL SCREAMING) and (JULIA SCREAMING). **Media:** http://ReadingSounds.net/chapter2/#paralanguage.

Captioners are fully responsible for creating nonspeech descriptions. They do not import descriptions so much as invent them. Matt Seidel (2014), a freelance closed captioner, tells a story of how he decided to caption the "deafening clamor of birdsong":

After being inadvertently zapped by a gamma ray, a young woman finds herself acutely attuned to the sounds of the surrounding forest. She pauses, stunned by the deafening clamor of birdsong. I paused the video and thought about how best to convey her newly enhanced auditory perception of the natural world, toggling between [HEIGHTENED WARBLINGS] and the Keatsian [FULL-THROATED AVIAN SINGING]. Suddenly aware of how ridiculous either would look on screen, I settled on the distinctly un-poetic [LOUD BIRD NOISES] and went on about my business. I had already wasted enough time, and who knew what other agonizing choices lay ahead?

Seidel reminds us that captioners must balance creative description with accessibility. There are limits to flowery language. Captioning is not poetry. It is not writing for its own sake. But I can't help but feel a little underwhelmed at Seidel's decision to go with "noises" to characterize the active sound of birdsong (and not simply because "noises" is a noun instead of a verb). "Loud" is also vague, qualifying the meaning only by volume. Of course, before passing final judgment, I would need to have more information about the context. Knowing little else about this example, I can only wonder about the vast excluded middle between "full-throated avian singing" and "loud bird noises." Searching my corpus of DVD caption files, I find quite a few references to bird sounds that Seidel might find "ridiculous" but that nonetheless aim to be more descriptive than "noises":

(birds squawking in distance)—*Aliens vs. Predator: Requiem* (2007)
(plaintive birdcall echoes from distance)—*Aliens vs. Predator: Requiem* (2007)
(BIRDS CAWING)—*Avatar* (2009)
[BIRD CHEEPING]—*Beasts of the Southern Wild* (2012)
[BIRD CHURRING]—*Beasts of the Southern Wild* (2012)
(BIRDS TWITTERING)—*Oblivion* (2013)

(BIRDS CHIRPING)—*Skyfall* (2012)

[flutes fluttering birdsong]—*The Artist* (2011)

(BIRD SCREECHES IN DISTANCE)—*Zero Dark Thirty* (2012)

Missing from "noises" is a more nuanced sense of the quality of the bird-song. If the sounds are significant enough to caption, then they probably need to be presented as more than vague noise. The differences among birdsong sounds—cheeping, squawking, churring, chirping, twittering, cawing, and screeching—collapse under the generic weight of "noises." It's easy to forget that birdsong serves a number of functions. In his groundbreaking contribution to sound studies, *The Soundscape*, R. Murray Schafer (1977, 33) lists a number of functions of birdsong, including calls of pleasure, distress, territorial defense, alarm, flight, flock, nest, and feeding. These functions are lost on most listeners, for "[o]nly those who live close to the land can distinguish birds by the sounds of their wings in flight. Urban man [sic] has retained this facility only for insects and aircraft" (33).

Should Seidel have identified the function of the birdsong in addition to or instead of describing its musical qualities? I don't have enough information to know. What I do know, and will argue in more detail throughout this book, is that *function* plays a crucial role in captioning. While there is a need for phonetic spellings in certain contexts—for example, Superman yelling "Ragh!" in *Man of Steel* in a supportive visual context in which the meaning of his angry scream is clear—the need to describe the function of a sound usually trumps any need to provide an irregular phonetic spelling. We need to know what the sound means, not just how it sounds. This claim takes us well beyond the standard pronouncements about closed captioning as a "transcription" or a "text version" of the soundtrack. Readers are better equipped to interpret the meaning of birdsong when it is described as cheeping, for example, than simply as loud. Cheeping also has the advantage of being onomatopoetic. The word itself is a description of the sound: the parakeet goes "cheep," as the children's book might say. We don't know if a cheep is a pleasure call or distress call. But we will probably have a better sense of the size of the bird, and perhaps even its age and the name of the bird family, when a sound is described as a high-pitched cheep, which might suggest a small, weak, or young bird such as a sparrow, as opposed to a squawk, which might suggest the landing call of a large bird such as a Great Blue Heron (Cornell Lab of Ornithology 2014). **Media:** http://ReadingSounds.net/chapter2/#birdsong.

At the risk of getting caught in the weeds—bird-watching herons, you might say—I wa[...] [...]l (2014) to raise some questio[...] [...]t more importantly, I wante[...] [...]respon- sibility are placed in[...] [...]sounds are significant and h[...] [...]ited to nonspeech sounds, si[...] [...]es deci- sion making. The cap[...] [...]r's edi- torial oversight, serve[...] [...]ducers and the millions of pe[...] [...]of the perspectives offered in[...] [...] types and the close-up shot[...] [...]er. In the remaining chapter[...] [...]ber of places, but I do so in o[...] [...]nd up and because, quite frar[...] [...]kinds of close analyses that a[...] [...] beyond the dominant view of captioning as simple transcription.

[Handwritten note:] "captioning is about meaning, not sound per se." p. 34
Subtitles + closed captions p. 35
culture bumps - 36
p. 45 - woman singing
46 - background music - mood

Captioning Style Guides and Prescriptions

Style guides for closed captioning are plentiful on the web. They run the gamut from informal suggestions posted to personal websites to formal documents written by large captioning firms. Style guides cover roughly the same ground, even if they do not agree on specific guidelines (on the problem of standardization, see Clark 2004). My own analysis of the style guides is limited to what is publicly available on the Internet. I am keenly aware that captioning vendors and individual captioners make use of a wide range of learning and training methods that are not typically available to the outside researcher (e.g., in-house style manuals, formal training sessions, informal conversations, etc.). However, my goal is not to provide a comprehensive review of in-house captioning style guides but rather to sample briefly a small number of authoritative, publicly available guides in order to suggest something about the contours of these texts and the terrain they cover. This review will allow me to situate this book's own approach to closed captioning alongside of, and in some cases in opposition to, the information that is currently available on captioning style. My review is based on the following style guides:

· *Captioning Key for Educational Media: Guidelines and Preferred Techniques* (DCMP 2009, 2011). A major resource on captioning style, and one I will refer to repeatedly in this book, this guide is authored by the Described and Captioned Media Program (DCMP), which is funded by the US Department of Education and administered by the National Association of the Deaf. The DCMP maintains a list of approved captioning vendors (including CaptionMax, National Captioning Institute, The Media Access Group/WGBH, and VITAC). *Captioning Key* mandates style guidelines for vendors who produce captioned content for the DCMP/ Department of Education. *Captioning Key*'s style guidelines are based on "captioning manuals . . . from major captioning vendors in the United States" (DCMP 2009, 2).

- *The CBC Captioning Style Guide* (Canadian Broadcasting Centre 2003). The Canadian Broadcasting Centre released this in-house captioning style guide in response to an information request filed by Joe Clark (2008), a well-known expert on closed captioning.
- Gary D. Robson's *The Closed Captioning Handbook* (2004). Chapter 3 of Robson's book covers "Captioning Styles and Conventions." Other chapters of Robson's book focus on caption timing and caption placement, two issues that are usually discussed as part of captioning style.
- WGBH's "Suggested styles and conventions for closed captioning" (Media Access Group 2002a, 2002b). Caption technology for television was developed at WGBH, a PBS station in Boston, in the early 1970s (Earley 1978). The first captioning agency—The Caption Center—was established at WGBH in 1972 (Robson 2004, 10). WGBH has arguably been thinking about captioning style longer than any other agency, thus making their style recommendations of particular value.

Standard topics in the style guides include: methods of captioning (pre-recorded or live), styles of captions (pop on or roll up), accuracy, verbatim vs. edited captions, screen placement, line breaking, timing, typeface and type case, grammar and punctuation, and nonspeech information (speaker IDs, music, sound descriptions). Style guides are light on theory; individual guidelines are typically offered up as truths in no need of justification. While readers should not necessarily expect style guides to provide a lengthy explanation of each best practice, the lack of good reasons is troubling for those practices that seem counterintuitive, such as the requirement to style captions in upper case letters only, using centered alignment, in the shape of an inverted pyramid, in no more than two rows, with new sentences always starting on a new line, and with speaker IDs and sound descriptions set in mixed case.

Of these requirements, the all-caps guideline is undoubtedly going to be the most troubling and confusing to rhetoricians, document designers, and typographers. While an all-caps style may have been necessary in the early days of TV captioning, it is unnecessary today and at odds with the most basic rules of good typography. With printed texts, lowercase letters are "more legible than those in upper case" (Kostelnick & Roberts 1998, 144). The same is true for electronic letters produced in the high resolution TV and Web environments of today. Joe Clark (2008) refers to the preference for uppercase caption styling in his review of *The CBC Captioning Style Guide* as "1980s nonsense." Uppercase styling is "a mistake, an archaism. It only ever came about because the original decoders' fonts were so lousy that all-upper-case captioning was deemed *less illegible* than mixed-case. CBC doesn't even know the reasons why it is using capital letters, or that such reasons are no longer in effect" (Clark 2008). Whereas CBC offers no explanation for why offline captions need to be set in uppercase style ("All text shall be presented in upper case, except for . . ." [CBC 2003, 9]), WGBH offers conflicting advice about type case. Examples in WGBH's captioning style guide are set in upper case, with lower case reserved for nonspeech information and speaker IDs. But WGBH's style guideline for type case stipulates that "caption text is generally rendered in uppercase and lowercase, or sentence case, Roman font" (Media Access Group 2002b). Readers are thus presented with a WGBH guideline that WGBH itself does not follow, a guideline that seems better suited to a preweb, low resolution, analog world. Moreover, if an all-caps style is interpreted as screaming for viewers who grew up with texting and instant messaging conventions, then how can upper case captions convey whispering?

(whispering):
PLEASE OPEN THE DOOR!

In this example from WGBH's (2002b) style guide, a screaming uppercase style comes into conflict with the intent to convey the opposite of screaming. Uppercase captions, in addition to being less legible than standard sentence case, are troubled by an association between all-caps and screaming. Is it even possible, in a postweb world shaped by email and texting conventions, to whisper in all-caps? For that matter, is it possible to whisper with an exclamation point? This example from WGBH is thus troubling for multiple reasons and begins to raise serious questions about the continued reliance on all-caps styling. While I have become accustomed to reading and identifying speaker IDs in all-caps (e.g., "JACK: Mission, we have a situation"), I am not able to justify or understand why only closed captions, among all of the types of extended copy we read daily on digital screens, are set in all capital letters.

A second issue likely to be of particular interest to rhetoricians and language experts is verbatim vs. edited captioning—whether it is possible to caption every word of dialogue and every audible sound and whether captioners should edit captions for certain readers (i.e., children), to meet maximum presentation rates, or to clean up a speaking style marked by presumably irrelevant verbal fillers like "um" and "uh." The first regularly captioned show on television—a nightly rebroadcast on PBS stations of *The ABC News* with open captions—was edited for content and for reading speed (Earley 1978). The producers at WGBH who created the open captioned version of *The ABC News* decided to edit the audio content for two reasons:

Captioners at WGBH recognize two needs: (1) to reduce the amount of text in the average television program (roughly 180 words per minute) to allow time to read the captions and watch the program video, and (2) to adjust the language itself so that comprehension of the captions can be rapid and accurate (Earley 1978, 6).

Today, reading speed continues to guide decisions by captioners to edit content. According to WGBH, "[A]ny editing that occurs nowadays is usually for reading speed only." The authors of *The Captioning Key* agree and have specified maximum presentation rates for captions (ranging from 130 words per minute [wpm] for children to "near verbatim" or 235 wpm for theatrical productions for adults [DCMP 2009, 12]). Some deaf and hard-of-hearing viewers may benefit from slower caption presentation speeds: "Where reading speed data are available, they suggest that the reading speeds of deaf and hard-of-hearing viewers are typically slower than those for hearing viewers" (Burnham et al. 2008, 391). At the same time, deaf and hard-of-hearing viewers have made it clear that they prefer verbatim or near-verbatim captioning because they want the same access as hearing viewers. According to Pablo Romero-Fresco (2009, 111–12), "the reason is not financial, but political. There is among these viewers a great deal of sensitivity and antagonism towards the idea of editing." Romero-Fresco (2009, 112) makes a distinction between "scholars who support editing to provide full access for the deaf" and deaf viewers who "line up with broadcasters to push for verbatim, which may not give them full access after all." Verbatim is cheaper to produce than edited captions, which explains why European broadcasters would push for it (Romero-Fresco 2009, 111). Reading speed is perhaps the most political and contested topic in caption studies. Reading speed guidelines also directly impact a captioning company's bottom line, because verbatim captions/subtitles "require less effort" and "are thus more economical" (Romero-Fresco 2009, 111). Regardless, the original practice of adjusting the language of captions for comprehension is no longer advocated in the North American captioning manuals. The DCMP equates editing with censorship: "Extreme rewriting of narration for captions develops problems, such as 'watered-down' language and omitted concepts. Language should not be censored" (DCMP 2009, 1). With the exception of CBC's style guide, which defines verbatim captioning as "difficult" to achieve (CBC 2003, 7), the style guides embrace verbatim captioning as standard practice that only presentation rate has the power to alter: "Editing is performed only when a caption exceeds a specified rate limit" (DCMP 2009, 12).

Captioning scholars, especially in Europe, continue to debate the question of edited vs. verbatim captioning (e.g. see Schilperoord, de Groot, and van Son 2005, Szarkowska et al. 2011, Ward et al. 2007). For example, Szarkowska et al. (2011, 363) begin their eye-tracking study with a pronouncement that "one of the most frequently recurring themes in captioning is whether captions should be edited or verbatim." The question remains relevant because some deaf people have been found to be "slower readers" (Szarkowska et al. 2011, 364). Josélia Neves (2008, 136) goes so far as to call the preference for verbatim captioning a fallacy. She offers at least three reasons. First, "the often mentioned fact [is] that deaf adults tend to have the reading ability of a nine-year-old hearing child." Second, it is more important to have subtitles that are "enjoyable to read, easily interpreted and unobtrusive" (136). Caption readers should have enough time to "tak[e] in the whole audiovisual experience" (136). Third, caption readers do not "make up one cohesive group," and "hard of hearing, deaf and Deaf viewers are, in reality, different audiences who may require different subtitling solutions" (131). This third reason resonates most strongly with me, particularly given the tendency among some captioning advocates to group everyone together in the name of universal design. We need to be reminded that audience members are different and have different needs. Audience analysis is central to rhetorical studies, after all. A verbatim solution may not always be the best choice. Nevertheless, verbatim captioning remains the method of choice in the United States for adult programming.

When speech is edited across multiple captions as part of a general captioning strategy, it becomes obvious right away to US listeners raised on or expecting verbatim captioning. US listeners are used to reading along with verbatim captions. When speech is not mirrored in the captions, the disconnect between what one hears and what one reads can pull listeners out of the narrative. *The Internship* (2013), starring Vince Vaughn and Owen Wilson, contains numerous examples of edited and summarized speech captions. Vaughn is a notorious fast talker in his films, a slick ladies' man and smart-ass who charms with his silver tongue. Editing tames Vaughn's high-velocity speech. Consider a scene in which Billy McMahon (Vaughn) is trying to motivate his teammates at halftime by invoking the 1980s movie *Flashdance*. I've italicized the words that were deleted from the captions:

This reminds me of a *little* girl who had to *get her head right and* start believing./ A *little* girl from a steel town who had the dream . . . / . . . to dance./ No one believed in that *little* welder girl, but *thank god* she believed in herself.

By removing six words from the first sentence, the captioner reduces the reading speed from 385 to 249 words per minute (wpm). In the third sentence, three words are excised ("little," "thank god"), bringing the reading speed down from 271 to 217 wpm. What is lost or gained, if anything, by stripping some of the color from Billy's description of the "little girl" in *Flashdance* who believed in herself, "thank god"? While the edited version preserves the basic meaning, some of the absurdity and chauvinism of the story is lost. Captioners also make substitutions when editing for reading speed, as when Billy is working the call-center phones at Google and tells a customer: "Here's what I want to do. I want to help you clear that up before the wife gets home—am I making sense?" The caption replaces the first fourteen words with three words: "Let's fix that before the wife gets home—am I making sense?" As a result, the reading speed drops dramatically from about 401 to 209 wpm. The edited version gets to the point much faster, but at the cost of fully conveying Billy's slick salesman vibe. (Billy's "Here's what I want to do" sounds like a version of the car salesman's "Here's what I'm going to do for you today.") When speech is summarized rather than rendered verbatim, captioners need to be careful not to strip the subtlety, color, and nuance from speech, especially in comedies where wordplay and creative turns of phrase not only enliven the dialogue but constitute the genre of comedy itself. *The Internship* is one of the most speech-intensive movies in my collection of DVDs. Whereas the average feature-length movie contains about 1,500 closed captions, *The Internship* contains 2,405. For comparison, the largest caption file in my DVD movie collection is *This is 40* with 2,747 captions. *The Internship* is the second-largest caption file in my collection. The smallest caption file is the silent movie *The Artist* with 208 captions. In other words, *The Internship*, which clocks in at just shy of two hours, delivers its content principally through dialogue, as opposed to physical sight gags or heavy action sequences devoid of dialogue. Because verbal comedy depends on both content and form (not just what you say but how you say it), captioners must see to it that their edits, when they are necessary to meet reading speed guidelines, preserve as much of the color, form, and delivery of speech as possible. **Media:** http://ReadingSounds.net/chapter2/#verbatim-edited.

The distinction between edited and verbatim captioning is misleading. Verbatim captioning is not an objective, neutral practice of channeling speech sounds directly. Captioning is always about choices. Speech is transformed when it is transcribed and prepared for the caption track, whether for the purposes of verbatim or edited captioning. Almost all traces of dialect and manner of pronunciation are scrubbed from every speech caption. Hesitations and verbal fillers like "um" and "uh" are also routinely eliminated, unless they play a significant role in a scene (e.g., drunk speech, nervousness). Moreover, verbatim and edited captions can coexist in the same caption track, as when a decision is made to mark background speech as (INDISTINCT CHATTER) that could have been rendered verbatim. In short, verbatim captioning is assumed to be a simple process that is readily understood but requires choices, although perhaps not as many time-consuming choices as edited captions require (cf. Romero-Fresco 2009, 111).

Verbatim captioning and editing for reading speed are complicated by nonspeech information (NSI) and, in particular, the variety of approaches to captioning it. How NSI is handled by the style guides is a third issue likely to be of interest to rhetoricians and others interested in language. Captioning NSI is a creative and at times complex rhetorical act that involves careful attention to the context and nature of the video text. The amount, quality, and variety of NSI can vary wildly from DVD to DVD. Movies and TV shows are teeming with sounds that either cannot be captioned due to space or presentation rate constraints, or are not deemed significant enough to warrant being captioned. It is simply not possible to caption every sound in a movie or TV show. Discussions of NSI in the style guides dutifully list which categories of nonspeech sounds need to be captioned and how, at a basic level, to caption them, but they do not offer suggestions for helping captioners identify which sounds are significant or how to develop and hone the captioner's rhetorical powers of description. The style guides seem to assume that it will be obvious which nonspeech sounds need to be described. Sometimes it is obvious, as when, for example, we agree that music lyrics must always be captioned verbatim (e.g., DCMP 2009, 23). But too often it is not obvious. We see a wide range of approaches to captioning NSI that the style guides cannot account for or reconcile.

The style guides assume that the captioner already knows how to describe nonspeech sounds. A style guide may direct the captioner to "convey all the concomitant non-dialog information that the hearing audience takes for granted" (Media Access Group 2002b), but the guide will not explain what challenges are involved in making captioners aware of information they usually take for granted or how to caption that information. A style guide may warn the captioner "not to congest a show with unnecessary descriptive captions" (CBC 2003, 15), but the guide will not help the captioner understand where to draw the line between too much and not enough, with the exception of providing examples of how to edit content to achieve a set presentation rate. (It is worth noting that examples of editing in the style guides draw solely on speech captions; they never draw on NSI to provide examples of how to edit content [see DCMP 2009, 12–13]). For rhetoricians, then, NSI is of interest for at least three reasons: (1) NSI is often the most distinctive and subjective aspect of a caption track, highlighting the captioner's rhetorical agency; (2) NSI calls attention to the relationship between sound and writing because similar sounds may have different captions and different sounds may have the same caption; and (3) NSI raises awareness of the challenges of distinguishing *significant* from *insignificant* sounds.

The question of *significance* cuts to the heart of a rhetorical approach to captioning. Definitions of captioning often make a distinction, sometimes implicit, between significant and insignificant sounds. But they seem to assume that the question of *significance*, *importance*, or *essence* is a straightforward one, easily answered. The *Captioning Key* (DCMP 2009, 21) states that "background sound effects" should only be captioned "when they're essential to the plot." Background music "that is not important to the content of the program" (24) only needs to be captioned with a music icon. And "background audio that is essential to the plot, such as a PA system or TV" (17), needs to be captioned in italics. These directives assume either that the captioner already understands implicitly which sounds are important or essential, or that the style guide authors do not believe that the question of significance is a significant one. As a result, the style guides are weighed down by the microdetails of presentation and design, some of which have resulted in guidelines that are misguided, arbitrary, or inconsistent from one manual to the next:

·   *Alignment.* Caption alignment remains an ongoing source of inconsistency. Should caption lines be center-aligned at the bottom center of the screen, as CBC (2003, 3) and CaptionMax recommend, or "left-aligned at center screen on the bottom two lines within the safe zone," as *Captioning Key* (DCMP 2009, 6) recommends?

Why have captioners not drawn more heavily on the convention in written English of using a strong left alignment with ample margins? What reasons continue to implicitly support center alignment in an era of high-resolution digital captioning?

- *Type case.* Why does uppercase text style persist as a viable and popular option for captions? The technical challenges of the early days of TV captioning (i.e., low resolution) have been resolved (Clark 2008). Given the superior readability of sentence case for all other types of extended reading (both on the printed page and on the computer screen), why have not more major captioning vendors changed their practices? The persistence of all-caps styling for offline, prerecorded captioning on television has to be the most troubling and baffling aspect of the visual rhetoric of closed captioning.

- *New sentences.* Two major style guides advocate starting new sentences on their own line for offline, pop-on style captioning (DCMP 2009, 10; Media Access Group 2002b). According to *Captioning Key* (DCMP 2009, 10), captioners should "never end a sentence and begin a new sentence on the same line unless they are short, related sentences containing one or two words." According to WGBH (Media Access Group 2002b), "A period usually indicates the end of a caption, and the next sentence starts with a new caption." Nowhere else in the world of English sentences do we find such an unusual guideline. New readers of English learn to move their eyes from the end of one line to the beginning of the next without needing each new sentence to begin on a new line. Captioners should always break lines for sense (e.g., at the end of a clause), but guidelines for breaking lines need not include breaking lines after periods. Granted, reading static text is not equivalent to time-based reading, but the new sentence rule seems arbitrary.

- *Caption shape.* Readability should always take precedence over the desire to create an allegedly pleasing shape out of a multiline caption. In other words, breaking caption lines for sense should always take precedence over the desire to "present a block of captions that are similar in size and duration" (CBC 2003, 8) or to present "captions [that] appear in a two-line pyramid or inverted pyramid shape" (CBC 2003, 17).

- *Maximum number of lines.* According to CBC's (2003, 8) style guide, "Captions should always be one or two lines in length. Three line captions are acceptable when time or space is limited. Four line captions are unacceptable." This prohibition against four-line captions seems arbitrary. If a captioner is trying to honor the viewer's preference for verbatim captions and avoid editing captions unless they exceed a specified presentation rate, then why deny the captioner an extra line, especially for no reason, if the presentation speed and nature of the text allow for it?

The style guides focus on microlevel issues of text presentation and avoid higher-level issues that seem crucial to a full account of how captions make meaning: how readers interpret the text, how context and purpose shape the production and reception of captions, and the differences between reading and listening. Style guides also fail to account for captioning as a creative act of rhetorical invention. Instead of focusing on how to rhetorically analyze the soundscape and create descriptions for important non-speech sounds, style guides focus on how to style non-speech descriptions that are presumed to already exist.

Caption Readers and Writers

This book is an extended rhetorical analysis, a close reading that foregrounds the critic's relationship with closed-captioned texts. But any discussion of captioning must also be responsive to the needs and preferences of deaf and hard-of-hearing viewers. It must also attempt to understand the professional captioner's perspective, including the economic constraints that shape closed captioning. To address these two perspectives, I conducted two small studies: a survey of regular caption viewers and one-on-one interviews with ten professional closed captioners. The results of these studies are situated against the backdrop of the captioning style guide recommendations.

Surveying Regular Viewers of Closed Captioning

Entitled "Surveying the Preferences and Attitudes of Regular Viewers of Closed Captioning," the survey ran in late 2013 and early 2014 on Survey Monkey. I advertised the survey on Twitter and a couple of high-traffic mailing lists focused on web accessibility and captioning (e.g., WebAIM and CCAC). The survey generated 114 responses to a range of questions about demographics, assistive devices, customizing the display, experiences with different delivery devices, Internet viewing habits, and live captioning. The survey also asked participants for assessments in a number of categories (visual, type case, style [pop on or scroll up], reading speed, comprehension, overall satisfaction). A final question asked respondents, "If you could make one change to improve the quality of closed captioning, what would it be?" 64 percent of respondents were female, 36 percent male. 94 percent of the respondents were 24–74 years of age, with a fairly equal distribution among 25–34, 35–44, and 55–64-year-olds (21 percent, 24 percent, and 21 percent, respectively). Respondents tended to be very highly educated: 55 percent held graduate degrees and 25 percent held bachelor degrees. Hearing status was not correlated with whether respondents held graduate degrees. Most respondents were from the United States (87 respondents, or 77 percent). The rest were from a range of other Western countries: Canada (6), Australia (5), New Zealand (4), and the United Kingdom (3).

Respondents chose to identify themselves in a number of ways. 35 percent identified as hearing, 23 percent as hard-of-hearing, and 23 percent as deaf. But given the option to specify another term they preferred, respondents wrote in answers that equated roughly with hearing, such as:

Live with a hard of hearing mother
typical
no hearing issues
normal hearing
Hearing—so far
hearing, but I prefer watching TV with captions
respiratory disability
AT [Assistive Technology] professional w[ith] friends who are hard of Hearing
My wife is deaf in one ear and has noise sensitivity because of Meniere's disease
Hearing autistic with auditory processing disorder
Hear too much. Background sounds + voice = muddle

Or that leaned towards hard-of-hearing or deaf:

Hearing impaired
I have moderate hearing loss
late-deafened
Deafened
DeafBlind (Usher Syndrome)
often miss important dialogue
I have bilateral cochlear implants
I'm probably HoH, but I'm in denial, plus I have a deaf son

In terms of assistive devices, 42 percent of respondents said they wear or use one, roughly the percentage of respondents who identified as hard-of-hearing or deaf. When asked to specify, respondents selected hearing aid(s) twice as often as cochlear implant (67 percent to 35 percent), with a very small number of respondents who chose both responses.

Not surprisingly, respondents reported that they primarily access closed-captioned content using a television (57 percent) or computer (26 percent). But they have experienced reading captions on a range of other devices, including DVD players (67 percent), handheld devices such as iPads (48 percent), Blu-Ray players (24 percent), and gaming consoles (12 percent). Surprisingly, only about half of the TV viewers (48 percent) have ever customized the appearance of their television's captions. I had assumed that anyone who took the time to answer a questionnaire about captioning would also be more likely to have experience adjusting the appearance of the captions on their televisions. Or perhaps this finding means that the default settings on television sets are sufficient for viewers' needs. One respondent with experience customizing TV captions did prefer the default settings ("I generally stick with the defaults, but sometimes make the default font smaller when that's an option"). Another respondent made changes to the font color and the background color before reverting back to the default settings ("Didn't like the changes so I left them in the default position, white letters on black background"). But in general, customizers gave no indication that they were unsatisfied with the changes they made to the background style (mentioned nineteen times in forty-six responses), color (twenty-six times), font type (sixteen times), and font size (twenty times). Hearing status had no bearing on whether a participant was likely to have experience customizing the appearance of captions. About half of hearing, hard-of-hearing, and deaf respondents (48 percent, 48 percent, and 48 percent) claimed some experience with customization.

There were few surprises in the category of viewing habits. Respondents varied in the number of hours per week they watched closed-captioned content. 25 percent watched less than five hours a week, 24 percent watched five to ten hours a week, 17 percent watched ten to fifteen hours per week, 18 percent watched fifteen to twenty hours per week, and 16 percent watched more than twenty hours per week. The percentage of that time spent watching Internet captions is likewise distributed pretty evenly, although it should be noted that 15 percent of respondents spend 0% of their viewing time watching closed-captioned content on the Internet in contrast to 25 percent of respondents who spend 76–100 percent of their viewing time watching captioned Internet programming. Hearing viewers were less likely to watch live-captioned programming such as live news (e.g., CNN), which is not surprising given the delay of five to seven seconds in live captioning, which can create a jarring disconnect for hearing viewers between what people are saying and the captions printed on the screen. Whereas 33 percent of hearing respondents never read live captions, only 16 percent of hard-of-hearing respondents and 0 percent of deaf respondents answered "never" to this question. For deaf viewers, I suspect, that jarring feeling of trying to read live captioning while listening to the broadcast is absent or at least much less pronounced.

The remainder of the survey asked respondents to weigh in on a number of recurring questions about captioning style. Under visual assessment, respondents were asked whether they ever had difficulty perceiving or seeing the letters and words on the screen. Neither hearing status nor age was a factor. At least 75 percent of hearing, hard-of-hearing, and deaf respondents answered "rarely" or "sometimes" to this question. These were the most popular categories for each age range as well. Respondents were in wide agreement about type case, choosing sentence-case over uppercase styling, 69 percent to 9 percent. At the same time, however, a large percentage of respondents had no preference (17 percent). Hearing status did not play a role in respondents' preference for sentence-case styling. Hearing, hard-of-hearing, and deaf viewers preferred it (77 percent, 62 percent, and 68 percent, respectively) over upper case (10 percent, 8 percent, and 4 percent). Captioning companies that continue to promote uppercase styling in their prerecorded practices should take note. A small number of respondents selected "It depends," cutting intelligent paths through the choices. One respondent wrote, "All caps can be used sporadically for emphasis, as in ordinary prose, especially since formatting options are limited." Another wrote, "Usually I prefer mixed case, but if the captions are poorly designed and too thin or have poor contrast, the easiest font to read might be 'all caps.'"

Respondents were also in wide agreement about pop-on vs. scroll-up styling. I defined the answer choices to participants this way:

Definition: In pop-on style, each 2–4 line caption appears on the screen all at once and may be placed underneath each speaker to help readers identify who is speaking. DVDs and most prerecorded TV shows use pop-on style.

Definition: In scroll-up or roll-up style, lower lines are continually moving vertically up a stationary two- or three-line content block to make room for new bottom lines. Live captioning is produced in scroll-up style.

Respondents preferred pop-on style to scroll-up style 62 percent to 13 percent, but just as many had no preference (13 percent) as said "it depends" (13 percent). Six of the fourteen respondents who said "it depends" leaned on a distinction between live and prerecorded programming. One explained: "It depends on the content. A DVD or finished program should only have pop-on style. But a live program needs scroll-up style captions." Another simply restated the status quo: "Pop-on for pre-recorded videos and scroll-up for real time programming." Rationales for this distinction were hard to come by. Only two of the six respondents gave reasons. One referred to pop-on style as "good for programs where you see two characters talking to each other." Another referred to the technical constraints—"pre-prepared block captioning [pop-on style] is not possible" for "true live content." But what if the presumed technical constraints were removed? Is pop-on style superior to scroll-up in terms of usability? Unfortunately, I wasn't able to press respondents on this question. Still, the findings show clear support for pop-on styling over scroll-up. **Media:** http://ReadingSounds .net/chapter2/#popon-scrollup.

Under the topic of reading speed, 75 percent of the respondents answered "never" or "rarely" to the question of how often they have difficulty keeping up with the speed of the closed captions. At the same time, a large percentage (21 percent) answered "sometimes," which suggests a continuing need to study the contested issue of reading speed to explore when it is appropriate, if ever, to replace verbatim speech captions with edited-for-speed captions. When broken down by hearing status, the percentage who answered "sometimes" increased from 18 percent for hearing respondents to 23 percent for hard-of-hearing respondents and 28 percent for deaf respondents. Similar percentages were found with the assessment of comprehension. 75 percent of respondents answered "never" or "rarely" to the question of how often they had difficulty comprehending the captioned words they were reading. 22 percent answered "sometimes," a high enough percentage to give us pause. When broken down by hearing status, the percentage who answered "sometimes" to the comprehension question was highest for hearing respondents (26 percent), followed closely by deaf respondents (24 percent), and finally hard-of-hearing respondents (19 percent). These findings are somewhat counterintuitive, given claims (and some stereotypes) about the poor reading skills of deaf children and adults. For example, Denis Burnham et al. (2008, 392) invoke the "known literacy difficulties of deaf and hard-of-hearing people":

In general, hearing status and literacy tend to covary. In a recent U.S. study, the median reading comprehension level (scaled scores on the Stanford Achievement Test, 9th edition) of deaf and hard-of-hearing students aged 15 years was comparable to the reading comprehension level of hearing students aged 8–9 years (Karchmer & Mitchell, 2003). In Australia, reading comprehension levels among deaf and hard-of-hearing students have previously been shown to be considerably lower than those for the hearing population. Walker and Rickards (1992) found that 58% of school-age deaf students in their Australian sample were reading below grade level.

These literacy difficulties and differences are not borne out by my survey data. Survey respondents were very highly educated, on average, which most likely muted any differences in reading comprehension that could be attributed to hearing status.

When asked to give an example of something that has caused comprehension difficulty, sixty-nine respondents gave a range of answers. Not surprisingly, typos and missed, garbled, or incorrect words were mentioned most often (thirty-seven times):

typos, mangled transcripts
misspellings, phonetic spelling
Sentences that make no sense, or captions that are letters that don't form real
    words
when the captioner [is] not accurately capturing the [dialogue]
garbled totally due to transmission errors
typos and captions that don't match aural content
Typos/poor captioning are a prime example
The correct word or phrase was not used
Misspelled or incorrect words
misspelled words, phonetically incorrect words, missing words, etc.

Mismatches between audio and captions were also noted repeatedly (ten times). Sync problems sometimes caused confusion for hearing and hard-of-hearing respondents:

If the text is behind audio that I can hear and it doesn't match it takes longer for me
    to understand what I have just read.
I rarely find comprehension difficult, but when I do, it is usually because the captions
    are not matching with what the speaker is saying.
Sometimes Netflix or other captions are not synched with the audio, which causes
    trouble.
Close[d] captioning is often in shorter sentences and I will notice that there is a dif-
    ference in what I hear and what I read—I have to rethink to make sure I follow
    and understand.

These examples might be considered a subset of the typos or incorrect captions category. But in this case, a sync problem is not limited to typos but reflects a perceived tension between the audio track and the caption track. Of the ten respondents who mentioned sync problems, six identified as hearing and four as hard-of-hearing (none identified as deaf), which should remind us that closed captioning, even though it is intended to be a substitute for sound, continues to be evaluated against the soundtrack by readers who have at least some hearing. Captioning attracts a diverse audience who employ a wide range of expectations and evaluation criteria. Another source of confusion for respondents, related to synching problems, was reading speed and timing (mentioned fifteen times):

when the timing is too short
relayed a lot of important information at once and then left the screen [too] quick
    to digest
When captions fall behind the spoken words
Sometimes, following the fast talking is lost on me
Sometimes it's gone before you can read it all . . .
Captions disappearing too quickly
when captioning is too fast
If captions are garbled or pop up faster than I have a chance to read
Rapid appearance of captions
moving too fast

Given that 75 percent of respondents answered "never" or "rarely" to the question of whether they have ever had trouble keeping up with the captions, it's notable that reading speed was mentioned as often as it was. Following up on reading speed studies (e.g., Jensema 1998), then, should remain an important item on the agenda of captioning researchers.

Autotranscription and live-captioning errors were also identified as sources of comprehension problems (ten times). Respondents called out YouTube's autocaptioning software (Google 2009), which is designed to provide closed captions automatically using speech recognition technology but is more famous, to date, for producing a large number of errors or "caption fails" (see Rhett and Link 2011). Respondents were confused by:

Captions that use voice-recognition programs, like the ones used on YouTube
cruddy auto-caption, via YouTube auto-caption, or when the timing is too short, or
    when the caption lines are epic long
Auto-captioning or poor quality captioning in the Internet is my biggest barrier

> Transcription errors in YouTube autocaptioned videos. Sometimes they are
> hilarious—for hearies, but if you need captions, they must be maddening.
> However, the YT autocaptions have been improving steadily since their introduc-
> tion [circa] 5 years ago. They'll probably never be accurate enough for con-
> sumption, but they are on the verge of being editable in less time than doing
> them from scratch. And now the admin interface of YouTube videos also offers a
> caption editing tool.

In this last response, we are presented with more than a critique of
"cruddy" autocaptioning. The respondent is both hopeful ("improving
steadily") and realistic ("probably never be accurate enough for consumption"), offering a solution that positions autocaptioning as the first
step (instead of the only step) in a process that relies on a human editor to clean up the crud. These were the most popular answers to the
question about sources of comprehension difficulty. Other less popular
answers included inaccessible visual design (small fonts, low color contrast), inadequate speaker identification, confusing nonspeech descriptions, unfamiliar historical references, "fancy" words, confusing line
breaks, long lines, and captions that covered sports scores and on-screen
news program text.

Respondents also offered an overall assessment of the quality of
closed captioning. Most felt "somewhat satisfied" (58 percent), with a
considerable percentage feeling "somewhat dissatisfied" (20 percent).
Less than 3 percent were "very dissatisfied," while 7 percent were "very
satisfied." More women (65 percent) than men (46 percent) felt "somewhat satisfied" but more women (23 percent) than men (16 percent)
also felt "somewhat dissatisfied." In terms of hearing status, fewer deaf
respondents (48 percent) than hearing respondents (66 percent) were
"somewhat satisfied" with the quality of closed captioning. More deaf
respondents (24 percent) than hearing respondents (16 percent) were
also "somewhat dissatisfied." When asked to explain what contributed
to their feelings of satisfaction and dissatisfaction (two separate questions), ninety-two respondents gave a range of answers. What satisfied
respondents, not surprisingly, were things like accuracy (twenty times),
precise timing (nine times), foreign language accessibility (three times),
captioned music lyrics (three times), legible and clear typefaces (three
times), being able to understand thick accents (four times), being able
to follow a program with the sound low or in a noisy environment (four
times), and having an additional mode of information (three times). But
respondents also wrote more generally about the value of closed captioning to provide access, enable participation, and equalize the playing field:

When captioning is done well—with good quality, pre-prepared block captions—I can access the content and more importantly enjoy and engage with that content, just like anyone else, and notwithstanding my profound hearing loss.

Satisfaction? INCLUSION—cannot follow without quality cc

without captions I often miss words don't understand jokes or get distracted for a second—it's nice to be able to see everything that was said

Can't do without them

It helps me keep up with important dialogue

It allows me to watch TV and DVD movies. Without captioning I cannot understand most of speech.

At least you have a chance to pick up something you are watching.

Captions or subtitles supplement my listening. I feel I hear better. Without them, I feel I strain more and am certain I miss nuances. So captions make me feel I get a more complete sound/hearing experience.

It generally allows me access to verbal communication for TV and movies that I do not have without it.

Grateful we have them. I enjoy music and love it when songs are closed captioned. Love how some of my apps allow closed captioning so I can catch up on favorite television programs. Love going to the movies with my older kids—something we missed out on when they were growing up.

Professionally executed and proofread captions are a delight because they mean I can fully understand dialogue.

That it exists, and that it is mostly usable.

Allows me to watch TV.

These comments are sobering reminders—not that readers of this book need to be reminded—that captioning is the difference between understanding and misunderstanding, inclusion and being left out. A number of respondents referred to access, inclusion, and being grateful that captioning exists at all. Given the alternative of not having any captions, participants were generally satisfied with the quality of closed captioning:

I'm happy to have cc/subtitles when watching a tv show/DVD. I can't enjoy watching anything without subtitles.

There are areas of misunderstanding, but given the alternative of no sound, I'm generally happy.

When provided, closed captions for pre-recorded programs are almost always high-quality, and the majority of live captions are at least acceptable when allowances are made for the inherent challenges of the enterprise.

Having it is better than not having it which is still the majority including advertisements.
Most shows are captioned and done well

Finally, respondents invoked a narrative of progress to explain the improvements to quality and satisfaction they have witnessed over time:

Becoming increasingly available, legible, and accurate.
More and more cc content available now than previously.
Availability is better than it was in the past. More devices supporting CC is great too.
well hell I can understand what the characters are saying . . . more and more
    captioned content is available, Netflix leading the pack, Hulu doing ok, YouTube
    mis[e]rable, Amazon lousy.
I've used a captioner to help me in my job since it was first invented. It's come a
    long way. To those complainers, I would only say that, in comparison to twenty
    years ago when hardly anything was captioned and what was captioned was
    slow, we are lucky to have what we do on TV and in first run movies these days.

In terms of satisfaction, then, respondents seemed to be generally happy
with closed captioning considering the alternative (no captions) and
given the increasing availability of captioning on the Internet and with
multiple digital devices.

A number of issues spurred feelings of dissatisfaction for the ninety-
two respondents who answered this question, the most popular of
which were transcription errors (thirty times) and timing/synching errors (thirty-two times). In a number of cases, these two issues were mentioned together in the same response:

Inaccurate words, delayed timing and placement over show information (example:
    date and address appearing on screen in Law & Order)
periods of poor transcription, captioner doesn't keep up with dialog
latency, gaps of info, misspellings, bursts of words all at once
Captions that are too far behind the action to be of any use—that just makes it more
    confusing to follow. Captions that have many errors may be funny but too many
    makes the viewing less than satisfactory.
Text not always in sync with speaker. Voice to text inaccurate. Sometimes blocks
    faces or other important things.

Respondents also repeatedly mentioned poor screen placement (sixteen times), inaccessible visual design ("when the captions are hard
to see . . .") (thirteen times), and incomplete or absent caption files

(nineteen times). In other words, the most popular topics mentioned by respondents also intersected with the common criteria for assessing captioning quality: accuracy, timing, placement, completeness, and visual design and typography (see FCC 2012). A number of other issues were also mentioned, including error-prone live captioning (eleven times), poor Internet captioning and autocaptioning (seven times), and captions that are paraphrased instead of rendered verbatim (six times).

Interviewing professional captioners

Entitled "Exploring the Rhetorical Agency of Captioners," this study set out to learn about the practices of professional captioners who specialize in prerecorded captioning. I conducted a total of ten one-on-one interviews: eight in fall 2011 and two in spring 2012. Each phone or Skype interview lasted about one hour. Eight participants were women, two were men. They were diverse in terms of years of experience with offline captioning, from one to twenty years. I did not ask any demographic questions, although I did surmise that all ten participants were based in North America (nine in the United States and one in Canada). I also assigned pseudonyms to all names, including company names. I asked a range of questions, focusing especially on the captioning of nonspeech sounds, which I assumed to involve a high level of creativity and freedom: How do captioners choose which nonspeech sounds to caption and how to caption them? Under what constraints do they work? How much editorial oversight do they receive? What is easy about captioning and what is hard? What kinds of training did they receive? Which skills are important for captioners? How would they describe the workflow or production cycle in their company, including division of labor with other captioners and supervisors? My interview participants provided me with a mountain of rich data. I want to report here on a small number of issues related to their experiences. I also want to focus on the participants with the most captioning and supervisory experience: Claire (fourteen years of professional captioning experience), Emma (six years), Grace (thirteen years), Mary (ten years), and Steve (twenty years). The majority of my participants fell into captioning as a way to make ends meet, perhaps with the exception of Grace, who was trained as a court reporter in 1980 and moved into live captioning before doing offline work starting in the late 1990s. Emma holds a music degree but could type fast, "answered an ad," and learned on the job. Steve was a self-described "starving actor" barely scraping by as a proofreader when he also answered an ad. Mary also started out as an actor but holds a law degree and has experience in the newspaper business. Claire was a film school grad who was looking for work in the film industry.

Given that most offline captioners don't anticipate doing this kind of work for a living when they are in college and thinking about career options, retention is a problem in the offline captioning industry. Claire reports that her department of four full-timers has been lucky to find "one or two people who stay on for several years, but generally speaking . . . three years seems to be the maximum for people who are interested in doing other things with their lives—and most of them are. And one year is more the norm in many cases." Contributing to the problem of poor retention is the low salary paid to entry-level offline captioners. As Claire put it, only new college graduates will accept the low starting salary in order to get some work experience on their resumes. Youth and inexperience can work against the goal of quality captioning. If captioners need a high level of cultural literacy, then the best captioners of prerecorded programming will be those with the most life experience and the greatest knowledge of popular culture, those who can recognize the sonic allusions in the soundscape. It's hard to teach captioners (or anyone, for that matter) about popular culture in a short amount of time. Even a lowbrow show like Fox's *Family Guy* requires a high level of cultural literacy spanning at least four decades of pop culture references. But captioning companies can mitigate this problem by hiring captioners with strong research skills who hold college degrees that emphasize reading widely and writing research-oriented papers.

What makes a good captioner? When I asked each interviewee about the skills required to be an effective offline captioner, they tended to list the obvious things first—fast typing speed, strong grammar skills, ability to spell, good listening skills, good text editing skills—but then transitioned to broader, less quantifiable skills. For example, Claire began her answer with a list of basic skills: "good typing skills, but even above that I would put spelling, knowing how to spell things, especially homonyms and knowing which homonym is . . . appropriate in [each] case." But then she considered the deeper importance of having a wide knowledge base and college-level research skills: "As far as knowledge is concerned, the wider the knowledge base of the person captioning, the better. We usually ask people [to have] at the very least a university or college degree . . . We figure at least somebody who's done three or four years of college will have done research, will have done readings and will have expanded their knowledge base up to a point." She summarized her answer like this: "So attention to detail, speed, and good spelling are important, but a good knowledge base is very very important too."

Emma and Steve also discussed the importance of having a wide knowledge base. While Emma reports that she "got the job because [she] was a fast typist," she also stressed the importance of having "knowledge of lots of different things," including "pop culture and different references people make." Emma summed up this knowledge in terms of having a "good ear" (a term that Mary also used). Likewise, Steve discussed cultural literacy (but didn't use this term) using hip-hop music videos as examples. For example, captioners can't always or simply rely on the music lyrics provided by online sites or even the artists' own record labels. As Steve explains, there's no guarantee that the record label's lyrics will be accurate, even when the captioning department requires official lyrics be sent with the program to be captioned. The lyrics may have been written by a record label employee who was ordered to type up some lyrics for the captioning company. Steve discussed a specific music video assignment that involved a slang term for a nine-millimeter gun, "nina." The music artist was "not talking about a girl named Nina. He's talking about nina in the context of a nine-millimeter gun." Even though the artist was singing "I left your nina on the table," the context made it clear that nina was a gun. If you aren't familiar with this slang term and assume that Nina is the name of someone's partner or lover, then you might think that the artist is singing metaphorically about leaving a woman. Steve used this example to call attention to the importance of having a wide knowledge base, which I will discuss in more detail in terms of cultural literacy in chapter 7.

Cultural literacy goes hand in hand with having common sense. Steve used the term "smart ear," which, like Emma's notion of a "good ear," is hard to quantify and can't be reduced to traditional grammar and spelling skills. "If you're listening to a show, and someone is droning on and on and on, or some piece of dialogue is going back and forth, if you hear and type something and it makes absolutely no sense, then you probably heard it wrong." Steve's point probably goes without saying, yet it also reminds us of the importance of context and common sense in captioning. He encourages captioners to listen to challenging or confusing passages of dialogue a second time: "If you do that, you'll realize, 'oh, in the context of what [the speakers] were just talking about before, this is the direction that this conversation is going.'" Context informs the captioner's understanding. Steve makes a helpful distinction between "the syllables that [the captioner] was hearing," or thinks she is hearing, and the context in which those sounds are situated. Context is crucial to making sense of the syllables. Having a smart ear, according to Steve, involves a "judgment element" that requires the captioner to "listen intelligently" and "rely on context to make sense." Smart ears can't be quickly learned or taught. Mary says something similar about the difficulty of training someone to listen and says she thinks it's a good sign when a job applicant has experience playing a musical instrument. Smart ears and listening skills must be developed over time, according to Steve, through increased awareness, self-criticism, and knowing when to make good use of one's fellow captioners. In his supervisory role, Steve says he can tell which captioners have a smart ear and which don't (or have yet to acquire one).

The smart ear learns to distinguish significant from insignificant sounds. Claire, Emma, and Steve leaned on a notion of "relevance"— what I have called significance—to explain how captioners know which sounds to caption. In Claire's department, they "don't describe absolutely every background sound, like I was led to believe we should do when I started. Some people were training me who said, 'Oh, everything you hear you have to describe.'" But that approach became problematic for Claire as she gained more experience: "Every time you heard a bird chirping, you put it in. And every time you heard an airplane overhead, you put that in, even though it has no relevance to the plot . . . . If I put that in, that's just going to mislead the viewer, the nonhearing viewer, into thinking that [the irrelevant airplane sound] has any relevance." This point about relevance is crucial for our reconsideration of the nature of closed captioning and the differences between sound and writing. Steve made a similar point. "It starts with relevance," he explained, using as an example the stock soundscape that often accompanies the establishing shot of a suburban home: "You've got a scene in the suburbs and . . . the generic soundtrack music is there, but you've got birds chirping and dogs barking. You might indicate that to give a flavor, because you want to indicate that it's a lovely suburban scene. The birds are chirping and the dogs are barking and everything's bright and cheery." But here's the problem:

. . . if the dog is barking in the distance [i.e. the sound is captioned] and the dog is still barking in the distance later [i.e. captioned again] but it's not relevant, it can become misleading. Because then you start worrying. Why is the dog barking? Why is the dog barking? If there turns out to be no answer to that question—it's just a sound filler that they're using—you're kind of taking the user in a particular direction by overemphasizing that. So, you need to be judicious when you type stuff that's less than immediately relevant to what's going on, because you can start taking them in the wrong direction and then you have to bring them back again.

Steve's example perfectly illustrates one of the challenges of captioning nonspeech sounds and the considerable power that captioners possess over the meaning of the text. The captioner is a driver, using Steve's metaphor of "taking" viewers in a certain "direction." In the hands of an inexperienced or careless captioner, the viewer can be shuttled to a destination that isn't on the map. The dog barking may just be filler, a stock sound and nothing more. Too much background or ambient captioning can lead viewers to assume that something is more important than it really is. Captions tend to equalize and flatten sounds. The irrelevant dog suddenly seems more relevant when captioned. Not all of my interview participants shared Steve's view, however. Grace reportedly attempts "to caption everything" in order "to become somebody else's ears." She qualified this view by explaining that speech in the foreground takes "precedence" over background sounds. But does becoming somebody's ears mean alerting the reader to every sound in the environment? Just as hearing is selective, so too should captioning be. The ears of hearing viewers do not experience the multilayered film soundscape as a series of equivalent sounds, which is why captioners need to be thoughtful and use judgment when translating the soundscape into accessible writing. How long it takes for new captioners to develop the ability to distinguish relevant from irrelevant sounds is unclear. Claire, a supervisor, says it takes "about six to ten weeks" to get new captioners "up to speed" and "semiautonomous" but longer to train them to intuit a sense of sonic relevance.

Knowing when and how to use one's fellow captioners for support and feedback is also an important skill. In a captioning department, regardless of its size, captioners should ideally have easy access to each other. In small departments of four or five captioners, the team can work in the same office space, at individual workstations and with headphones. Swivel chairs allow captioners to turn quickly and face each other at those times when one captioner, for example, wants to consult with her colleagues about a challenging sound. As Mary put it, "It can be a real team effort when something unusual shows up and you have a sound effect and you don't really want to use onomatopoeia unless you absolutely have to." (Steve also cautioned against using onomatopoeia due to its ambiguity, a topic I take up in chapter 8.) For a hypothetical example, Mary made a high-pitched whooping sound that could not be explained with a conventional nonspeech caption such as "tires screeching." "We work in a bullpen. It's an editing space. We're all on swivel chairs. We're all working with headphones. The room is set up so when somebody turns, your attention can be caught. It'll just be thrown out in the open, because we're all working on a different thing, and somebody will say, 'How would you describe . . .' And we'll all talk about it." In her department of five captioners, Mary reports that collaborations of this kind occur about once a week. When I asked her to recall the last time one of these discussions had taken place, she told me it was during a discussion of the "quality of surf hitting the shore on the beach. And it had some sort of strange reverb on it. And the question was, number one, was it 'surf crashing,' or is that too violent? Is it 'waves lapping'? And do you include the sound effects afterwards? I think what we went with was 'surf crashing echoes.'" Working together in the same physical space also allows captioners to collaborate on big projects and rush orders. For example, a movie might be carved up into thirty-minute segments and assigned to four captioners. This team of captioners will need to coordinate with each other, and, ideally, with a final reviewer/supervisor who can ensure the continuity across their respective segments.

How much freedom do captioners have? In response to this question, captioners claimed near total freedom over every aspect of the captioning process. While a captioner may be overruled by her supervisor during an editorial review, in-house negotiating over the final product doesn't diminish the captioning industry's freedom and independence, generally speaking. Let me put it this way: if a team of captioners decides to caption the waves as "surf crashing echoes," their decision is final. No film or television producer will ever overrule their decision, because producers don't work with captioners to ensure that the captions honor their creative vision. Captioners are rhetorical proxy agents because they work on behalf of the content creators to express their intentions as faithfully as possible, but they usually do so without receiving any direction or feedback from the clients they are serving. Claire put the matter this way in an email she sent to me before I asked her for an interview:

The vast majority of clients do not care what the captioning looks like, as long as it gets done in time for the stations to receive their captioned masters. The vast majority of them do not watch shows with the captions on, and they don't know anyone who's deaf or requires captioning for one reason or another. Clients don't have first-hand experience of bad versus good captioning, therefore it doesn't matter to them if we use roll-up or pop-on, whether we caption verbatim or edit, or even whether we know how to spell!

The captioning department will receive a script, at least some of the time, but the script will always be of limited help and guidance to the captioner. Claire very helpfully framed the question of freedom in terms of percentages: "The script would only be about 60–70 percent any good. We have to rewrite about 30 percent of it." She summarized what's involved: "fixing all of the dialogue that has changed [i.e., the ad libs that aren't scripted], adding all the sound effects, and then positioning everything, and then timing everything. . . . A script, if it's any good, will save us, maybe, I would say, 40 percent of the work. If it's not so good . . . it will save us only about, maybe, 25 percent of the work." Even so, the captioners, not the producers, have the final say over how the script is made accessible. No one from the production team is ever likely to verify that the captions have faithfully and accurately represented the script. There are exceptions, such as the producers who, as reported in Claire's email to me, "demand to see the captioning before it's shipped out" to make sure that the captioners have "follow[ed] the as-recorded script exactly word-for-word, including punctuation." But when I asked Steve how much freedom captioners have over nonspeech sounds, he didn't mince words: "quite a bit." Producers are more likely to be absent than demanding.

On the other side of this equation are the viewers who depend on quality captioning. The captioners are positioned between the producers and the viewers but have few to no interactions with either side, at least in the case of the captioners I interviewed. These captioners also received almost no feedback from deaf and hard-of-hearing viewers, with the possible exception of Grace, who referred in her interview to the importance of understanding Deaf culture and employs a deaf engineer to assist with file formatting issues and some captioning tasks. Grace also works in both prerecorded and real-time captioning, having been trained originally as a court reporter. She talked about receiving "immediate" feedback from the users of her real-time captioning services. For the other captioners I interviewed, who work exclusively in prerecorded contexts, feedback from the primary audience is neither immediate nor readily forthcoming. I asked Emma if she ever received feedback from deaf viewers. She replied, "We didn't, and I always wondered about that." As Steve put it, feedback and focus groups are a luxury that captioning departments, especially small ones, can usually neither afford nor justify financially. "We don't have a really formal mechanism," Steve explained, for soliciting feedback from deaf and hard-of-hearing viewers. "A lot of the judgments about the quality of what we do [are] just based on my experience." Granted, Steve's experience is extensive, spanning two decades. But the lack of regular, formal contact with members of the primary viewing audience should raise questions for captioning advocates about their ability to shape and improve captioning quality except in the usual and limited ways (e.g., filing an FCC complaint or emailing a broadcaster to advocate for better captions). Further, the captioners I interviewed were not regular viewers of closed captioning at home. Emma told me that she watches captioning on TV "occasionally," while Claire watches "from time to time."

Freedom is thus double-edged. These captioners were, for the most part, like ships floating between two land masses they could, at times, see only dimly in the distance. On one side are the producers who offered no instructions or feedback to the captioner, outside of a production script or music lyrics that had limited value. The captioners were thus free to describe the soundtrack in whatever ways they saw fit. Editorial oversight and regular discussions in the bullpen contributed to a more productive environment, despite the creative team's permanent absence. (It should be clear that the captioners I interviewed worked for independent captioning companies and had no direct lines to movie producers or music artists.) On the other side are the deaf and hard-of-hearing audiences whose opinions were not regularly sought out. The interviewed captioners also didn't regularly watch television or DVDs at home with the captions turned on, so they were cut off from the work of their peers at other captioning firms.

Complicating the notion of freedom further are the economic constraints under which captioners work. I've already mentioned the low entry salary for captioners, which tends to attract fresh college graduates who will be hired for their fast typing and strong research skills (among other things) but will probably be lacking in cultural literacy and knowledge of pop culture history. Steve also referred to the real-world constraints that prevent captioners from spending an inordinate amount of time worrying about how to caption a single nonspeech sound:

One thing that's missing from so many discussions of captioning is a real-world understanding of the demands, not just of getting the words right but of getting the job out the door when the client needs it. Over the years I've taken exception to some of the criticisms that I've heard because we frankly don't have a lot of time to agonize over whether it's a "wet squishing" or a "loud squishing." You can't spend an indefinite amount of time.

Steve didn't discuss the economics of closed captioning explicitly, but the implications are clear. Time is money; deadlines must be met. In her email to me, Claire was more direct about the negative influences of the market economy on the quality of captioning:

In my 14 years in this field, I've come to the conclusion that the main factor that drives captioning quality is what clients are willing to pay for it. I regret to say that most clients—whether it's for commercials, series, films or whatever else—see captioning as that mandatory last step that has to get done as a condition of their materials going on air. The vast majority of them do not care what the captioning looks like, as long as it gets done in time for the stations to receive their captioned masters. Clients will often choose to go to cheaper captioning houses who promise to get their feature film captioned in a day. And when a captioning company charges low prices on high volumes of work, it's because they hire lots of people at low wages.

Captioners should be responsive to the needs of both the viewers and the clients, but these two forces do not pull equally on the captioner. As Claire writes: "We always strive to balance the needs of the deaf and the hearing-impaired with clients' requirements; but because clients pay, and users do not (at least not directly), if a client decides to have us do an awful mess of their captions, they win. The best we can hope to achieve before this happens, though, is to convince them to let us do it our way." The trend towards using scroll-up styling for prerecorded programming is an example of captioning companies yielding to market/client forces over the needs of viewers. Scroll-up styling is cheaper than pop-on styling because it's faster to produce. Scroll-up style for prerecorded shows costs about half to three-fourths of pop-on style (FCC 1999), which explains its growing popularity. Pop-on captioning requires more time to shape, place, and time individual captions. The extra time taken in production creates a more usable experience for readers. Lines that are constantly on the move, and timed with only approximate accuracy, are less usable. The economic pressures on the captioning industry represent perhaps the biggest challenge to advocates who demand thoughtfully produced and carefully reviewed pop-on style captions.

In this chapter, I sought to lay the groundwork for the detailed analyses that follow. I approached this large task from a number of perspectives, starting with a detailed breakdown—the first of its kind—of major NSI types. Three small studies followed—a review of major style guides, a survey of regular viewers, and interviews with professional offline captioners—that, when taken together, paint a picture of closed captioning as complex, subjective, and contextual.

THREE

Context and Subjectivity in Sound Effects Captioning

Captioners don't caption sounds. They caption shows. They interpret and convey contexts. They make meaning. They mediate authorial intentions. This chapter explores how sounds acquire meaning in specific audiovisual contexts. Even if we know, in raw technical terms, what a sound *is*—where it comes from, who or what made it, how to reproduce it precisely—we still won't have enough information to caption it. To explore this claim, this chapter analyzes a single recurring sound—the Hypnotoad's drone on *Futurama*—through the variety of captions attached to it over the course of nine years (2001–2010). The sheer variety of captions for a single nonspeech sound points to a deeply subjective practice in which captioners rhetorically invent meaning that hasn't quite existed before within the universe of the show. This same variety of captions also raises questions about the need for consistency in episodic television captioning. This chapter is rounded out with five additional examples from *Family Guy, Twilight, Sunshine Cleaning, Hick,* and *Aliens vs. Predator: Requiem.*

Unlike most speech sounds, nonspeech sounds do not always have clear linguistic meaning attached to them. In other words, it's not always obvious how some non-speech sounds should be described. Even sounds that might seem obvious, such as a telephone's busy signal, can only be effectively captioned by attending to the needs of the audience. What US viewers know as a busy signal, viewers in

(LOUD BUZZING DRONE)

3.1 **How do you caption an ambiguous nonspeech sound in a novel context?**
In this frame from an episode of *Futurama* ("Rebirth"), the Hypnotoad sits on four legs surrounded by a background of undifferentiated whiteness. The Hypnotoad is a large cartoon
horny toad with unusually large round eyes. These eyes vibrate and change color when
animated, giving the toad his powers of hypnosis. He is also wearing a dog collar. His head
is positioned to the left of his bumpy body, allowing us to see his multicolored brown
body in profile while his eyes gaze directly at us, hypnotizing his audience just as television
hypnotizes its viewers. Caption: (LOUD BUZZING DRONE). Twentieth Century Fox Television,
2010. Blu-Ray. http://ReadingSounds.net/chapter3/#figure1.

Great Britain call an engaged tone. Moreover, many nonspeech sounds
can have multiple meanings when taken out of context. For example,
how do we know whether a breathy sound expressed by Bella Swan
(Kristen Stewart) in *Twilight* (2008) should be called a gasp, groan, grunt,
pant, wheeze, exhale, inhale, sigh, or ignored by the captioner? Out of
context, the same gasping sound can denote fear, surprise, desire, an-
ger, pain, and so on. Broadly speaking, "[t]wo sounds may be identical
but have different meanings and aesthetic effects," just as "two sounds
with quite different physical characteristics may have the same meaning
and aesthetic effect" (Schafer 1977, 149). When faced with multiplicity and uncertainty, the captioner must show sensitivity to context and
purpose. But because nonspeech sounds can also easily accumulate and
overlap in a scene, especially in an action movie, the captioner must be
selective about which nonspeech sounds to caption.

We can explore these ideas in the context of the rich variety of official
closed captions associated with one recurring nonspeech sound on a
long-running animated TV show. The sound and its myriad captions
complicate many of our definitions of closed captioning and expand our
understanding of what it means to make sound accessible.

Context and Subjectivity in the Hypnotoad Captions

Sound can only be effectively captioned in context. Listening to a discrete sound out of context may not provide enough information about what it is or how to caption it. Hearing readers might take a moment right now and listen carefully to this sample sound: http://Reading Sounds.net/chapter3/#sample. How should it be captioned? Out of context, the sound could be described as a violent rumbling, an industrial engine, or angry white noise. The source is unmistakably mechanical and unnatural, the output of some kind of large engine or factory motor, perhaps. Visualized in an audio editing program such as Audacity, the sound wave appears uniformly jagged, with only small differences in the heights of the sharp peaks of the wave. In other words, there are no characteristic breaks in the wave that one would find in visualized speech, breaks indicative of pauses between words. The wave's uniformity—the absence of a clear start or end point—means that the sound could easily be set to play on an endless loop. It also means that one second in the wave could most likely stand in for any other second. The rumbling engine sound is continuous.

When I've played this sound out of context and asked hearing people at conferences how they would caption it, I have received a variety of responses, including [electric guitar] from a professional captioner who was providing real-time captioning for the audience. (The captioner's output was displayed on a second projection screen alongside my slides.) I've asked audiences to play along so I can make a point that may not have been obvious: the meaning of every sound *only* becomes secured in context. But how much and what kind of context do we need? What if we also know what this mystery sound *is*—where it came from or who/what produced it? Let's say we know that the mystery sound is a turbine engine played backwards. Do we now have enough information to caption the sound? Probably not. We can't know for sure (or even know whether the sound needs to be captioned at all) without access to a visual context in which the sound functions. A "turbine engine played backwards," as it turns out, is the Hypnotoad's trademark sound on *Futurama* (*Futurama Wiki* 2012b), an animated TV series created by Matt Groening of *The Simpsons* fame. New episodes of *Futurama* ran on Fox from 1999–2003 and on Comedy Central from 2008–2013. The series is set in the year 3000, which is 1000 years after Philip J. Fry (voiced by Billy West), a New York City pizza delivery boy, was accidentally frozen cryogenically. When he is revived, he gains employment at a space courier service with a cast of regular characters—human, robot, and alien— providing an endless supply of episode ideas as the crew travels around the galaxy. Put simply, *Futurama* is a workplace comedy, set in the future.

None of the main characters know the Hypnotoad personally or have any face-to-face interactions with it. Still, the Hypnotoad is popular among audiences. The Hypnotoad is a large, alien toad with unusually round eyes that warble and change color. The animated toad resembles the Amazon horned frog with its large, sloped head, raised eyes, wide mouth, and spiked, multicolored back. It's hard not to anthropomorphize the Amazon horned frog. With its long mouth sloping down on both sides, it looks like it's having a bad day. The animated version almost always appears as a fictional television personality within the *Futurama* universe. The Hypnotoad is host of "America's most popular show," *Everybody Loves Hypnotoad* (a not so subtle nod towards the 1990s sitcom *Everybody Loves Raymond*). The content of the Hypnotoad's show-within-a-show seems to be limited to the Hypnotoad staring at the viewer with flashing, warbling, multicolored eyes while emitting an engine-like noise. The sound emitting from the toad never varies across episodes, except in volume when the Hypnotoad's television show is playing in the background. The Hypnotoad does not speak and seems to have only one purpose: to hypnotize viewers who will, in turn, worship him. We learn very little about the Hypnotoad or its TV show. On its TV show, the Hypnotoad performs only one action as it sits on four legs staring straight into the camera. Outside of its TV show, The Hypnotoad makes only one appearance. In "The Day the Earth Stood Stupid" (2001), the Hypnotoad is a contestant at a stock show who, by emitting the hypnotic drone and flashing its mesmerizing eyes, hypnotizes sheep to move into their pen during a "sheep herding" contest. The judges are also hypnotized to award the Hypnotoad the grand prize, uttering the catchphrase now associated with the Hypnotoad: "All glory to the Hypnotoad." The stock show audience, which includes the main characters, is similarly hypnotized, clapping in unison. In paying homage to the Hypnotoad, online fans will sometimes interrupt their own written statements to indicate that they have fallen under the Hypnotoad's spell: "Why is this Hypnotoad video so popu—ALL GLORY TO THE HYPNOTOAD" (Know Your Meme 2012). While the Hypnotoad appears to be all powerful, it also wears a collar, suggesting, perhaps, that it can be tamed.

What makes the Hypnotoad so popular? The character is absurd, even for *Futurama*. It appears rarely on the series and never plays any significant role in the plot or engages with the main characters, except through a fictional TV screen. In short, the Hypnotoad is a flat character that never exhibits any real complexity or movement, except for those wild eyes. The hypnosound is constant too, never wavering in its sonic intensity. For these reasons, the Hypnotoad functions as a satire. The presence of a television show-within-a-show and hypnotized sheep suggest a satire of mindless television worship, with audiences serving as pliant sheep who worship people and characters they never engage with directly. We sit in front of the screen hypnotized by the mesmerizing colors and enthralling but repetitive situations. Our critiques of television culture are then appropriated by the objects of critique themselves, turning a potentially serious issue (viewers as passive consumers) into a running gag and Internet meme divested of critical power. When Fry complains in season four that *Everybody Loves Hypnotoad* has "been going downhill since season three," he may as well be reflexively referencing *Futurama*, which was cancelled at the end of the fourth season (before being revived again on a new network a few years later).

The critique of sitcom culture as mindless and hypnotic is suggested in the DVD special feature of *Bender's Big Score*, a twenty-two-minute uncaptioned "episode" of *Everybody Loves Hypnotoad*. The episode has an opening theme song, an up-tempo piece with rhythm guitar and a strong drum beat. The establishing shot is a suburban home, but one that seems to have been subtly reinforced with surveillance or security technologies. The "content" of the episode is more mindless and continuous hypnosis from the Hypnotoad, who is set against an undifferentiated white background, albeit with a few twists sprinkled throughout: canned studio reactions (e.g., laughter, "awww" sounds, applause), borrowed theme music, animated establishing shots from other sitcoms (e.g., the *Seinfeld* bass line and diner shot), commercials (e.g., an ad for "Calorie Substance"), and an announcer ("sponsored by"). At the end of the episode, the announcer counts down from fifty to bring us out of the hypnosis, "remembering nothing." In these ways, the episode spoofs mindless sitcoms that provide momentary entertainment but no lasting knowledge.

The closed captions associated with the Hypnotoad turn the meaning of the hypnosound away from its *actual* origin (a turbine engine actually produces the sound) and towards the hypnosound's *apparent* origin (the Hypnotoad apparently produces or controls the sound). This turn is typical for film sound insofar as the "phenomenon of synchresis" leads us to identify not "the real initial causes of the sounds, but causes that the film makes us believe in" (Chion 2012, 49). For example, when the hypnosound is captioned as [Eyeballs Thrumming Loudly] in season four, the caption reinforces the visual context for the sound. The sound apparently emanates from the toad's eyes, so that's how the sound is captioned. Again, it's not surprising that film sound would direct us towards constructed as opposed to real causes. What's of interest here is that, as we will see in a moment, captioners don't agree on the causes and nature of the hypnosound. In other words, synchresis operates both at the level of film sound and in the minds and practices of captioners who interpret causes and caption them accordingly. **Media:** http:// ReadingSounds.net/chapter3/#hypnotoad-eyeballs.

The Hypnotoad captions do more than describe the hypnosound. They rhetorically perform and produce the sound. While it would be technically correct to caption the Hypnotoad sound as a "turbine engine played backwards," it wouldn't be contextually correct. The meaning of a sound is always contextual, "for no sound has objective meaning" (Schafer 1977, 137; see also 149). Captions do not simply convey the origin of a sound—or even what the sound *really* is in any objective, technical, or decontextualized sense—but what the sound is doing in a scene. That's why, as I explain in chapter 8, it doesn't work to simply describe a nonspeech sound objectively with onomatopoeia ("rrrrrr" or "engine roar") or some decontextualized phrase such as "turbine engine in reverse." Caption readers need to understand how sounds function in specific contexts, not merely what they sound like.

In addition to helping us recognize the importance of context in closed captioning (what sounds are doing, not just how they sound), the Hypnotoad also disabuses us of any lingering belief in the objectivity of sound. Over the course of nine years (2001–2010), the hypnosound has been captioned a variety of ways on *Futurama*, even as the sound and its contexts have remained constant across episodes. Indeed, consistency across Hypnotoad appearances is part of the running gag. The Hypnotoad seems to be limited to making only one sound and taking one action. With context held fairly constant, then, we can attribute differences across captions to the preferences of captioners as opposed to fluctuations in episodic contexts. Moreover, because the exact nature of the sound is unclear on the show—the Futurama Wiki (2012b) says that the Hypnotoad "emits a droning hum" but doesn't say where that hum comes from—we can also explore how captioners have drawn their own conclusions about the nature of the sound (see table 3.1). **Complete collection of Hypnotoad media:** http://ReadingSounds .net/chapter3/#hypnotoad.

Table 3.1 Closed captions for the Hypnotoad's sound on DVD, cable TV, and Netflix streaming, 2001–2010.

Futurama episode title	Hypnotoad caption	Source
"The Day the Earth Stood Stupid" (2001)	[Electronic Humming Sound] [Electronic Humming] (low humming) None	Netflix and DVD Cable TV
"Bender Should Not Be Allowed on Television" (2004)	[Eyeballs Thrumming Loudly] (sustained electrical buzzing)	Netflix and DVD Cable TV
Bender's Big Score (2007 DVD movie)	(droning mechanical sputtering) (MECHANICAL GRINDING)	Netflix and cable TV DVD
"Everybody Loves Hypnotoad" (DVD special feature on Bender's Big Score)	None	DVD
Into the Wild Green Yonder (2009 DVD movie)	None	Netflix and DVD
"Rebirth" (2010)	(LOUD BUZZING DRONE) (deep, distorted electronic tones blaring)	Netflix and DVD Cable TV
"Attack of the Killer App" (2010)	None (mechanical humming)	Netflix and DVD Cable TV
"Lrrreconcilable Ndndifferences" (2010)	None (electronic static)	Netflix and DVD Cable TV

Source: Twentieth Century Fox Television. http://ReadingSounds.net/chapter3/#table1.

Analyzing the Hypnotoad Captions

The hypnocaptions present a remarkable variety of possibilities for the same sound, despite the relative stability of the surrounding visual and rhetorical contexts. This variety raises a number of questions that relate directly to caption quality. For this analysis of the hypnosound, I used the official captions on DVD, Netflix streaming, and cable TV. I relied on the Futurama Wiki (n.d.) to gather a list of episodes and DVD movies in which the Hypnotoad character appeared. I streamed episodes on Netflix. I viewed DVDs. I compared the Netflix captions with the DVD captions and noted any differences. I also set my TV to record all episodes of Futurama. I identified two Hypnotoad appearances not listed on the Wiki because I also watched the show regularly and became culturally literate in the visual and sonic landscape of Futurama.

Should the hypnosound be captioned at all? At a basic level, we can distinguish between uncaptioned and captioned instances of the hypnosound in the same episode. Whether the sound should be captioned at all is apparently open for debate. In "The Day the Earth Stood Stupid," "Attack of the Killer App," and "Lrrreconcilable Ndndifferences," the

Table 3.2 Is it captioned? Differences between formats in the
Futurama episode "The Day the Earth Stood Stupid."

	Hypnotoad's sound	Audience's reaction
DVD/Netflix	Yes [Electronic Humming]	No
Cable TV	No	Yes (clapping in sync)

Source: Twentieth Century Fox Television, 2001. http://ReadingSounds
.net/chapter3/#table2.

same sound is either captioned or not depending upon the delivery format. For example, in "The Day the Earth Stood Stupid," the Hypnotoad
is the grand prize winner at the stock show. Upon hearing the hypnotized judge make the announcement, the Hypnotoad responds with its
trademark drone, which hypnotizes the audience. On the DVD/Netflix
version, the hypnosound is captioned as [Electronic Humming]. On the
cable TV version, the same sound is not captioned at all. Instead of the
hypnosound, the TV captions emphasize the sound of the hypnotized
audience: (clapping in sync). On DVD and Netflix, the audience's clapping is not captioned at all. In other words, the caption files for this
episode offer reverse images (see table 3.2). Granted, both sounds are
significant to the scene as cause to effect: The hypnosound causes the
audience to become hypnotized and clap in unison. But each sound is
contextualized differently. The act and sound of clapping have visual
and cultural salience over the hypnosound. We not only see the audience clapping but quickly recognize the gesture they are making as
clapping. It seems redundant to be told in the captions that the audience is clapping in sync. The hypnosound doesn't have the same visual
or cultural salience for viewers, which is precisely why it needs to be
captioned in this scene. Its low visibility (only the toad's eyes alert us to
something going on) doesn't make up for its lack of familiarity (what do
those warbling eyes mean?). An audience that immediately recognizes
the sight of clapping can't also be counted on to know the concept of a
toad that hypnotizes people.

In other words, captioning is not simply about the sounds themselves
but about the relationships among sounds, images, and the audience's
presumed knowledge (or cultural literacy). A sound should be analyzed
not only in terms of its sonic and contextual salience within a scene but
also in terms of its visual and cultural salience to the audience. Can the

sound be clearly seen and will audiences be familiar with it? The example of clapping reminds us that some sounds are visual and highly familiar. We can see the audience clapping; it is an action, a visual performance, as much as it is a sound. Clapping is also an action that most babies learn to perform (along with waving and pointing) at a very young age, making it an abiding cross-cultural gesture with a high level of recognizability. (Compare the sign for applause in American Sign Language, waving the hands in the air, which suggests that it may be more accurate to say that applause, not clapping per se, is cross-cultural.) But just because a sound is visible or familiar doesn't negate the need to caption it. Clapping sounds may still need to be captioned. Clapping and applause can take many forms, although the standard form is the familiar boilerplate or default option: [applause] or [audience applauds].

(SCATTERED CLAPPING)—*Lincoln* (2012)

[clapping in rhythm]—*The Master* (2012)

(LIGHT APPLAUSE)—*Zero Dark Thirty* (2012)

[CROWD CHEERING AND APPLAUDING]—*Argo* (2012)

Captioners may need to distinguish, for example, a sarcastic slow clap from a rousing ovation, especially if the sound can't be clearly seen or fluctuates over time (a slow clap turns into cheering). When there's not enough time to fully caption two consecutive and significant nonspeech sounds, the captioner should give more weight to the one with less familiarity and/or less visibility (in this cast, the hypnosound). While it would appear as though the captioners of this episode were forced to make such a decision, the fact is that there was plenty of time to caption the two-part announcement of the contest winner (106 words per minute [wpm] and 57 wpm), the Hypnotoad's response (63 wpm), and the audience's reaction (96 wpm). These speeds are well below 145 wpm, which was rated as the "most comfortable" reading speed by participants in Carl Jensema's (1998) well-known captioning speed study. I will have more to say about reading speed in chapter 5. **Media:** http://ReadingSounds.net/chapter3/#applause.

Why aren't the DVD extras captioned? One of the uncaptioned hypnosounds occurs throughout a special DVD feature of *Bender's Big Score*. The twenty-two-minute episode of "Everybody Loves Hypnotoad" features almost nonstop hypnosound action, but the entire episode is uncaptioned, which is not uncommon for DVD special features. An agreement by the major studios to close caption DVD releases has not extended to the special features of those DVDs. Some special features of DVDs are closed captioned; others aren't. This example reminds us of the need to continue to advocate for universal access for all multimedia content.

Should the sound's volume be identified? With the exception of the hypnosound placed into the background of a scene from *Into the Wild Green Yonder*, every instance of the hypnosound is approximately the same volume. Yet three hypnosound captions make a point of referring to the loud volume of the sound: [Eyeballs Thrumming Loudly], (LOUD BUZZING DRONE), and (deep, distorted electronic tones blaring). Moreover, as one of the readers of my blog pointed out, "humming" is by definition a softer sound (see "electronic humming," "low humming," and "mechanical humming"), which implies the sound's volume without stating it explicitly. (My dictionary offers a number of examples of quiet humming sounds: computers and insects hum, people hum a tune with closed lips, and a room can hum "with an expectant murmur.") But the hypnosound isn't soft by nature, which is reason enough to think twice about the appropriateness of "humming" as a descriptor. In addition, "low" (see "low humming") can mean deep or soft, providing a potential source of ambiguity that can mistakenly reinforce the quietness of humming. The hypnosound is deep but it is not soft. When a sound dominates the soundscape and the environment, captioners should be mindful to avoid descriptors such as "humming" and "low" that might work against the nature of the sound. Describing a sound as "loud" is always a popular option:

(LOUD MUSIC PLAYING)—*21 Jump Street* (2012)
(flies buzzing loudly)—*Aliens vs. Predator: Requiem* (2007)
(LOUD PROTESTS)—*Lincoln* (2012)
(LOUD RUMBLING)—*Moonrise Kingdom* (2012)
(GROANS LOUDLY)—*Skyfall* (2012)
[Loud Blubbering]—*Star Wars: Episode 1—The Phantom Menace* (1999)
(LOUD WHOOSH)—*Twilight* (2008)
(BLADES THRUMMING LOUDLY)—*Zero Dark Thirty* (2012)
[loud clattering]—*Silver Linings Playbook* (2012)

But "loud" can also become an easy fallback option when a more creative or distinctive term might be called for (e.g., "blaring"). **Media:** http://ReadingSounds.net/chapter3/#loud.

How do conventions and linguistic affordances shape the description of sustained sounds? Sounds that play continuously—*sustained* as opposed to *discrete* sounds—are often indicated in closed captioning with a present participle: *verb+ing* (see chapter 2). The Hypnotoad captions tend to follow this convention: humming, thrumming, buzzing, droning, sputtering, blaring, and grinding. "Drone" is an exception, but then again "drone" already suggests the sense of being continuous. "Drone" is also paired with a present participle, "buzzing": (LOUD BUZZING DRONE). "Static" might also count as an exception, but it too gives the sense of having a sustained, ongoing quality: (electronic static). One hypnocaption uses the word "sound": [Electronic Humming Sound]. But viewers already know that captions describe sounds, so adding the word "sound" to any caption is usually a waste of precious caption space. Putting "sustained" next to a present participle (as in "sustained electrical buzzing") might also qualify as unnecessarily redundant.

Which of the verb+ing *options best illustrates the hypnotic function of the hypnosound?* Are the present participle options mostly synonymous— humming, thrumming, buzzing, droning, sputtering, and grinding—or does one option suggest a greater degree of hypnosis? "Droning" might be the best candidate, since it appears to be paired with "hypnotic" much more often in a search of Google books (e.g., "hypnotic drone"). By this logic, "sputter" would appear to be the worst choice to be paired with "hypnotic." These results make sense to the extent that droning, by definition, can have hypnotic effects. The droning sound of a car engine or washing machine can lull a newborn to sleep. People who speak tediously and monotonously are said to drone on. A drone is a pilotless aircraft or worker bee that is controlled by or in the service of an external agent, just as hypnotized people are controlled by a hypnotist. In short, the most hypnotic of the action terms would appear to be "droning." **Media:** http://ReadingSounds.net/chapter3/#ngramviewer.

When multiple descriptors are used in the same caption, how well do they attune to each other? When multiple action words are yoked together in the same caption, we should ask whether the actions are compatible. In (droning mechanical sputtering), one word meaning *continuous* grinds up against another word meaning *sporadic* (i.e., not continuous). Contrast (LOUD BUZZING DRONE), which combines two roughly synonymous words (but see avoiding redundancy above). Most likely, the captioner of (droning mechanical sputtering) was combining the hypnotic qualities of droning with the engine-like qualities of sputtering. When building descriptions out of multiple action terms, captioners need to consider their connotations and how descriptors acquire meaning from being paired with (or attuned to) other terms.

Are individual captions accessible? Putting the hypnosound captions side by side also productively complicates our understanding of accessibility.

Usually, when we talk about accessibility for deaf and hard-of-hearing viewers, we refer to the entire TV program, movie, or web video. But what if we also asked whether specific captions were accessible? For example, "thrumming," the main action term in [Eyeballs Thrumming Loudly], is a somewhat obscure term. Thrumming is defined as a "continuous rhythmic humming sound" that an engine or musical instrument might make (cf. strum). The hypnosound is technically the sound of a turbine engine, so the descriptor fits. But is it an accessible option for viewers of *Futurama*, an animated series enjoyed by kids and adults? Thrumming only appears in one other movie in my corpus of DVD captions. In *Zero Dark Thirty* (2012), "thrumming" appears four times. All four instances are associated with the sound of helicopter blades whirring, while two are paired with a second simultaneous nonspeech caption, (ENGINE WHINING) and (ENGINES WHINING):

(BLADES THRUMMING)

(BLADES THRUMMING LOUDLY)

(THRUMMING FADES)

(BLADES THRUMMING)

Assuming we know what thrumming means (a big assumption), it is probably easier for audiences to process the concept of copter blades thrumming than eyeballs thrumming. What does it mean for eyeballs to thrum, anyway? Should thrumming be replaced by a more accessible, potentially more familiar and common term such as humming, whirring, or droning? I'm not suggesting that nonspeech captions need to be dumbed down or reduced to default placeholders or overused boilerplates devoid of creativity. In fact, I find [Eyeballs Thrumming Loudly] to be the most provocative and inspired Hypnotoad caption of the available choices. Rather, I'm suggesting that we need to reframe accessibility around the needs of audiences in the context of specific programs. Accessibility is more than a summary judgment applied to an entire program. One commenter on my blog put the problem this way: "I know most of my former students would not understand humming, thrumming, buzzing etc. and it would require creative teaching to bring these ideas into the deaf experience. I have experienced electronic noise but from what I understand about frogs they don't buzz, thrum, hum etc. I'm lost." How do captioners ensure not only that the sound is captioned accurately but that the words don't leave viewers lost? Viewers may know that a frog makes a croaking sound but be confused by the notion of a fictional frog that sounds like an airplane engine. Again, captioners must be mindful of which sounds and contexts are likely to be familiar to viewers and how to translate with the needs of audiences in mind. **Media:** http://ReadingSounds.net/chapter3/#thrumming.

Captions should also reflect the spirit of the content being captioned. Nonspeech sounds are embedded in specific contexts and captions must reinforce those contexts. Let me offer a crude but particularly compelling example. There are a large number of euphemisms in English for flatulence, from the more polite (passing wind, passing gas), to the medicalized and less familiar (flatulence), to the popular and somewhat profane (farting), to the strictly comedic (crop dusting, stepping on a duck). Captioners have a wide range of options for captioning flatulence, although probably only a small number of terms are potential candidates for nonspeech captions. Fart jokes are pretty common on *South Park*, a long running animated program on Comedy Central that is known for being crude, if not downright offensive, but also politically keen and thought provoking. Fart captions on *South Park* should reflect the juvenile contexts in which the farting sounds occur. In "Ass Burgers" (season 15, episode 8, 2011), Stan (voiced by Trey Parker) becomes cynical and jaded after his tenth birthday. One character offers this explanation to Stan: "You see everything as shit, don't you?" The episode interprets this explanation literally: Stan sees feces in everything and hears fart noises everywhere. As the episode opens, Stan is waking up to a pair of radio shock jocks who are "talking about the new hit movie," but Stan only hears [flatulence] interspersed throughout:

BIG HARRY AND MIKE
IN THE MORNING

TALKING ABOUT THE NEW
HIT MOVIE [flatulence].

IT'S A GREAT COMEDY.
IT'S STARRING

[flatulence]
AND [flatulence].

To be clear, the DJs aren't actually making these fart sounds, but that's how Stan interprets what he hears and what the audience actually hears. While [flatulence] is technically accurate, it's also overly formal and not appropriate in this situation. In terms of accessibility, [flatulence] is also likely to be less familiar to viewers. It sounds like an outsider's description, not a description of what Stan hears. To Stan, these are fart noises, not flatulence sounds. The *South Park* kids would never refer to farting as flatulence. A juvenile term is called for, a term that reflects Stan's attitude towards his shitty world and the childish antics of the radio DJs, a term that also reflects the crude and often scatological humor of *South Park*. By inhabiting the situation, captioners can channel their descriptions of nonspeech sounds through the characters who experience them. These sounds belong to Stan—he is the only one who hears them—and they should be captioned in a way that resonates with Stan's experience. We could perform a similar analysis of (sotto voce), which occurs occasionally in nonspeech captioning as a less accessible way to note an aside or under the breath remark that's not intended to be heard by others. **Media**: http://ReadingSounds.net/chapter3/#flatulence.

What is the nature of the hypnosound? Where does it originate? The nature of the sound is described as either electronic, mechanical, or biological (i.e., emanating from the eyeballs). Every caption, except perhaps for (LOUD BUZZING DRONE), describes the sound using one of these three options. Electronic/electric is used four times (five if you add the second occurrence of "electronic" in the same caption file for "The Day the Earth Stood Stupid"). Mechanical is used three times. The biological (eyeballs) option appears once and only in an implied form. But the biological option is perhaps the most compelling for caption studies, for only in [Eyeballs Thrumming Loudly] does the sound find a specific location in or around the toad. The different options cannot simply be chalked up to the nature of the sound or what the producers of the show have said about the sound (because, as far as I can determine, the producers have given no indication of how the hypnosound should be described or where it comes from). We must appeal to the captioner's subjectivity—the subjective practice of closed captioning itself—to fully account for the differences among the options. In this case, captioners have offered a range of competing explanations for the location and origin of the hypnosound.

The hypnocaptions remind us that the concept of a single, authoritative caption file does not exist in closed captioning. For any TV show or movie that has been subject to redistribution (e.g., a movie that has been rebroadcast and recaptioned for network TV), multiple caption files will be available, and each one will be official. In analyzing closed captions, we have to give up the practice, common in rhetorical analysis, of finding and analyzing the single most authoritative or "final" text. The

Table 3.3 Build your own hypnocaptions.

First term	Second term	Third term
Deep	Buzzing	Blaring
Droning	Distorted	Drone
Eyeballs	Electrical	Buzzing
Electronic	Grinding	Loudly
Loud	Humming	Sound
Low	Mechanical	Sputtering
Mechanical	Static	Tone(s)
Sustained	Thrumming	

Note: Terms in the table have been culled from the available options. First terms have been placed into column one, second terms into column two, and third terms into column three. For the sake of efficiency, I've reduced (deep, distorted electronic tones blaring) to make the terms fit into three columns. Use any combination of terms. Move back and forth among the columns to create longer descriptions if desired. Send your user-generated captions to the author. Try the interactive version: http://ReadingSounds.net/chapter3/#table3.

notion of an *urtext*—the original or earliest version—is foreign in closed captioning. While we might be able to locate the earliest caption file for a program, it doesn't make sense to treat it as the version to which later versions should be compared. Captioners don't typically consult the work of their predecessors anyway (making comparisons problematic), and later versions are no less authoritative than earlier versions.

Even something as seemingly minor as the Hypnotoad's sound has the potential to raise important questions for caption studies. I've avoided being overly critical of specific hypnocaptions because I wanted to account for (and revel in) the impressive range of options that captioners have offered to describe the same sound (and because there isn't one correct hypnocaption). These options vividly show how captioning is a deeply subjective and context-dependent practice. In table 3.3, all of the possible options have been placed into a grid in the spirit of a mix-and-match game.

A New Method of Analyzing Closed Captions

The style of analysis I've offered here is new to caption studies. Or rather, the development of caption studies depends on our willingness to dig into the data in ways we haven't before. Captioning advocates have never before gathered together different captions for the same

sound across different episodes and media formats of the same show—
and then evaluated the various options. This new method of caption
analysis can be summarized as follows:

Take a significant sound that recurs across episodes of a TV show or the scenes of a
movie, preferably a nonspeech sound. Chart that sound and its captions across the
cinematic landscape. Use only the official closed captions. Note any similarities and
differences across episodes but also within the same episode on different formats (e.g.,
DVD, cable TV, Netflix streaming, etc.). Analyze and evaluate.

Despite generating a wealth of insights, this method of analysis is limited. It can only speculate about why the captions are different. We
know that WGBH has been the longtime captioner of the *Futurama* TV
episodes, because WGBH is listed as the captioning company at the end
of every episode broadcast on TV. But we don't know how many different captioners have worked on the show or what practices, if any, they
followed to ensure continuity across episodes (what I refer to as "series
awareness" in chapter 7). Moreover, we can only speculate, given the
agreement between the Netflix and DVD captions, that the Netflix caption files were generated from the original DVD caption files, whereas
the TV captions were most likely produced separately. (Only in the case
of the straight-to-video production, *Bender's Big Score*, do the Netflix
captions for the Hypnotoad depart from the DVD captions.) While
we can interview captioners about their experiences and practices, we
can't recreate the original contexts that produced a ten-year-old caption or hope to track down the captioner(s) originally responsible. We
don't know whether the same captioner worked on multiple episodes
or whether subsequent captioners consulted the work of their predecessors. Outside researchers and advocates who weren't there when the
captions were produced will most likely have limited access to production scripts and other written materials, beyond what is widely available
online (although scripts will not typically include textual representations of nonspeech sounds anyway). Ethnographic studies from inside
captioning firms are needed to explore the decision making processes,
organizational values, client demands, and economic constraints that
shape the production of closed captions. I encourage other scholars in
caption studies and web accessibility to take on these projects in concert
with the method of analysis I've outlined here.

The flexibility of this method can be applied to one recurring sound
on the animated TV series *Family Guy*: the so-called hurt knee gag that
extends across a number of episodes (figure 3.2). In fact, this chapter

3.2   Peter hurts his knee in a recurring gag on Fox's *Family Guy*.
A side by side comparison of two identical frames from an episode of *Family Guy* ("Wasted Talent") that is part of the recurring hurt knee gag on the show. Peter sits on the sidewalk outside his house, grabbing his left knee with both hands. His eyes are closed, his mouth is open. He looks like he's in pain. The left frame shows the plain text caption track (Caption: "AHH!") and the right frame shows the bitmap caption track (Caption: [Howling in pain]). Twentieth Century Fox Television, 2000. DVD. http://ReadingSounds.net/chapter3/#figure2.

could have been organized around the hurt knee gag instead of the Hypnotoad. But the contexts for the hurt knee are more varied, involving multiple characters and situations into which the gag is inserted. The context for the Hypnotoad tends to be stable, akin to a cutaway gag when an episode shifts to the self-contained and seemingly unchanging universe of the Hypnotoad's TV show. A stable context allows us to analyze the differences in the captions on their own terms. Regardless, the hurt knee gag provides another opportunity to chart the differences in captions for the same sound. In the hurt knee gag, a character, usually Peter Griffin (voiced by Seth MacFarlane), falls down and grabs his knee in pain, repeating the same sequence of sounds—sucking in air through his teeth and exhaling it with an open-mouthed "ahhhhh" sound—as he rocks back and forth. (In Internet memes, examples of which are included on the book's supplemental website, the hurt knee gag is written as some version of "sssss-ahhhhh," where "sssss" is the intake of air through the teeth and "ahhhhh" is the pained exhalation of air.) The sequence usually goes on for too long, which is part of the gag. For example, in "Wasted Talent" (season 2, episode 20), the first appearance of the gag in the TV series, Peter rocks back and forth seven times, sucking in air and exhaling each time. The entire twenty-seven-second hurt knee sequence, complete with fourteen sounds plus the original "ahhhhh" when he falls, receives only one non-speech caption: [Howling in pain]. This caption appears with Peter's second "ahhhhh" sound and stays on the screen for two seconds. Viewers can see that Peter is audibly expressing his pain. Caption readers don't need to be hit over the head with fifteen captions that essentially repeat the same two-sound sequence seven times. **Media:** http://ReadingSounds.net/chapter3/#hurtknee.

In [Howling in pain], function and meaning take precedence over phonetic description. This example comes from the bitmap track on one of the official *Family Guy* DVDs for season 2. Bitmaps are images composed of bits of color arranged on a map or grid. Individual characters or letters are rendered as bitmap images and decoded with a DVD player. Because the letters are pictures, they can't be resized or otherwise reformatted by the user. DVDs will usually include multiple caption and subtitle tracks, although only one track can be displayed at a time with a DVD player. Different formats satisfy different specifications and allow different devices to process the caption or subtitle data. On some DVDs, the bitmap track contains subtitles (speech) only, while a separate text track carries the closed captioning (speech and nonspeech). On other DVDs, multiple caption/subtitle tracks may include the same information. So-called "Subtitles for the Deaf and Hard of Hearing" (SDH) contain both speech and nonspeech information, troubling the distinction between subtitles and captions (Neves 2008). The *Family Guy* DVDs include a bitmap track and a plain text track, but both contain speech and nonspeech information (though not identical information). When we look at the same scene of "Wasted Talent" through the lens of the text captions, we see the influence of a captioning style not based on describing the meaning of the sound (Howling) but on channeling its phonetic qualities (Ahh!) (see table 3.4). Instead of a single nonspeech caption, the text captions describe all eight exhalations in the hurt knee sequence: "AHH!" is repeated eight times and interrupted by three [Whimpers], each of which accounts for one inhalation sound. Why only three of seven inhalations are captioned, even though all eight exhalations are accounted for, is unclear. Also unclear is why the exhalations are captioned phonetically while the inhalations are described in terms of their function. Finally, what should we make of the gulf in meaning between howling and whimpering? If "Ahh!" is ambiguous because it can refer to pleasure or pain, then any term paired with it will inform its meaning. In other words, the meaning of "Ahh!" is inflected differently when it is paired with howling than when it is paired with whimpering. Whimpering and howling are closer to antonyms than synonyms. The pained expression on Peter's face suggests that howling is a more faithful interpretation of what Peter is feeling. Regardless, this examples shows how different captioners can arrive at divergent interpretations of the same situation.

Table 3.4 A comparison of the DVD text captions and DVD bitmap captions for the five instances of the hurt knee gag in *Family Guy*.

Episode	Bitmap captions	Plain text captions
"Wasted Talent" (2.20, 2000)	Run home, Peter! Run as fast as you can! [Cheerful instrumental music] [Howling in pain]	RUN HOME, PETER! RUN AS FAST AS YOU CAN! AHH! [Whimpers] AHH! [Whimpers] AHH! [Whimpers] AHH! AHH! AHH! AHH! AHH!
"Wasted Talent" (2.20, 2000)	CHUMBA WUMBAS: [Singing] "Chumba Wumba gobble" [Howling in pain]	♪ CHUMBA WUMBA GOBBLE ♪ AHH! AHH! [Whimpers] AHH!
"Brian Goes Back to College" (4.15, 2005)	(WINCING) (GROANING)	Same
"Fox-y Lady" (7.10, 2009)	Now run home, Lois! Run as fast as you can! (*THE GOLDEN TICKET* PLAYING) (EXCLAIMING IN PAIN)	Same
"Something, Something, Something, Dark Side" (8.20, 2010)	(EXCLAIMING) (GROANING)	Same

Source: Twentieth Century Fox Television, DVD. http://ReadingSounds.net/chapter3/#table4.

Placing caption tracks side by side generates some productive tensions for caption studies. We not only can compare the choices made in each instance but also can ask questions about those choices and the approaches to captioning that inform them. When describing the repetitive, two-part hurt knee sequence, which term is best: exclaiming, howling, groaning, whimpering, wincing, or something else? Or rather, which term captures the pained expression on Peter's face in "Wasted Talent"? Should a more phonetic substitute be used instead, such as "Ahh!" or even "Sssss," while bearing in mind the ambiguity of phonetic description? Does every instance of a repeating sequence need to be captioned, or is one caption sufficient? When do captions need to cite the cultural touchstones or "prior texts" that inform them? For example, "I've Got a Golden Ticket," a song made famous in *Willy Wonka and the Chocolate Factory* (1971), connects two instances of the hurt knee gag together. In

"Wasted Talent," an instrumental version of the song is captioned in the bitmap track as [Cheerful instrumental music], but not captioned at all in the text track. In "Fox-y Lady" (2009), the extended gag is duplicated almost frame for frame but this time with Lois Griffin (voiced by Alex Borstein) running down the street and falling in front of their house. On both caption tracks for this episode, the song is captioned as (THE GOLDEN TICKET PLAYING). While "The Golden Ticket" is not technically the title of the song, this caption, unlike the vague reference to cheerful music in "Wasted Talent," sufficiently alerts viewers to interpret the sequence as a spoof of the scene in *Willy Wonka* when young Charlie Bucket (Peter Gardner Ostrum) finds the last remaining Golden Ticket hidden inside a chocolate bar wrapper. After a crowd forms around Charlie, a passerby yells: "Run for it, Charlie. Run straight home, and don't stop 'til you get there." *Family Guy* streamlines this imperative as: "Run home, Peter [Lois]! Run as fast as you can!" Without access to the name of the song, it's possible that caption readers will miss the deeper significance of the scene as a parody of a classic movie scene. Recognizing the song as more than cheerful music requires cultural literacy, a topic I take up in chapter 7. **Media:** http://ReadingSounds.net/chapter3/#golden-ticket.

Beyond the Hypnotoad

The Hypnotoad isn't an isolated or special case. Captioners are regularly confronted with sounds that demand subtlety and creativity. Such examples demonstrate the extent to which captioning is deeply contextual and subjective, less like science and more like art. A single nonspeech sound can be captioned in multiple ways. Consider "breathy" sounds such as gasps, scoffs, pants, sighs, grunts, and heaving breathing. The differences between a gasp and a pant may come down to more than just the objective qualities of the sounds themselves. Facial expression and context influence how a sound should be described. Similar "breathy" sounds may warrant different captions, just as markedly different sounds may rely on the same caption. With this in mind, consider a compilation of breathy, nonspeech sounds from *Twilight* (2008), a movie I only half-jokingly call the "gaspiest" movie in the world. Dramatic breathing, particularly from Bella Swan (Kristen Stewart), plays a recurring, visible, and captioned role in the film. Bella gasps, sighs, grunts, screams, yells, giggles, and pants her way through the narrative. For example, of the 190 nonspeech descriptions in *Twilight*, thirteen involve gasping, and twelve of these are associated with Bella. Gasping is used in the caption track as an all-purpose placeholder for the audible intake of air when Bella is *scared* (during the attempted sexual assault in Port Angeles), *startled* (when she wakes to find Edward [Robert Pattinson] watching her sleep), *aroused* (when she and Edward are kissing), *excited* (when she is lifted by Edward high into the trees), and *dying* (when Bella gasps three times within a span of thirty-five seconds following the nearly fatal vampire attack at the end of the film). In fact, when we analyze Bella's character through the nonspeech descriptions associated with her, we have to conclude that gasping is one of her key personality traits, at least as far as the captioner was concerned. Bella gasps. We could continue this line of analysis, identifying the nonspeech sounds associated with Edward, Jacob Black (Taylor Lautner), the evil vampires, and so on. We could also analyze any nonspeech descriptors that are associated with and bond together multiple characters. At one point, for example, when Edward and Bella are kissing for the first time, Bella (GASPS SOFTLY), but Edward breathes heavily too during this scene. In fact, it's possible that the caption was meant for him. It's hard to tell because the DVD captions for *Twilight* do not use placement to indicate who is speaking. In this moment, they are bound together around a single caption spanning multiple breathy sounds between them as they embrace. Finally, we might also explore whether some descriptors, like gasping, giggling, and growling, are gendered because they are overwhelmingly associated with men or women. If captioned gasping is feminine in *Twilight*, it may have different connotations in another movie. At the end of *Return of the Jedi* (1983), for example, Darth Vader lays dying with [GASPING BREATHS THROUGHOUT]. **Media:** http://ReadingSounds.net/chapter3/#gasping.

Captions also create contexts; they don't simply respond to them. Patterns and themes which might otherwise be latent in a film become manifest when captioned. Speech and nonspeech sounds can become indelibly linked together on the caption layer. I use the term *captioned thematics* to refer to the potential of closed captioning to make themes manifest. Consider *Sunshine Cleaning* (2008), a movie starring Amy Adams and Emily Blunt as two sisters who start a business cleaning up crime scenes and biohazard waste. The movie opens with a suicide. Sitting in his car, an unnamed man prepares to enter a gun store, where he will purchase a gun and shoot himself before leaving the store. The first caption of the movie is (spritzes), which is the sound of the man spraying breath freshener into his mouth, no doubt foreshadowing another more gruesome kind of spraying when bullets and blood fly inside the gun store and spray the man's body. The nonspeech sound of "spritzing" is explicitly linked in the closed captions to a later scene involving the same word uttered by one of the main characters. While cleaning up a grisly bathroom, blood everywhere, one sister commands the other sister to "Spritz." The sisters are linked together in this moment—one sister says the word and the other performs the action by spraying some cleaner on the shower wall for the first sister to scrub away. And both, of course, become linked to the unnamed man who freshened his breath at the beginning of the movie. Even though they are not cleaning up the spritzing man's blood, they are cleaning up another similar crime scene. While this thematic link is also available to hearing viewers who are watching without closed captioning—assuming they are already attuned to the subtle thematic potential in every film—it comes to the fore on the caption layer. I don't think I would have quickly identified the breath freshener sound as spritzing without the captioner identifying it for me. The captions forge a clear line of connection leading from the spritzing man to the spritzing sisters. All three are engaged in acts of cleaning. He cleans his breath and they clean up the mess he leaves behind. His spritzing is directed inward, towards the self—just as suicide is—and theirs is directed outward, towards restoring order in the world following tragedy and death. Closed captions bring these thematic patterns to the surface and increase their intensity for viewers. **Media:** http://ReadingSounds.net/chapter3/#spritzing.

Another example of a thematic pattern that is activated on the caption layer is found in *Hick* (2012), a movie "about a damaged 13-year-old girl who runs away from an alcohol-soaked home and encounters only hateful, cruel or moronic people" (Ebert 2012). The girl, Luli (Chloe Grace Moretz), meets Glenda (Blake Lively) on the road after she runs away. Glenda introduces Luli to what is likely powdered cocaine. After Luli snorts a small amount, we see the world through Luli's buzzed eyes. The first caption after Luli's (Snorting) is a song lyric, "♪ Sweet dreams ♪," which is the first line in Patsy Cline's "Sweet Dreams" and a reflection, perhaps, of Luli's dreamy mental state. (The song title and artist are not captioned but should have been; Cline's song is a country standard and perfectly accompanies the pair's travels down a country road.) The road rushes by from Luli's (and the viewer's) vantage point in the passenger seat. To give us a sense of how Luli is feeling under the drug's influence, the scene is shot with a high contrast lens filter, blurred effects, and adjustments to chronological time (i.e., the footage is sped up and slowed down). Later, when her buzz is presumably wearing off, Luli asks Glenda, "you got any more of that stuff?" Glenda responds: "No kid, fresh out." The scene then shifts to the outside of a bowling alley, "Blane's Lanes." The neon B in the sign flickers as it prepares to burn out, hovering between life and death, on and off. The sound of the flickering light is captioned as (Buzzing), which immediately calls to mind Luli's buzzed state (even though a drug-induced buzz doesn't make any sound). The thematic connection between Luli and the buzzing neon light suggests that Luli's altered state is similar to the light's. Both are in the process of turning off or going out. Because all of the captions in this movie are mistakenly timed to appear about 1.7 seconds early, the (Buzzing) caption appears on the screen before the new scene shifts to the bowling alley and the flickering sign. In other words, we see the (Buzzing) caption while Luli is reacting to Glenda being out of the "stuff." While the caption is visually associated much longer with Luli (1.49 seconds) than the neon sign (0.21 seconds), it still hovers between two signifieds, one human (Luli) and the other nonhuman (the neon sign), just as both signifieds also hover between two states (on/off, high/low, buzzed/still, etc.). This caption was clearly intended to be associated with the neon sign, but the premature timing unintentionally associates it with Luli too—and thus creates the condition for the term's double meaning. Perhaps the director intended the audience to make this connection. But that would require a pretty big interpretative leap without the caption, especially when the term "buzzing" or "buzzed" is never used in the film. The caption makes this thematic connection possible by bringing it to the surface, while the poorly timed caption compels us to try to make sense of why Luli might be the source of (Buzzing). **Media:** http://ReadingSounds.net/chapter3/#buzzing.

When the same caption is repeatedly associated with a specific character or recurring context, it comes to serve as a kind of leitmotif for that character or context. In classical music, especially the operas of Richard Wagner, a leitmotif refers to a musical phrase, melody, or flourish that is associated with a specific character or situation. The leitmotif can (re)introduce characters and situations as well as support recurring emotional contexts. In *Star Wars* (1977), for example, composer John Williams created musical themes for a number of different characters (Luke, Leia, Yoda, Darth Vader, etc.) and different settings (battles, fanfares, and marches). In literary works, leitmotifs are word phrases or patterns that are thematically linked to specific characters or settings. Similarly, nonspeech captions can link specific words or descriptions to specific characters at a level that is only available within the captioned viewing experience. For example, in *Aliens vs. Predator: Requiem* (2007), (guttural croaking) is repeated twenty-two times over the course of the movie and always in the context of the Predator's animalistic noisemaking. No one in the movie utters the word *guttural* or *croaking*; these words only exist in the captions. Sometimes, viewers encounter the captioned description before they see the Predator. Given that the Predator is usually couched in darkness and sometimes resembles little more than a foreboding dark outline, (guttural croaking) becomes an identity marker as much as a description of a nonspeech sound. The recurring caption is the Predator's leitmotif, as it repeatedly announces his presence in the narrative. **Media:** http://ReadingSounds.net/chapter3/#gutturalcroaking.

These thematic patterns, which are suggested through the captioner's process of interpreting the soundscape, or through poorly timed or poorly placed captions, come to serve as the film's permanent, official record. Spritzing is thematized on the caption track and becomes part of the meaning of the movie. The caption track is not incidental or an add-on for the millions of Americans who depend on quality captioning. The captions provide the official meaning of the text. They travel with the text and, in the case of popular media such as movies and TV shows, cannot be changed. They may seem insubstantial and fleeting because they are mere text and can be turned on and off at will. But they are just as enduring and permanent as the movies they support.

Epilogue

The Hypnotoad remains a popular, albeit minor, recurring character on *Futurama*. As I was finishing up the first draft of this chapter in the summer of 2012, the Hypnotoad was making yet another appearance on *Futurama*. In "Decision 3012" (season 7, episode 3, 2012), the Hypnotoad appears twice. Neither appearance is closed captioned on the first-run version on Comedy Central. The first appearance is akin to an inside joke, a tease, a gift for hardcore fans of the show. While I'm not a hardcore fan, I am nevertheless intimately familiar with the nature of the hypnosound and can recognize it—the sound is always the same—even when I'm not watching intently. While working on my laptop, I heard the sound first and then looked up at the TV to catch a glimpse of the toad. A flurry of activity followed as my family kindly reversed the DVR so I could see exactly what was going on. The Hypnotoad is easy to miss in its first appearance on "Decision 3012." The sound is initially loud, but only momentarily, as the scene opens with the crew gathered in the break room. We can't immediately see the source of the sound on the TV or the TV itself. Unless we are familiar with the hypnosound (associating it with the Hypnotoad's fictional television show), we don't even know for sure that the crew is watching TV. No captions alert us to any sound coming from the television. When Bender (voiced by John DiMaggio) comes in two seconds later and initiates a dialogue with the other crew members, the hypnosound is immediately reduced to a quiet ambient rumble for the rest of the scene. We then catch a glimpse lasting 3.3 seconds of the Hypnotoad on the TV screen behind Bender. That's our only visual clue, but it comes five seconds after the loud hypnosound has turned to a quiet, almost inaudible background whirr. In other words, the sound is detached from its source. We hear the hypnosound without seeing it and then see the source without hearing it. The same sense of detachment is characteristic of some ambient and environmental sounds that can be heard but not seen, or heard but not easily reduced to a single source or single location (e.g., diffuse, complex city noise). It's easy to miss the hypnosound in this quick glimpse unless you know what to listen for. Drawing on prior knowledge of the toad's importance to the series, the captioner must recognize the significance of the hypnosound in the absence of any synchronized visual clues. Even a quick tease may need to be captioned. The second appearance of the Hypnotoad later in this episode fulfills the promise of the initial tease: the classic Hypnotoad now fills the frame (akin to figure 3.1) but with a twist in keeping with the episode's focus on political campaigning. According to the gag, a popular political candidate, Chris Travers, has purchased air time on *Everybody Loves Hypnotoad*. The candidate's face flashes numerous times within the Hypnotoad's mesmerizing eyeballs. The hypnosound is characteristically loud when the bit opens but not captioned. Instead, a news report fills us in on the details: "REPORTER: Chris Travers wins the South California primary handily, thanks to a series of well-placed cameo appearances." **Media:** http://ReadingSounds .net/chapter3/#hypnotoad-decision3012.

As "Decision 3012" is an episode from the current season, it was not yet available on Netflix or DVD at the time of writing. Hence, no comparisons to other captioned versions of the episode could be made. Nevertheless, this most recent appearance of the Hypnotoad raises questions that other chapters will take up again, especially chapter 6 on captioned silences and chapter 7 on series awareness: How should ambient, quiet, and background sounds be captioned? What if a particularly quiet background sound is in fact significant to the series itself, as the Hypnotoad's drone has become? These two facets of captioning—how to squeeze a multilayered soundscape into a highly constrained space, and how to measure the importance of sounds on the basis of their global relevance across episodes, not merely on the basis of their volume or visual prominence—remain key challenges for captioners. The Hypnotoad doesn't produce just any sound on *Futurama*. As a cultural touchstone on the show, the Hypnotoad's sound—like other significant sounds that span episodes on other TV series—cannot be dismissed as just another ambient sound, even when it competes with presumably more relevant speech sounds or cramped caption spaces. Let the Hypnotoad's popularity as an Internet meme—"All glory to the Hypnotoad" (Know Your Meme 2012)—be a reminder to captioners that some sounds are more important than others within a show's sonic universe. If you aren't a fan of the show, it's easy to miss the importance of a fleeting or background sound. Ideally, captioners should be die-hard fans of the shows they caption. Otherwise, they risk passing over, not understanding, or downplaying significant sounds.

Although it hasn't been previously discussed in terms of rhetoric, closed captioning is not unlike other rhetorical or compositional practices that demand sensitivity to audience, context, purpose, and genre. Captioners filter and interpret sounds; they don't simply transcribe them. Captioners reconstruct contexts; they don't simply record them. Captioners negotiate the needs of audiences, work within the space-time constraints of alphabetic literacy in motion, and interpret the intentions of content producers. The preceding analysis of a single recurring sound raises questions for caption studies that lead us away from accessibility as a summary judgment (e.g., is the movie captioned or not?) and towards accessibility as a situated, rhetorical assessment. The example of the Hypnotoad reminds us not only that sounds are subjective but that closed captions are too. Both become laden with meaning only in specific contexts.

FOUR

Logocentrism

Closed captioning encompasses the entire range of significant sounds, both speech and nonspeech sounds. Speech sounds are not inherently more important than nonspeech sounds, even if the latter (including speaker IDs) may comprise only a fraction (anywhere from 10 to 40 percent) of the total number of captions in a two-hour movie (see chapter 2). But in some cases, captioners privilege speech sounds at the expense of nonspeech sounds. Despite a growing awareness of the importance of closed captioning among the general public, closed captioning is still routinely confused with speech-only subtitling. For example, Google's automatic speech recognition technology, available on YouTube, has increased the public's awareness of closed captioning, even as this technology has defined captioning in terms of speech sounds only. In short, Google's automated captioning technology does not transcribe nonspeech sounds. It is essentially a subtitling service masquerading as a captioning service. Moreover, as we've seen, a number of definitions of closed captioning also reduce captioning to speech sounds. Narrow definitions that simplify captioning by equating it with the "spoken part" need to be countered with examples of robust, full-spectrum captioning for deaf and hard-of-hearing viewers.

We need to remember that closed captioning is not subtitling. Foreign-language subtitling provides access to the spoken language for viewers who are already presumed to be hearing. The spoken part of a French movie, for example, is subtitled in English for hearing viewers who read English

4.1 **Please maintain visual contact.**
A close-up of Roy Miller (Tom Cruise) wearing sunglasses in a frame from *Knight and Day*. His head fills the entire height of the frame. Caption: "Ladies and gentlemen, please maintain visual contact." What's significant and ironic about the accompanying clip is that these words spoken over the airport PA are impossible to make out without captions, regardless of one's hearing status. In other words, the only way to access these words is to maintain visual contact with the caption track. Twentieth Century Fox, 2010. Blu-Ray. http://ReadingSounds.net/chapter4/#figure1.

but not French. While English subtitles provide some level of access for deaf and hard-of-hearing viewers, they do not provide a sufficient level of access. Subtitles only account for speech sounds, leaving out nonspeech sounds and information about sounds that hearing viewers are assumed to have access to already. For example, the sounds of wolves howling would not be included in the subtitle track because subtitle viewers are assumed to be able to hear and understand nonspeech sounds. Definitions that reduce closed captioning to the "spoken part" confuse subtitling and captioning. Josélia Neves's (2005) PhD thesis on "subtitling for the deaf and hard of hearing," for example, starts from the assumption that captioning is an inferior variant of subtitling. As Joe Clark (2006a) puts it in his critique of Neves's thesis:

If you start from the vantage point that captioning is really subtitling, except that it isn't *really* subtitling, it's subtitling *for the deaf and hard-of-hearing*, then you immediately start out with a patronizing attitude based on translation theory. Captioning is not translation, and, while "audiovisual translation" as a field of study may occasionally intersect with captioning, it is a fatal mistake to act as though [captioning is a variant of the "Platonic activity" of subtitling].

When equated with or treated as a subset of foreign-language subtitling, captioning is reduced to speech within an economy organized around translation. Captioning advocates need to challenge methods of captioning that downplay the distinction between speech-only subtitling and captioning for deaf and hard-of-hearing viewers.

This chapter is organized around two approaches to closed captioning—what I call *undercaptioning* and *overcaptioning*—that reflect major confusions over what it means to provide full access for deaf and hard-of-hearing viewers. Whereas undercaptioning equates closed captioning with subtitling by leaving out or minimizing nonspeech sounds/information, overcaptioning artificially and intrusively elevates speech sounds, even when such sounds only serve as background, ambient, or "keynote" noise.

Undercaptioning: Muting Nonspeech Sounds

Closed captioning is more than a written record of speech sounds. But when captioning is confused with the technologies or practices of subtitling, nonspeech sounds can be squeezed out. Consider the current state of speech recognition technology, which, thanks to Google, is now easily confused with closed captioning. Google's so-called "automatic captions," which debuted at the end of 2009 and have since that time been the subject of much criticism, "combined Google's automatic speech recognition (ASR) technology with the YouTube caption system to offer automatic captions, or auto-caps for short" (Google 2009). But speech recognition is not the same thing as closed captioning. By conflating the two practices, Google reduces captioning to the transcription of speech sounds. Captioning becomes aligned with "machine translation" (e.g., see Cutts 2009): English subtitles become simply one of many language options for viewers within an economic system in which the only remaining barrier to access (now that space has been conquered) is assumed to be linguistic rather than auditory. In short, people are assumed to be separated only by language and not hearing status.

Speech recognition technology enthusiasts hold out the promise of a fully captioned web, but their assessments of quality have focused on "caption fails" (speech accuracy) at the expense of asking larger questions about access. Even if or when Google improves their speech recognition technology to 100 percent accuracy, it still won't be sufficient to provide full access to people who can't hear. Google's autotranscription technology on YouTube doesn't recognize or account for nonspeech sounds. Also absent are nonspeech identifiers for speaker, language, and media type (e.g., sounds coming over the radio, public address announcements, etc.). Nonspeech sounds need to be added manually to a YouTube video that's been transcribed automatically by Google's technology. (See Wald and Bain [2007] for an approach that uses human editors to correct speech recognition transcripts and supplement them with non-speech captions such as [Laughing]). Automatic speech recognition researchers have been focused on contexts in which speech dominates and nonspeech sounds such as [laughter] and [applause] are rare and limited in type: the classroom lecture (Papadopoulos and Pearson 2008), business meetings and conferences (Wald and Bain 2007), and news broadcasts (Pražák et al. 2012). Because nonspeech sounds play a minor role in these contexts—for example, the classroom lecture mainly presents a speech transcription problem—these contexts allow researchers to make strong claims about accessibility. These claims tend to hold up if we assume that (1) the researchers' speech recognition technology produces an accurate transcription of speech (not yet possible), and (2) the context is speechcentric, such as a classroom. If there are no (or very few) significant nonspeech sounds to be accounted for, then speech recognition technology can indeed be claimed to provide an accessible solution. But we need to be careful not to mistake what speech recognition technology can do, even at 100 percent accuracy, with what deaf and hard-of-hearing people require, particularly on captioned TV shows and movies, which rely on significant nonspeech sounds. An automated transcript of a YouTube video does not produce closed captioning, despite what that little "cc" button in the lower corner of the YouTube interface might suggest.

Research on automatic recognition and classification of nonspeech/ environmental sounds is still in its infancy, unlike the research on spoken language processing which "has achieved significant advances" (Liao and Lin 2009, 2695). According to Michael Cowling (2004, 2), "The research that has been done into computer hearing revolves around the recognition of speech and music, with little research done into the recognition of non-speech environmental sounds." Kim, Moreau, and Sikora (2004, 716) put the problem this way: "Because the environmental sounds consist of multiple noisy and textured components as well as higher order structural components such as iterations and scatterings, they are generally much harder to characterize than speech and music sounds." Despite these challenges, researchers have developed a number of classification schemes and recognition algorithms for nonspeech sounds (see Casey 2001). For example, Guodong Guo and Stan Z. Li (2003, 213) used "an audio database of 409 sounds" divided into sixteen classes—altotrombone, animals, bells, cellobowed, crowds, female, laughter, machines, male, oboe, percussion, telephone, tubularbells, violinbowed, violinpizz, water—to train their sound recognition technology and test its effectiveness. Roman Jarina and Ján Olajec (2007, 3) developed software for detecting applause sounds, testing their approach on "more than 9 hours of audio sounds, of which about 50 minutes is an applause." Michael Cowling's (2004, 60) PhD dissertation on environmental sound classification for autonomous surveillance explores the development of a system that can recognize a large number of environmental sounds associated with security (e.g., a computer system that can recognize the sounds of home burglary): jangling keys, coins dropping, footsteps (close and distant), footsteps on leaves and grass, wood snapping, and glass breaking. Kim, Moreau, and Sikora (2004, 723) consider the problem of film/video indexing, focusing on a number of sounds in three classes: animal (bird, dog), foley (bell, gun, horn, motor, telephone, water), and people (baby, laughter, male speech, female speech). Wen-Hung Liao and Yu-Kai Lin (2009, 2695) are interested in "smart living spaces" and classify environmental sounds as cough, scream, sneeze, snore, and laugh, with a special emphasis on snore sounds. One application of this research on snore sounds focuses on "whether the subject suffers from sleep apnea syndrome" (2698). In short, there's a lot of interest in sound classification and autonomous nonspeech recognition, but this research lags behind speech and music recognition technology.

Speech sounds dominate subtitling efforts on the web as well. What drives crowdsourcing efforts and subtitle sharing sites is an assumption that people around the world are principally divided by language differences rather than hearing differences. This should give accessibility advocates pause. Within a translation economy, true full-spectrum closed captioning may become less common or be deemed less important. Companies like Dotsub.com are organized around concepts such as "transcription" and "translation." Built on a crowdsourcing model, Dotsub encourages users to contribute to the site by transcribing videos into their source languages and then translating those videos into different languages. In the official Dotsub tutorials (2013a), the terms "captions" and "subtitles" are interchangeable. While the Dotsub site stresses the importance of making videos accessible to deaf and hard-of-hearing people with information on captioning laws (Dotsub 2013b) and blog posts about hearing loss (Dotsub 2012), the pull towards a narrow view of captioning as foreign language subtitling is strong, beginning with the company's motto: "Any video, any language." In this model, language and speech trump nonspeech sounds when the primary barrier to access is assumed to be language, not sound.

Nothing prevents Dotsub's users from producing full closed captions in any language. Closed captions are created using the same semiotic resources as subtitles. But the tutorials put a premium on the transcription and translation of speech sounds. The official tutorial video on "how to transcribe a video on Dotsub" (2009a), which has been translated into four languages, describes the subtitling process as one of listening for and transcribing "talk" (i.e., speech). The narrator describes part of the transcription process using the web-based application: "Now, if you notice that there's no talking for a few seconds . . . you can just play [the video] until the talking starts again." In the official tutorial video on "how to translate a video on Dotsub" (2009b), the sample video being translated opens with a non-speech caption: [♪ CBS News theme music ♪]. But the company representative who is narrating the screencast video misses an opportunity to discuss the importance of nonspeech sounds and captions for deaf and hard-of-hearing viewers. He faux-translates this music caption as "The first line" while reminding viewers that "If I was typing in French that would be in French." Clearly, then, nonspeech sounds and the needs of deaf and hard-of-hearing viewers have a role to play in the Dotsub universe, but they are obscured in the interest of promoting Dotsub.com as a language translation service. Given that the technology behind Dotsub, including their interactive transcript technology, powers the subtitling and translation of TED talks (ted.com), the reduction of closed captioning to foreign-language subtitling extends beyond the Dotsub site.

On subtitle-sharing sites like SubtitleSource.org and Subscene.com, subtitle files are available for thousands of movies and TV shows in a number of languages. Each site's content is generated by users, so it's not unusual to find a dozen or more different subtitle files in the same language for the same movie. Users of the subtitle-sharing sites are primarily interested in speech-only subtitle files. While subscene.com provides access to closed caption files, access is limited and uneven. For example, Subscene.com hosts seventy-nine English-language subtitle files for the final *Harry Potter* movie (*Deathly Hallows: Part 2*), as well as hundreds of additional subtitle files of this movie in other languages. Of the subtitle files in English, twenty-four are marked with an ear icon, which means they are presumably intended for viewers who are deaf and hard-of-hearing. The quality of all of the files varies. Finding a suitably accessible file can be frustrating. Just because a file is marked with an "ear" icon doesn't guarantee that the file will make appropriate or sufficient use of nonspeech captions and nonspeech information. I was unable to find true closed captions in English for *The Artist* (2011) on any of the subtitle sharing sites. Of the thirteen files in English on subscene.com, three files were marked as accessible to deaf and hard-of-hearing viewers. But none included any nonspeech captions. In short, subtitle-sharing sites participate in the same translation economy as subtitle-creation sites. While language translation services play a vital role in a world flattened by globalization, they aren't necessarily inclusive.

When captioners for whatever reason leave out an entire class of non-speech sounds, the results will likely be counter to the interests of deaf and hard-of-hearing viewers. Consider *Country Strong* (2011), a movie starring Gwyneth Paltrow about a country singer's efforts to return to the stage following a stay in rehab for alcoholism. The movie features close to two dozen instances of singing from Paltrow and others. While the movie includes some nonspeech captions—indeed, the first caption of the movie is (AUDIENCE APPLAUDING)—none of the music lyrics are captioned on the DVD version. Music is identified in the captions by song title such as (BEAU BEGINS SINGING *FRIENDS IN LOW PLACES*) and, occasionally, by short, unhelpful descriptions such as (SINGS LINE OF A COUNTRY SONG), (RESUMES SINGING), (SINGING), and (MUSIC PLAYING). But lyrics are never captioned in *Country Strong*, even though the movie is, ironically, about music. The movie reaches the height of irony during the song "Words I Couldn't Say," which might as well be referring to the captioned lyrics. While lyrics can't always be accommodated in the closed captions due to time and space constraints, every effort must be made to include them when context demands it. **Media:** http://ReadingSounds.net/chapter4/#country-strong.

It's not clear why all of the lyrics were left out of *Country Strong*, but we do know one thing: US copyright should have had nothing to do with it, despite the claim to the contrary that captioners need special permission from music owners to publish lyrics. While lyrics themselves are copyrighted, closed captions merely provide access to them. There's a difference. Closed captions do not license content but mirror it so a broader audience can access it (see Peltz Strauss 2006, 252). Special permission from the copyright holder is not required to close caption any content, including lyrics. Intellectual property rights still apply, of course, but professional captioners who are hired by producers to caption their content have the legal authority to provide full access to that content, including full access to the music lyrics. As attorney Jodie Griffin (2011) puts it, "Copyright does not trump disability rights law." In an effort to curb "copyright overreach," Griffin argues that captioning should be construed as an example of "fair use":

> The nature of the copyrighted work (here, video programming) is probably highly creative, and so it enjoys a fair amount of copyright protection. However, captioning is a non-commercial use that is simply intended to make the programming accessible to individuals with disabilities, and to legally comply with the CVAA [21st Century Communications and Video Accessibility Act].

In the context of the CVAA, then, Griffin argues that the Federal Communications Commission (FCC), which was tasked with implementing the provisions of the CVAA, has the authority to create a statutory exception in the case of closed captioning:

> Even if captioning infringed copyright, the CVAA explicitly orders the FCC to "revise its regulations to require the provision of closed captioning on video programming delivered using Internet protocol. . . ." If captioning does indeed violate copyright, then the FCC has statutory authority to create a limited exception to copyright protection for the purposes of implementing the CVAA. Copyright law is not a shield against all other legal obligations.

A similar case for fair use has been made in the name of digitizing books to accommodate people with print disabilities, such as blind and low vision users of text-to-speech technology. According to Section 107 of US Copyright Law, "the fair use of a copyrighted work . . . for purposes such as criticism, comment, news reporting, teaching (including multiple copies for classroom use), scholarship, or research, is not an infringement of copyright." In the lawsuit brought by the Authors Guild against HathiTrust, the Authors Guild argued that the HathiTrust digital library, which comprises millions of digitized titles through a partnership of academic and research institutions, violated copyright and authors' rights to control their own intellectual property. In the appeal of this case to the Second Circuit of the US Court of Appeals, the judge upheld the lower court's "finding that digitization for full-text search and access for print-disabled readers is fair use under US law" (HathiTrust 2014). In another case, *Sony Corp v. Universal City Studios, Inc.*, the court found that "making a copy of a copyrighted work for the convenience of a blind person is expressly identified by the House Committee Report [on the Copyright Act] as an example of fair use" (quoted in Reid 2014, 19).

The copyright issue is far from settled or simple, however. What right, if any, do users have to create and share translations of copyright-protected movies? The Swedish police's 2013 takedown of Undertexter. se, a movie subtitle fansite populated by contributions from site members, on the grounds that the site's owners were profiting from the intellectual property of others through advertisements placed on the site, suggests that the copyright industry takes a narrow view of subtitling as unauthorized copying: "The basic argument is that creating captions effectively 'copies' the protected dialogue and soundtrack in a video program by transcribing them in a nearly verbatim fashion" (Reid 2014, 9). Moreover, when fans share translations online ahead of a movie's release in foreign countries, movie distributors are denied the right to "delay the release of a video in foreign countries to maximize profits in a process known as 'release windowing'" (Reid 2014, 13). But fans and those who defend fansubbing have suggested that fansubs are more like interpretations, transformations, or remixes than copies—and thus should be protected, at least in the United States, by the fair-use exception.

Consider as well the problem of poor or nonexistent YouTube captions. Do users have the right to improve inaccurate, incomplete, or automated speech-to-text captions, even if they do not own the video? While only the video owner can upload YouTube captions, users—third parties—can embed videos on other sites using an interface that supports captions or interactive transcripts. Amara.org and Dotsub.com work from a crowdsourcing model in which anyone can add and edit subtitle tracks for any video on the major video sharing sites. According to Blake E. Reid (2014, 3), a law professor who authored a policy white paper on third-party captioning and copyright, "the need for third-party captioners" is "ever growing" to address both poor captioning (e.g., Google's automated captioning) and nonexistent captioning. The CVAA does not require the captioning of user-generated Internet video, thus necessitating the need for third parties to step in and provide accommodations for millions of user-generated YouTube and Vimeo videos. Right now is a "critical moment for captioning," according to Reid, because "[w]ell-meaning third-party captioners striving to improve video accessibility face potential liability for infringing the copyright of video creators" (4). Third-party captioners, whether creating new caption files or correcting errors in existing captions, can infringe the copyright holder's rights under US law to create both derivative works (e.g., adaptations) and reproductions (see Reid 2014, 10–11). Reid (2014, 15) offers a number of "potential workarounds" that would allow third parties to caption videos while complying with copyright law. One is contract work. The official DVD captioners on *Country Strong* were contracted by the movie producers and thus had the full legal authority to provide a complete account of both speech and nonspeech sounds such as music lyrics. In cases where a contract can not be negotiated between the copyright holder and the captioner, Reid (2014, 18) suggests the possibility of leveraging "several more general statutory limitations and exceptions in U.S. copyright law that might exempt captioning activities under specific circumstances," including Section 108 (exemptions for libraries) and Section 110 (exemptions for classroom use and certain performances). But the fair use exemption (Section 107) holds the greatest potential for justifying the legality of third-party captioning: "Under the principles articulated in *Sony* and the House Committee Report, it is reasonably likely that most third-party captioning—at least captioning undertaken strictly for accessibility purposes—constitutes a non-infringing fair use" (Reid 2014, 19). When third-party captioning is undertaken for nonaccessibility purposes that might alter a video's market value, such as for purposes of search engine optimization, captioners "should be wary of the viability of fair use" (20).

Overcaptioning: Unnaturally Elevating Speech Sounds

Whereas undercaptioning emphasizes speech sounds at the expense of nonspeech sounds, overcaptioning unnaturally elevates speech sounds. When indistinct speech sounds become distinct through verbatim captioning, captioners play god. Just because speech sounds can be discerned through the careful and repeated listening of a trained captioner does not mean that these sounds should always be captioned verbatim. Let me say this again, somewhat differently: captioning is not a game of listening to sounds repeatedly until they can be discerned clearly and distinctly. That's not how sonic contexts on TV and film are intended to be experienced (or actually experienced). Captioning is about listening to shows, attending to and supporting the contexts in which TV and film sounds circulate. Sounds cannot be divorced from the meaningful contexts in which they occur. The captioner's ear must be tuned to the show, not the sounds per se. When we shift our focus from sounds to sonic contexts, the limitations of our definitions of captioning come into sharp focus. Captioning from this broader perspective is concerned not simply with representing sound in writing but with providing access for deaf and hard-of-hearing viewers. To provide sufficient access, captioners must attend not to the sounds themselves but to the rhetorical work that sounds perform.

What should we do about speech sounds that can only be heard clearly in hindsight or with the assistance of external support such as a production script? Even if the captioner has advance knowledge of which words are actually being spoken, gibberish still needs to be captioned as gibberish if that's its purpose. The same goes for rapid-fire speech and inscrutable background speech. Overeager, well-meaning captioners can deflate a scene or ruin a joke when they caption verbatim what was intended to be indistinct. Just because speech sounds can be discerned through a process of slowed-down or repeated listening doesn't mean that they should be. The captioner's job is not to reveal a set of de-contextualized truths about what is being said but to interpret the soundscape in specific rhetorical contexts.

Captioning Contextually: The Public Address Announcement

The opening to *Knight and Day*, a 2010 action flick starring Tom Cruise and Cameron Diaz, provides a dramatic example of how closed captioning can mistakenly privilege speech over nonspeech, even when the speech sounds are barely audible and/or insignificant. I use the term *logocentrism* to describe this audacious appetite for speech and discuss it in more detail below. First, consider the uncaptioned version, which I pulled from the official DVD. If you are a hearing viewer, think about how you would caption it. Which sounds are significant? Every sound cannot be captioned, so which ones are most important here? And how would you convert those sounds into words? If you are a deaf or hard-of-hearing viewer, think about how this scene visually establishes a context or mood without relying on speech from either of the main characters. In this opening scene, we see the world through Cruise's eyes. His flight is delayed and he's killing time. He seems harmless enough while eating ice cream, playing video games, and shopping for knickknacks. But we glimpse a darker side, too. He seems to be looking for someone—a young blonde woman pulling a wheeled carry-on, perhaps? He scopes out two targets before settling on Cameron Diaz. **Media:** http://ReadingSounds .net/chapter4/#knight-and-day-DVD-uncaptioned.

What sounds do we hear in this clip? While the first dozen seconds of the scene rely solely on indistinct airport noise (footsteps, crowd sounds, an indistinct public address [PA] announcement), the rest of the scene is dominated by an instrumental music track, which is timed to start on the beat of Cruise's first dramatic step. The music conveys a light, expectant, even playful mood. Indistinct airport noise continues faintly in the background as a dull echoey hum or keynote, becoming more distinct only when Cruise grabs his ice cream cone and plays video games. As the shot of Diaz comes into visual focus at the end of this clip, the music tells us that she's the one he's been looking for: six musical notes—three visually associated with Cruise and three with Diaz—build to a pleasing plateau. Cruise's three notes are mirrored in Diaz's, connecting them together. The notes are nicely balanced around a small knight figurine that Cruise is eyeing in a gift shop. The knight is an appropriate object given that Cruise plays a character named Matthew Knight. The knight figurine is visually associated with the first note (and with Cruise) and serves as a transition to the fourth note (and to Diaz) through a rack-focus, point-of-view technique in which Diaz comes into focus as the knight fades. Clearly, this entire sequence, while short, is intricate and was painstakingly planned. The musical notes are meaningful. They sustain and interpret the visual sequence, telling us implicitly that Cruise has found his mark. How should such a musical sequence be captioned, if at all? Captioners have yet to develop approaches to captioning ambient music at the level of the individual note or musical phrase. **Media:** http://ReadingSounds.net/chapter4/#knight-and-day-music.

If there's a PA announcement playing throughout this scene, it's little more than background noise. If I strain unnaturally, I think I can make out the PA announcer saying "Welcome to Wichita" and maybe "attention, please." While background noise may be important as a way to establish the scene's "key," it is not so important that every (inaudible) word the announcer says needs to be captioned. In this scene, wordless music reigns, not the irrelevant specifics of the PA announcement. With that in mind, consider the captioned version of this scene as it appears on the official DVD. The captions focus exclusively on rendering the PA announcement verbatim:

WOMAN ON PA:
Welcome to Wichita Mid-Continent Airport.

Convenient,
friendly, affordable.

Fly with us.

Please note,
the Kansas Clean Air Act

has designated
this airport as non-smoking.

Your attention, please.
Baggage claim is located

on the east side
of the terminal.

Hotel and ground
transportation information boards

and public telephones
are located opposites the baggage carousels.

Ladies and gentlemen,
please maintain visual contact

with your personal
property at all times.

Please do not leave
your baggage or other items unattended.

Thank you for
your cooperation.

The official DVD captions miss the point entirely. They are distracting and contribute little to our understanding of how sound functions in this scene. While some might argue that it is useful to know the name of the airport (Wichita Mid-Continent) and to appreciate the ironic way in which Cruise is performing the PA announcement (maintaining "visual contact" and looking for "unattended" baggage), the scene (and the movie more broadly) has little to do with the specific information contained in the PA announcement. The announcement is only intended to add ambience, to convey the underlying "key" for the scene. The scene is not intended to convey information about the announcer's gender, the "Kansas Clean Air Act," or where to find the baggage claim area or the public telephones. It's a safe bet that no hearing viewer watching for the first time without the aid of closed captioning has ever made out or cared to make out such details. They don't matter. They are insignificant as details but significant as keynote or ambient sounds. The largely indistinct PA announcement contributes to the sonic construction of "airport." Not only do we *not* need to know the specific words the announcer is saying, but it is potentially distracting and confusing to be subject to the specific minutiae of the announcement. Like the repeatedly captioned barking dog in an exterior set-up shot of a suburban home (see chapter 1), the PA announcement threatens to turn our attention to irrelevant details. **Media:** http://ReadingSounds.net /chapter4/#knight-and-day-DVD-captioned.

These clips from *Knight and Day* were pulled from the official DVD. The broadcast television captions for *Knight and Day* are substantially different. That's not surprising, since the television captions for most movies usually depart noticeably (and sometimes dramatically) from their DVD counterparts. As part of the process of preparing movies for redistribution on TV, movies are often recaptioned, instead of merely retimed, to account for commercial breaks. We only need to compare two official caption files from the movie of our choice—the DVD captions and the broadcast TV captions—to begin to appreciate just how radically interpretative the practice of captioning is. Contrast the thirteen DVD captions in the opening scene with the five captions in the broadcast TV version:

WOMAN (over P.A.): Welcome to
Wichita Mid-Continent Airport.

Convenient,
friendly, affordable.

Fly with us.

(woman speaking indistinctly
over P.A. system)

(electronic blipping,
whooshing)

The TV captions come closer to approximating the experience of the airport PA as ambient noise (rather than distinct speech), but they are no

Knight and Day (2010) opening scene: DVD vs. TV captions

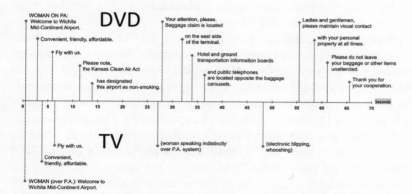

4.2 **A timeline comparing the DVD captions with the TV captions for the opening scene of *Knight and Day*.**

The DVD captions have been placed above the timeline. The TV captions have been placed below the timeline. The DVD captions present the PA announcement verbatim, whereas the TV captions attempt to convey something of the ambience of the scene. Neither caption track accounts for the ambient music that provides an emotional register for the scene. Twentieth Century Fox, 2010. Data in HTML table: http://ReadingSounds .net/chapter4/#figure2. This timeline was adapted from a Microsoft Excel template: http://www.vertex42.com/ExcelArticles/create-a-timeline.html.

less partial, selective, or incomplete. When the two streams are visualized on the same timeline, as in figure 4.2, a number of differences stand out (along with one similarity). Both the TV and DVD clips begin with the same three basic captions, which are also the loudest and clearest portions of the PA announcement. After that, all bets are off. The DVD captions interpret the scene exclusively through the mostly inaudible speech of the PA. The captioned speech falls roughly into three clusters (0–14 seconds, 28–36 seconds, and 55–66 seconds). When visualized on a timeline, these clusters stand out more clearly. If the DVD captions exaggerate the prominence of the PA announcement, the TV captions adopt a minimalist approach. After the first three PA captions, two nonspeech captions round out the scene in the TV version, one of which appropriately characterizes the PA speech as indistinct, and the other describes the arcade game sounds. By representing the PA speech as indistinct after the opening three captions, the TV captions dramatically change the meaning of the text. Interestingly, neither set of captions accounts for the music, even though the music provides the controlling mood,

121

beginning on the beat of Cruise's first step and reaching a crescendo as it draws the two main stars together. **Media:** http://ReadingSounds.net /chapter4/#knight-and-day-TV.

PA announcements are tricky to caption. The tendency has been to caption them verbatim regardless of the contexts in which they appear. Sometimes, PA announcements do need a verbatim treatment. But not always. That's why I would add the "public address announcement" to a list of sounds for caption studies to investigate more thoroughly. Let me share two more examples of captioned PA announcements in the movies, one that doesn't need a verbatim treatment and one that does. Consider the emotional family reunion at the end of *Taken*, a 2008 thriller starring Liam Neeson. The reunion at the airport following the rescue of Neeson's daughter from human sex trafficking is disrupted by a captioned, incomplete, and ironic announcement over the airport's public address system. As the family embraces and speaks to each other in person for the first time since their daughter was kidnapped and sold into sex slavery, a partially muffled PA announcement interjects on the caption layer:

[Man on P.A.]
Attention travelers, you are not required—

Shall we go?

This airport does not
sponsor their activities.

In these three consecutive captions, a question from the stepfather ("Shall we go?") interrupts the PA announcement just as the announcer is about to tell viewers and listeners what is "not required." The main point of the announcement is muffled as the stepfather speaks. It is not possible for hearing viewers to make out the uncaptioned spoken words of the PA. Ironically, the movie itself is about activities that are not officially "sponsored" (i.e., kidnapping, human trafficking). In this sense, then, the PA announcement is relevant to the larger themes of the movie, even if the announcement is most likely not a public warning about kidnapping or slavery. But irony should never be enough to trump a scene's thematic intensity. Because the announcement disrupts the emotional intensity of the triumphant reunion, and its main idea is inaudible (whose activities are not sponsored?), it should not have been captioned verbatim. A complete, verbatim rendering of the announcement is impossible anyway and only leads to confusion and distraction. **Media:** http://ReadingSounds.net/chapter4/#taken.

When we assume that only significant sounds should be captioned (rather than starting from the mistaken assumption that all sounds can be captioned), we begin to explore significance beyond volume level. The loudest sounds are not always the most significant, just as quiet and even partly inaudible sounds are sometimes in need of captioning. The stepfather whispers in his stepdaughter's ear in this scene but his words are hard to make out as clearly as other spoken words in this scene (including the PA announcement). A nonspeech manner identifier, [Whispers], modulates the spoken words "It's so good to have you back." The daughter audibly cries out while the whisper is being uttered, making it difficult for hearing viewers to determine precisely what is said without the aid of captions. Yet despite the low volume of the whisper, it needs to be captioned because it is a crucial component of the emotional reunion at the end of the film. Captioning should thus be driven by the scene's purpose. Volume levels alone may be helpful but not sufficient in determining which sounds should be captioned. Based on my reading of this scene from *Taken*, I offer five guidelines for thinking through the question of significance:

1. Captions should support the emotional arc of a text.
2. A sound is significant if it contributes to the purpose of scene.
3. Caption space is precious. It should never be wasted on superfluous sounds that may confuse viewers or diminish their sense of identification with the protagonist(s).
4. Sounds in the background do not necessarily need to be captioned, even if they are loud.
5. Every caption should honor and respect the narrative. While a narrative does not have one correct reading, it does have a sequence and arc that must be nourished.

By suggesting that the captioned PA announcement at the end of *Taken* might be edited down to a sound description such as "PA announcement" or "crowd talks indistinctly," I realize that I open myself up to charges of censorship. Indeed, I am acutely aware that deaf, hearing, and hard-of-hearing viewers do not want dumbed-down captions. Every viewer deserves sufficient access. The original practice at WGBH of editing TV speech for comprehension (Earley, 1978) is no longer advocated today in the United States. But what discussions of verbatim captioning leave out are (1) the differences between writing and speech, and (2) the limited space available for captions. Everything cannot be captioned. There are some pretty loud footsteps in the *Taken* scene, some loud but indistinct chatter, and at least one loud car horn. Someone had the good sense not to caption them. The footsteps seem to be almost as loud as the PA caption, but volume alone should not drive caption design. The captioner must contend with spatial and temporal constraints while being responsive to the rhetorical needs of the narrative. Countless sounds are left out of every caption file. Time and space are working against the captioner, but more importantly, viewers do not want to be burdened by a screen full of insignificant captions. Censorship is simply the wrong word to describe the selective and creative process of captioning. What I am describing is not censorship—far from it—but the art and rhetoric of captioning. Someone must make these decisions. Captioning is not an objective science; it is a highly interpretative practice. (By the way, it sounds as if there is a foreign language PA announcement at the beginning of the *Taken* scene. It is not significant and thankfully there is no reference to it—even a vague one like "Foreign PA announcement"—in the caption file.)

Decisions about the significance of any PA announcement must be made in the context of a narrative and not out of some preconceived sense of obligation to provide a verbatim record of every speech sound. Arguments about the importance of verbatim captioning need to be informed and tempered by a broader rhetorical awareness. As the miscaptioned example from *Knight and Day* suggests, the goal of verbatim captioning must be balanced with a respect for how speech sounds function in a scene. Properly captioned verbatim PA announcements, in contrast to the two examples from *Taken* and *Knight and Day*, provide crucial information to viewers within the context of a scene or narrative. Consider *Whiteout* (2009), a bloody thriller set in and around an Antarctic research base a couple days before the last plane is set to depart the base in advance of a series of blinding whiteouts and the quickly approaching long dark winter. Here's the premise: "The last plane out leaves in just three days, and just after that Antarctica will not see the sun again for another 24 weeks. Then, just 72 hours before [US Marshal Carrie Stetko] is set to escape the coming darkness, a body is discovered in the ice—prompting the first murder investigation ever to take place on the desolate Antarctic Circle continent" (Rotten Tomatoes n.d.). It is within this context of time running out that the recurring PA announcements serve an important expository function to remind us of this impending deadline. Here's just one example of many in the movie:

OPERATIONS TECH: *This is a 72-hour warning, people.*

That's three days until the last flight for
those of you mathematically challenged.

Two clips from the *Whiteout* DVD are relevant here. In the first clip, we are introduced to the Antarctic research base for the first time as the camera follows the main character (Kate Beckinsale) to her room amidst lively preparations for an end-of-season party. The PA announcement orients us to the impending deadline: "72-hour warning," "That's three days until the last flight," "tonight's station-closing party," "bag drag will start at 0700," "all your gear must be in the hallway." Other sounds compete with the PA announcement—the music, some crowd chatter, one loud party decoration at 0:00:15.81—which is perhaps why the announcement basically repeats the same information in different ways to ensure the audience gets the message: it's base closing time. Despite the competition from these other sounds, however, the PA announcement is the most significant sound in this clip. It's also foregrounded in the scene but has no visual corollary (i.e., the PA announcement can't be seen), unlike the sounds of the lively party preparations and end-of-season activities, which are conveyed both visually and aurally. **Media:** http://ReadingSounds.net/chapter4/#whiteout-intro.

This last point about PA sounds—that they are almost always invisible—should remind us that some sounds may need to be captioned *because* they are less visible. The meaning of visible sounds (i.e., sounds that have clear visual referents) can sometimes be roughly conveyed contextually without captions. We can see the hustle and bustle in the corridors as Kate Beckinsale makes her way to her room. We don't necessarily need to be told in the closed captions that the scene is filled with [indistinct chatter] or even [musicians practicing]. We can see these activities. But we do need to know what the PA announcer is saying, because his instructions and warnings are crucial to the plot and invisible. Even without the benefit of additional captions, the visual scene provides a rough substitute for the bustling party/chatter/music sounds, which acquire specific meaning in the PA announcer's reminder about "tonight's station-closing party."

At best, then, the PA is a version of an ancient Greek chorus that summarizes and interprets the dangers for the characters and the movie audience. At worst, it's a rather lazy form of film exposition in which the announcer *qua* narrator hits us over the head with some basic plot info. When the operations tech explains to the "mathematically challenged" residents on the base that 72 hours is the same as three days, it's like the movie is really talking to us, treating us as challenged and in need of some very basic knowledge about the world. Regardless of the relationship between the PA announcements and the overall quality of the film (*Whiteout* was universally panned by critics), the PA announcements in *Whiteout* need to be captioned verbatim when they engage in foregrounded plot exposition.

The second clip from *Whiteout* is a compilation of selected public address announcements (including the announcement in the first clip). These excerpted clips suggest how a verbatim approach is warranted when it is driven by the expository function of the announcements themselves, not by some external obligation to caption every speech sound verbatim. But mediated announcements (over radio, phone, TV, PA system) are typically quieter, sometimes barely audible sounds in the background, even when they are significant. Closed captions, in contrast, tend to *equalize* all sounds (everything is equally "loud" on the caption layer). When background sounds are captioned verbatim, they come forward and take center stage, which may conflict with the intentions of the content producers to place such context-informing sounds in the background. The soundscape is multilayered. Captions reduce a three-dimensional soundscape to two-dimensional print. Qualifying background speech captions with manner of speaking identifiers such as [quietly] or [sarcastically] (see chapter 8) is an imperfect solution to the problem of equalization and runs the risk of cluttering the screen. Captioners need to use their best judgment in identifying manner of speaking. A good approach is to simply indicate how the sounds are mediated—over PA, over radio, on the phone, on TV—and let caption readers interpret the relative loudness and clarity of such sounds. Channel identifiers can be neatly nested inside speaker IDs: "Operations Tech [over PA]: *Attention, the transport plane . . .*" **Media:** http://ReadingSounds .net/chapter4/#whiteout-compilation.

The PA announcement has served as a mini case study in this section. But the tendency to confuse indistinct sounds for distinct speech is not limited to PA sounds. Let me conclude this section with other examples that perform similar kinds of work. We might look to recent zombie movies for additional examples of expository (as opposed to ambient) PA-like sounds. In order to convey a sense of growing fear of global pandemic, some zombie movies rely on TV and radio broadcasts about strange illnesses, invisible toxins, and global unrest. These broadcasts situate the audience in a context in which zombie apocalypse is possible. The announcements tend to overlap too, suggesting a frenetic and overwhelming situation that has reached a breaking point. When the zombie virus begins wreaking havoc in one suburban neighborhood at the start of *Dawn of the Dead* (2004), for example, Ana (Sarah Polley) flees her home in a thrilling scene shot from multiple cameras mounted to and inside of her car. She turns on the car radio and flips through stations quickly as the neighborhood burns, people are running and screaming, zombies are eating people, and the streets are cluttered with objects she must avoid. The radio provides context and, just to make sure the audience gets the message, repeats the main warning to stay inside. Consider this sampling of radio announcements in this scene:

FOR LOCAL
EVACUATION CENTERS.

STAY INSIDE AND
LOCK ALL DOORS—

PLEASE STAY INSIDE AND
LOCK ALL DOORS AND WINDOWS.

MILLER PARK IS NO LONGER
CONSIDERED A SAFE HAVEN.

PLEASE AVOID THE STADIUM AND
PROCEED TO OTHER LOCATIONS.

MEANWHILE, CIVIL UNREST
IS STILL BEING REPORTED

PLEASE AVOID TRAVELING . . .

THAT SEVERAL MILITARY
PERSONNEL HAVE FALLEN ILL . . .

The radio sounds compete with the sounds of chaos: (police sirens approaching), (woman screams), HELP!, (Truck honking), (woman screaming), (screaming), and other sound effects not captioned, such as a helicopter drowning out the emergency broadcast announcements. Because Ana is a witness to the terrifying scene, the announcements take on an ironic cast. Ana is traveling (*PLEASE AVOID TRAVELING . . .*), she is outside

(*STAY INSIDE AND LOCK ALL DOORS*), and she is presumably looking for a safe haven (*MILLER PARK IS NO LONGER CONSIDERED A SAFE HAVEN, PLEASE AVOID THE STADIUM AND PROCEED TO OTHER LOCATIONS*). In short, because Ana is doing everything she is being told not to do, the announcements effectively add to the sense of terror, even though they all can't be heard clearly without the aid of verbatim captions. **Media:** http://ReadingSounds.net /chapter4/#dawnofthedead.

World War Z (2013), starring Brad Pitt, also opens with a montage of radio and TV broadcasts, but the broadcasts precede the zombie outbreak and serve as dire warnings. Some of the announcements are invisible, whereas others, such as the television broadcasts, are accompanied by supporting visuals and synchronized lip movements. The announcements pile on top of each other, zombie-like, creating tension through interruption and overlap. Here's a sampling:

if the virus changes in a way
that allows transmission
between humans.

another group of dolphins
became stranded.

CO2 emissions have
dramatically increased in . . .

Police say they've seen
similar cases recently

of people behaving strangely.

MALE REPORTER:
The subject growled at him

and continued
to maul the victim.

The music swells too, beginning as a simple and repetitive eight-note musical phrase played on the piano and reaching a crescendo as the name of the movie appears on the screen about two minutes later. Though it is uncaptioned, the music plays important roles in this scene: it is both hypnotic (again, zombie-like) and somewhat reminiscent of a slowed-down theme from the classic horror movie *Halloween* (1978). The announcements are mostly clear and distinct, despite being layered and strung together as a series of interruptions. But as the music swells and the visuals become more disturbing with images of rabid and frenzied animals and insects, the announcements become increasingly difficult to hear clearly as distinct words. Towards the end of the opening scene, (INDISTINCT RADIO CHATTER) initiates a series of captions that can only be heard as ambient noise:

MAN 1: Monitors have arrived,
even carrying them . . .

MAN 2: With that much
public at risk, we cannot . . .

MAN 3: They're trying
to move us on,

trying to tell
the journalists to get out.

Only caption readers have access to the specific meanings contained in these final captions. While the scene is expressly designed to thwart listeners' attempts to remain calm in the face of growing threats, the captions provide measured and calm transcriptions throughout. Captions clarify meaning, but in doing so they take away some of the rising tension from the scene. That we can no longer make sense of what we're hearing is conveyed in part through indecipherable announcements but belied by a series of verbatim and seemingly rational captions. **Media:** http://ReadingSounds.net/chapter4/#worldwarz.

A captioned phone call provides another opportunity to explore the tension between distinct and indistinct speech. Consider the opening scene from *Hyde Park on Hudson* (2012), starring Bill Murray as FDR, in which a phone call from one of the president's staff sets the plot in motion. The purpose of the phone call is a mystery. We only learn that Daisy (Laura Linney) has been invited to visit the president when she arrives at his mother's home. No one can hear what the person on the line is saying. The very low volume of the voice on the other end, in addition to the (DOGS BARKING), prevents Daisy and the hearing audience from hearing the full conversation. That's the point: Daisy is lost—she says "What?" and "Sinus?" during the call—and so are we. It is thus ironic that the full conversation would be captioned verbatim on the DVD version when the scene is intended to keep the reason for the call a mystery and the conversation can't be fully discerned anyway without the aid of a production script. While audiences can benefit from knowing that the person on the other end of the line is saying, "Now I would like you to come over, and I shall expect you this afternoon," the scene explicitly works against providing viewers with that knowledge. That's the mixed blessing of captioned irony, as discussed in chapter 5: caption viewers gain knowledge over hearing viewers who are watching with the captions turned off, but sometimes that knowledge comes at a cost to the film's integrity. **Media:** http://ReadingSounds.net/chapter4/#hydeparkonhudson.

In a similar example from an episode of the animated series *South Park* ("Margaritaville," 2009), Randy (voiced by Trey Parker) lectures his family on the US economic crisis. He criticizes those "PEOPLE WITH NO MONEY WHO GOT LOANS TO BUY FRIVOLOUS THINGS THEY HAD NO BUSINESS BUYING." Then he stands up to shovel ice into his own frivolous purchase, a Margaritaville blender. He turns on the loud blender but continues to lecture his son Stan. The machine is so loud that it almost entirely drowns out Randy's voice. His speech is intentionally obscured for humorous effect, but the closed captions display Randy's speech verbatim nevertheless. The captions continue as though nothing has changed, thus mistakenly revealing what the episode intentionally hides. We read verbatim what can not and should not be heard: "IT GOES BACK TO WHEN THE GOVERNMENT HAD THE IDEA / THAT EVERYONE IN AMERICA DESERVES TO OWN A HOUSE . . ." Just because some speech can be captioned verbatim doesn't mean it should be, even if the knowledge gained through access to the obscured speech might be perceived as beneficial to viewers. **Media:** http://ReadingSounds.net/chapter4/#southpark-margaritaville.

Because one of *South Park's* main characters, Kenny (voiced by Matt Stone), can only mumble through the orange-hooded parka stretched tightly around his face, a number of gags on the series revolve around Kenny making jokes, usually sexually explicit ones, that the other characters can hear clearly but the audience may or may not. To produce Kenny's mumbling voice, series cocreator Matt Stone reportedly speaks Kenny's lines into his hand. Sometimes, Kenny's words are partially discernable to the hearing audience. At other times, it is difficult to make out clearly what Kenny is saying, though listeners may grasp the explicit nature of his words from the context and the other characters' reactions. Kenny's words are usually captioned as either (mumbling) or onomatopoetically (e.g., some version of "mrph"). In the opening theme song, for example, Kenny's muffled line is captioned on cable television as "♪ MRPH RMHMHMRM! MRPHRMHMHMRM! ♪" The same line is captioned in the same episode ("The Jeffersons," 2004) by the official online distributor, Hulu.com, as ♪ (muffled mumble muffled mumble) ♪. An extended example of miscaptioned mumbling occurs in "Obama Wins" (2012) when Butters (voiced by Matt Stone) is in the hospital with a severe allergic reaction and can't speak clearly. Kenny tries to translate for Butters when the other kids pressure Butters to tell them what he knows. The gag is a timeless one, a translation joke in which the person translating makes no more sense than the person being translated. The hospital scene opens with Butters making nonsense sounds through his swollen cheeks, appropriately captioned as "UGHH—WAHGHGH" and "UGHH." His first and second replies to Kyle's threat to "TELL YOUR DAD YOU HELPED GET THE WRONG PERSON ELECTED PRESIDENT" can be heard clearly for the most part, and are thus captioned verbatim as "NO!, PLEASE, YOU CAN'T!" and "OKAY! OKAY!" The rest of the scene continues to offer full verbatim speech captions, but the speech sounds being captioned are not clear enough to warrant the verbatim treatment. That's the joke—we can't understand what anyone is saying. It goes like this: Butters makes some nonsense sounds, the boys don't understand what he's saying, so Kenny translates. The boys seem to understand Kenny (until the end, when Jimmy steps in to translate for Kenny), but the audience does not. When Butters's mumbling speech is captioned verbatim as "BECAUSE THE CHINESE ACTUALLY WANT TO PROTECT SOMETHING THAT'S IMPORTANT," and Kenny's mumbled speech is likewise captioned verbatim as "BECAUSE THE CHINESE WANT TO PROTECT SOMETHING THAT'S IMPORTANT," the captions have failed to capture the gist of the joke and the scene. While caption viewers may surmise that Butters's speech is muffled because his face is massively swollen, they won't necessarily realize that Kenny's speech is similarly obscured, or that the other characters can understand Kenny while the television audience cannot. Regardless, the captions have confused indistinct sounds for distinct speech without providing any clues about the manner in which the speech sounds are being produced. What the caption viewer gains in clarity she loses in understanding. The captions clarify but at the expense of the joke, which is the entire purpose of the scene. **Media:** http://ReadingSounds.net/chapter4/#southpark-kenny.

Back from the Future

Captioners overreach when they transcend the specific context of a scene to provide a transcription that stretches interpretation past the breaking point. The opening to *Knight and Day* falls into this category, because the DVD captioner swooped down, godlike, armed with a verbatim transcription of speech sounds that are mostly indistinct and inaudible. In this section, I provide an example of overreaching from an episode of *Arrested Development* in which a short but significant stretch of nonspeech is mistakenly replaced with a speech caption. In order to execute this magic, the captioner travels into the future, retrieves a speech caption from later in the episode, and offers it up as a caption for some gibberish earlier in the episode. This example is both remarkable and rare; I do not hold it up as representative of TV captioning but present it here as one dramatic and real example of what can go wrong when captioners overreach. It is also a reminder that nonspeech sounds need to be honored with appropriate nonspeech captions.

Captioners have a luxury that the typical moviegoer doesn't have, even with the pressures captioners routinely face to get the work done quickly. This luxury includes going back and listening to the same sounds again and again, searching the web for lyrics and other info, working from a detailed production script, captioning earlier scenes based on advance knowledge of what's to come in later scenes, and consulting with other captioners about puzzling sounds. This power comes with great responsibility. Captions are neither the annotations of a diehard fan nor a set of decontextualized truths about the soundscape. Just because a listener can decipher the truth of what's being said after repeatedly listening to a sequence of inscrutable background speech sounds—or after consulting the presumed truths contained in a production script, as the *Knight and Day* captioner has most likely done—doesn't mean that these truths should be captioned verbatim.

An episode from *Arrested Development* offers us an opportunity to explore this tension between verbatim and indistinct speech. In "Forget Me Now" (2005), Buster Bluth (Tony Hale) is reenlisted in the Army without his knowledge when the enlisting officer talks too fast for anyone to understand clearly. Buster, the entire Bluth family, and the episode's viewers (with the exception of the viewers watching the captions on broadcast TV, as we'll see) only learn the hilarious truth of Buster's reenlistment in hindsight at the end of the episode. In the initial reenlistment scene, Buster believes he is being awarded a medal. The officer is talking very quickly, but it's not only that. The middle portion of the officer's fast talking has been overwritten with gibberish during postproduction editing to prevent audiences from learning the truth too soon. The gibberish is captioned as [Rapid Muttering]. At the end of the initial reenlistment speech, the officer's speech slows down enough for listeners to hear him asking Buster, and for the captioner to write, "DO YOU AGREE? SAY THANK YOU." To which Buster replies, "THANK YOU. WOW." Towards the end of the episode, the scene is played again after Buster receives a letter from the Army telling him to "REPORT FOR DUTY TOMORROW." In the replay, the speech is slowed down for comedic effect so listeners can hear what the officer was really saying. Narrator Ron Howard (who is only credited as "Narrator" in the captions) introduces the slowed-down speech by explaining that "IF THAT SERGEANT HAD SPOKEN MORE SLOWLY," it would have been clear that he was reenlisting Buster. In truth, while the beginning and end of the speech are slowed down, the performative speech act in the middle is unmasked to reveal the truth (i.e., the gibberish sounds are replaced with the original audio). Instead of gibberish, listeners can now clearly hear the six words that perform the reenlistment: "HEREBY REENLIST IN THE ARMED SERVICES." **Media:** http://ReadingSounds.net/chapter4/#arresteddevelopment-dvd.

These DVD captions are far from perfect. It's hard to ignore the problematic all-caps styling or the unforgiveable failure to distinguish speakers at 0:10, 0:12, and 0:21 in the first reenlistment scene. Nevertheless, [Rapid Muttering] rings true contextually as well as technically. Buster's mother contextually prefaces the officer's speech in the first scene by identifying it as unclear ("OH, WHO KNOWS WHAT THEY WERE SAYING"), a point that she repeats when the same scene is slowed down at the end of the episode ("I DIDN'T HEAR THEM SAY THAT"). The muttering is extremely difficult to express phonetically, but it might sound something like this, spoken very quickly: hee buh la buh eee buh la muh. It's much easier to express gibberish visually as a sound wave (see figure 4.3). Technically, gibberish takes on a different form than speech when both are visualized as sound waves. Whereas speech is made visible through the clear demarcations in the sound wave—the breaks in the wave suggest pauses between spoken words—gibberish is visually closer to an undifferentiated mass. You can literally see gibberish in the sound wave, with its lack of distinct visual breaks. The sound wave suggests that, technically speaking, the first reenlistment scene and the second reenlistment scene are fundamentally different, even after we account for audio speed. The second scene is not a slowed-down version of the first scene; the first scene is actually overwritten in key places with gibberish.

The TV captioner, unlike the DVD captioner, plays god, jumping ahead into the future, grabbing the verbatim speech from the second scene, and dropping it into the first scene. As a result, the joke is ruined, a joke that depends on keeping the truth from everyone, viewers and characters alike, for nearly the entire episode. We aren't supposed

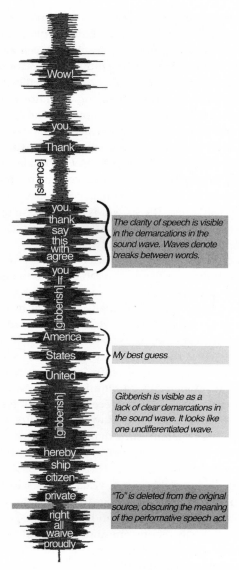

The clarity of speech is visible in the demarcations in the sound wave. Waves denote breaks between words.

My best guess

Gibberish is visible as a lack of clear demarcations in the sound wave. It looks like one undifferentiated wave.

"To" is deleted from the original source, obscuring the meaning of the performative speech act.

4.3   **The first reenlistment scene from *Arrested Development* ("Forget Me Now"), presented as an annotated sound wave.**
Audacity was used to create the sound wave, which was imported into Photoshop and marked up with text. Each word in the sound wave is identified. Clear speech is visible in the demarcations in the sound wave. By contrast, gibberish is revealed through the lack of clear demarcations in the sound wave. Gibberish appears as a single wave. Twentieth Century Fox Television, 2005. http://ReadingSounds.net/chapter4/#figure3.

to know, initially. The producers went to lengths to create uncertainty by technically overriding the performative speech act with gibberish and contextually having Buster's mother express her confusion. The TV captioner replaces nonsense sounds with verbatim speech—indeed, the captioner inserts the very words that comprise the performative speech act and are intentionally hidden from viewers ("hereby reenlist in the armed services"). Caption viewers are now armed with knowledge that non-caption viewers do not have and aren't supposed to have. Put simply, the TV captioner mistakenly captions speech sounds that don't exist. **Media:** http://ReadingSounds.net/chapter4/#arresteddevelopment-tv.

There are larger questions here for caption studies about what happens to the meaning and intent of fast talk (not gibberish per se but distinct speech) when it is captioned verbatim and read off the screen. Captions tend to tame and rationalize speech. The barely containable speech of an excited Vince Vaughn, or the Gilmore Girls, for example, is contained by the printed form. Words spoken so quickly that they can't be articulated fully (or, for hearing viewers, can't be understood fully) are corralled, tamed, and standardized when displayed in printed form.

A Logocentric Bias for Speech

Captioning nonspeech sounds is hard work. It involves a high level of care and creativity. There are no published style guides to help caption writers invent words and phrases for nonspeech sounds, beyond a few simple formatting rules. There are no published standards for assessing the quality of nonspeech sounds. Approaches to captioning nonspeech sounds vary wildly. The number of nonspeech captions in a typical film also varies quite a bit. At times, it feels like captioners are flying by the seat of their pants, inventing rules for captioning nonspeech sounds that make little sense from a usability perspective, such as prefacing nonspeech sounds with an inaccessible acronym like "SFX" (for "sound effects," presumably), as the captioner of a GEICO Insurance commercial has done. Not only is "SFX" less accessible (viewers can't be assumed to know what it means), it is also potentially distracting, cluttering up an already crowded, rapidly changing scene of captions. Another inaccessible acronym is "VO" and "AVO" for voiceover narration (see the Subaru and One a Day commercials), as an alternative to the eminently more accessible "Narrator." Caption readers don't need to be told explicitly which captions are sound effects or nonspeech sounds; they rely on a widely recognizable shorthand (brackets or parentheses) to distinguish speech from nonspeech. The brackets efficiently tell readers which sounds are sound effects. Acronyms can't be counted on to be widely accessible. SFX is redundant; VO and AVO are confusing alternatives to "Narrator" that should be avoided. **Media:** http://ReadingSounds.net /chapter4/#sfx-vo.

It should come as no surprise when captioners privilege or fall back on speech captions or when they fail to caption significant nonspeech sounds. A listening bias for speech over nonspeech, which has been measured in hearing infants, could even be considered innate for most hearing people. According to Athena Vouloumanos and Janet F. Werker (2007, 159), "humans are born with a preference for listening to speech," one that offers survival and behavioral advantages. Measuring sucking pressure rates with a "sterilized pacifier coupled to a pressure trans-ducer" (160, 162), they found that the infants in their study "sucked significantly more to listen to speech . . . than to complex nonspeech analogues" (162). (For a clarification and critique of this study, see Rosen and Iverson 2007.) Even if we don't claim a natural bias for speech in hearing people, speech sounds are still easier to identify and caption than nonspeech sounds for which meaning must sometimes be creatively in-vented by the captioner. If hearing people are more attuned to speech sounds, and if production scripts tend to privilege speech sounds, then a bias for captioning speech and reducing music to its lyrical content would seem to be strong if not inevitable. Moreover, in an industry with a high turnover rate and constant pressure to get the job done quickly in the name of preserving slim profit margins, speech sounds provide a quicker and easier route to a finished job, particularly for inexperienced captioners, than any route through nonspeech sounds that are highly interpretative and require more cognitive effort to describe. Instead of trying to convey the mood of a musical score, captioners sometimes default to the spoken aspects of music, reducing mood, ambience, and subtlety to a series of captioned lyrics. Granted, music lyrics will often express significant meanings in a movie or TV show, but music, like all film sounds, must be assessed in context. For example, consider how Iron & Wine's "Flightless Bird" is used as background music in the prom dance scene at the end of *Twilight* (2008), the teen love story between a human and a vampire. At the precise moment Samuel Beam sings of the "American mouth," Edward Cullen (Robert Pattinson) kisses the neck of Bella Swan (Kristen Stewart) with his American mouth. The lyric "Bleeding or lost you" takes on vampiric overtones—Edward is a blood-sucking vampire, after all—while also conveying Edward's fear of losing Bella. What is significant about these lyrics for captioning advocates is that they are so subtle as to go unnoticed by hearing viewers until captioned. When captioned, they powerfully reinforce the meaning of the scene. **Media:** http://ReadingSounds.net/chapter4/#twilight.

Sometimes a musical piece is chosen for a film because it supports or enhances a particular mood, not because there's some particular meaning in the lyrics. In other words, both lyrical content and ambient meaning may be sonically significant, and it is incumbent upon captioners to convey all aspects of a musical score's significant meanings instead of falling back on the easier option (lyrics and title only) or, worse, the easiest but least helpful option (musical note only *or just* [MUSIC]). While approaches to captioning music should vary according to context, some approaches to captioning music convey very little information to viewers. The music note caption [♪] is at the top of this list, for it fails to convey any information about the music except that it exists. Similarly, [MUSIC] is equally unhelpful and should not be used, since it tells us nothing about the way that instrumental music creates ambience for a scene. For example, in *The Lost World: Jurassic Park* (1997), all the instrumental musical sounds are captioned as [MUSIC] in the cable TV version broadcast in 2014 with scroll-up style captions. In the opening scene, instrumental music (combined initially with the ominous sound of a dinosaur growling) creates a sense of foreboding as a wealthy British family anchors their yacht off the coast of a presumably deserted island and enjoys their lunch on the beach surrounded by uniformed ship officers. The scene seems innocent enough, except for the music, which subtly alerts us to impending danger. (*The Lost World* is the second installment in the *Jurassic Park* dinosaur movie franchise, so movie audiences will expect an intense opening scene even before the first caption.) When the young daughter wanders off to play by herself, the music rises. When the girl is attacked by a pack of small dinosaurs, the music reaches a crescendo as her parents and the crew rush to save her. The scene is visually intense, but that intensity is muted by three distracting and unhelpful [MUSIC] captions: the first as the scene opens, the second as the daughter wanders off, and the third as everyone responds to the daughter's screams. With the exception of two additional and identical nonspeech captions—[INDISCERNABLE CHATTER]—no other nonspeech captions appear in this opening scene. The most significant nonspeech sounds are not captioned: the rustling of the bushes before the gang of tiny dinosaurs emerges and surrounds the girl and the dinosaurs' chittering, chirping, and screaming sounds. In other words, the scene is not only dominated by speech captions but seems to be driven by a speech-centric approach to captioning that has failed to account for some of the most significant sounds in a movie about dinosaurs: the sounds of the dinosaurs themselves. **Media:** http://ReadingSounds.net /chapter4/#thelostworld.

Captioning advocates need to take note when speech sounds appear to be mistakenly privileged at the expense of nonspeech sounds, especially when a scene has nothing to do with speech, when speech can't easily be heard but is captioned anyway, when captions fail to distinguish background (ambient) from foreground (musical) sounds, or when captioners have not properly analyzed how sounds function in a particular scene. *Just because words are spoken doesn't mean they need to be captioned.* This claim may be hard to take seriously, but that's only because of our cultural tendency to privilege the spoken word over other sounds, to make speech the center of our meaning universe even when speech sounds are irrelevant, ambient, or barely audible. When irrelevant speech sounds are captioned seemingly on the basis of their presumed privileged status alone, we need to call out the practice as yet another expression of logocentrism or logophilia.

Logocentrism is a new term for caption studies that asks us to reflect on the relationships between speech and nonspeech sounds. In the context of literacy studies, the term refers to the reduction of literacy to writing and "the neglect of other modes and their interconnectedness" (Archer 2006, 451). Applied to caption studies, logocentrism refers to the act of privileging speech and the neglect of (or misunderstanding of the complex roles played by) nonspeech. Logocentrism is rational. The teeming soundscape, when approached in logocentric mode, is corralled and tamed at the expense of its complexity and nuance (which, in an important sense, is what all captioning does). Nonspeech is irrational. Nonspeech sounds are hard to caption. They sometimes exceed the reach of language and understanding. They require from captioners a great deal of creativity and attention to context. The tendency to fall back on the pull of *logos*—the spoken word—is strong but not surprising. What is the quickest route for the captioner? Follow the speech sounds. Or: follow the production script, which will always promote dialogue over sound effects. Likewise, the preferred route within a global economy of translation is a speech-centered one, because people are assumed to (1) be hearing and (2) speak different languages. Accessibility for deaf and hard-of-hearing viewers becomes, as a result, a mere byproduct of a global economy driven by the need to reconcile linguistic, not hearing, differences. Accessibility is equated with speech translation. While examples of full spectrum, high-quality closed captioning

abound, particularly on DVDs, we need to be mindful of, and challenge
when necessary, logocentric tendencies by asking:

- How sounds *really* function in a scene and not which sounds we think are important because of our assumptions about the intrinsic value of speech;
- What important roles nonspeech sounds play, especially instrumental music;
- How hearing viewers experience scenes and how soundscapes were intended to
  be experienced;
- How caption readers, especially deaf and hard-of-hearing viewers, make sense of a
  scene's visual calculus and logic;
- How assumptions about the importance of speech can shape the captioned landscape and the resulting meaning of the text as it is expressed through the captions.

In this chapter, I've called attention to the tension between speech
and nonspeech captions through examples of the gravitational pull that
speech sounds sometimes exert on captioners and on what passes for
"captioning" in a global subtitling (speech-only) economy. But to what
extent is captioning itself logocentric? It usually goes without saying in
captioning discussions and style guides that every film soundscape—no
matter how complex, transcendent, or immersive—can always be translated into words. Is it problematic to assume that language is always up
to the task? Is captioning really just a simple matter of translating across
modes? Does every sonic event communicate semiotically? Can every
soundscape be sliced up into discrete layers or units, with each unit
communicating individually? What do we miss if we operate according
to the assumption that captioning is a process of extracting the most
salient sounds, where salience usually means speech (cf. Rickert 2013)?
In other words, we need to be mindful not to equate significance with
speech but to attend to the sonic ground that embodies and informs
the meaning of speech sounds. The captioner's job should not be to extract only the most salient sounds from the situation, "like a comedian
extracting what is comical from life's situations" (Rickert 2013). While
it is not unusual for captioning advocates to wonder about the fit of a
single word or an entire caption, no one has questioned the assumption
that informs captioning itself: namely, the logocentric tendency to assume that language is fit enough to represent the soundscape. While I
think this assumption is largely (but not wholly) justified, it needs to
be made explicit to encourage us to reflect more deeply on the limitations of language and in particular the different affordances of sound
and writing. Ambient sounds—the background "key" of a scene, such as

the low rumble of a commercial airliner's interior—are especially challenging to represent linguistically because of their duration (they can span entire scenes), stability (they don't change), low respective status as background noise, and complex psychological impact (i.e., they are not always consciously noticed, but they inform viewers' experiences nonetheless). If we remember that words and sounds are fundamentally different, that an immersive soundscape can exceed the reach of an accessible collection of captions, just as language can overpower a scene's subtle sonic ambience, captioners will be in a better position to invent a fitting rhetorical transcription, and we will be in a better position to evaluate it. Our collective faith in words—"the linguistic imperialism that subordinates the sonic to semiotic registers" (Goodman 2012, 71)— must be informed by a greater sensitivity to the affordances of language and a respect for the complex tensions between writing and sound.

FIVE

Captioned Irony

Under the right conditions and with the right readers, closed captions can manipulate time, transporting readers into the future or, in more general terms, providing them with advance or additional information. In this chapter, I coin the term *captioned irony* to account for this significant difference between captioned and uncaptioned texts. Just as dramatic irony accounts for situations in which the audience knows more than the characters, captioned irony accounts for situations in which caption viewers know more, sooner, or differently than noncaption viewers. This different experience is not dependent solely on poorly timed captions (although sometimes it is) but on the different affordances of reading and listening, and, in some case, the inflexible application of style guidelines to situations that require a more subtle approach. This chapter will explore the time-traveling potential of punctuation at the end of a line (when coupled with fast reading and/or slow speaking), speaker identifications (when they give away the identity of characters too soon), and ironic juxtapositions (when captions cross boundaries they shouldn't).

If we move beyond narrow approaches to quality captioning, we can explore quality in more nuanced, sophisticated ways. When quality is defined primarily in terms of accuracy and completeness, we have little room to talk about captions and viewing experiences that are not simply right or wrong but . . . *different.* The different affordances of sound and writing create different experiences of the text, and this is true for all readers regardless of hearing ability. Writing stabilizes, formalizes, clarifies, and names sounds

5.1   Premature captions and ending punctuation help readers predict the future.
In this frame from *Taken*, the final bad guy in the movie (Nabil Massad) holds a long knife to the throat of a young woman (Maggie Grace) who was kidnapped. The em dash helps us identify the exact moment when the bad guy will be shot because his utterance is interrupted midword. Caption: "We can nego—." EuropaCorp, 2008. Blu-Ray. http://ReadingSounds.net/chapter5/#figure1.

that are uncertain, ambiguous, or not yet ready to be named in the narrative. The untamed and teeming soundscape is corralled in the act of captioning it. As a result, savvy caption readers are presented with, and take advantage of, opportunities to read ahead, freeing up precious reading time to experience more of what's visually happening outside of the caption frame (e.g., reading characters' lips and facial expressions).

The Ironic Potential of Closed Captions

Irony is a complex term with a very long history. According to Jonathan Tittler (1985, 32), "Irony has meant and means so many different things to different people that rarely is there a meeting of minds as to its particular sense on a given occasion." In the simplest sense, "irony highlights a difference or a contrast" (Giora et al. 2005). The most common form is verbal irony, which refers to a difference or tension between what is said and what is meant. In Socratic terms: "expressing what we mean by saying something contrary to it" (Vlastos 1991, cited in Van Goor and Heyting 2006, 479). Penny Pexman et al. (2005, 259) distinguish between verbal irony for criticism and praise: "[Verbal irony] can be used to criticize, as in 'You look really gorgeous' said to someone who looks disheveled (an ironic criticism), or to compliment, as in 'You look really terrible' said to someone who looks stunning (an ironic compliment)." Because verbal irony "is prevalent in everyday discourse," very young children acquire the capacity to distinguish the literal meaning from the intended meaning, although this capacity continues to mature through middle childhood, particularly where grasping humor in irony is concerned (Pexman et al. 2005, 259). More broadly, irony is associated with negation, subversion, incongruity, contrariety, duplicity, doubling, multiplicity, uncertainty, humor, teasing, mockery, and impersonation. Irony is traditionally grounded in an awareness of difference: between appearance and reality, surface meaning and deeper intention, expectations and effects, contexts and statements, earnestness and irreverence. After postmodernism, however, these distinctions have been called into question:

Traditionally, irony has been a means to expose the space between what is real and what is appearance, or what is meant and what is said, revealing incoherence and transcending it through the aesthetic form and meaning of a work of art. The irony of postmodernity denies a difference between what is real and what is appearance and even embraces incoherence and lack of meaning. It claims our interpretations of reality impose form and meaning on life: reality is constructed rather than perceived or understood, and it does not exist separately from its construction (Colletta 2009, 856).

From this perspective, we could say that we live in a postmodern world of surfaces without depths, an endless play of signification and pastiche, style over substance, and reality subsumed by rhetoric (see Colletta 2009, 856–66). Postmodern irony is self-referential. Television repeatedly mocks itself as a "waste of time" and "not real" (857) even as it ironically needs to keep us tuned in and served up to advertisers. This postmodern view is valuable for caption studies because it rejects the notion that there's a single set of "real" meanings in the film that can be ferreted out by the captioner. The captioner isn't a decoder so much as an interpreter of meaning and a performer of intentions. The same film can be interpreted in wildly different ways, as we have already seen. Some of these ways will be criticized, but from a postmodern perspective, there is neither one true meaning of the text nor an objective means of translating from sound to writing. While the captioner ideally serves as a rhetorical proxy agent for the producers, that agency will always be a rhetorical performance.

Dramatic irony provides the inspiration for this chapter. In ancient drama, according to Claire Colebrook's (2004, 14) book on irony, "most of the plots were mythic. The audience watched a drama unfold, already knowing its desired outcome." In the case of Sophocles's *Oedipus the King*, for example, " 'we' (the audience) can see what Oedipus is blind to" (14). This same awareness on the part of the audience persists today: "We might say that we can get a sense of this tragic or dramatic irony today, either by the fact that we know the plot of *Macbeth* and can see Macbeth hurtling towards his end, despite his ambitions, or by the fact that we are aware of the forces of plot and genre" (Colebrook 2004, 14). Our familiarity with standard Hollywood plots creates our ironic relationship with the text. We know where the action is going and how, in broad terms, it will be resolved (happy endings and neat resolutions are usually the safest bet in Hollywood films). In short, dramatic irony accounts for what the audience knows that one or more characters do not. It operates on a global level when audiences activate their knowledge of forms and genres to see into the future and at the local level when we become privy to information that a character doesn't possess.

Captioned irony is a new term that calls attention to the differences between experiencing a movie through captioning and experiencing it through listening. What do captions bring to the table for readers that isn't available to hearing viewers who are only listening? Such a question hasn't really been asked before. We tend to focus our discussions of closed captioning on what is lacking—missing captions, inadequate technologies, "caption fails." Indeed, these kinds of differences can sometimes create an ironic awareness in audiences, as when programs about "deaf-related topics (for example, a news clip about cochlear implants)" are not captioned when delivered online (Berke 2009). But in this chapter I want to call attention to some of the productive differences between listening to a movie (with no captions) and reading a movie (with closed captions). These differences allow caption readers to know more or sooner than members of the hearing audience who are merely listening to the same movie or show. In some cases, differences can be chalked up to one or more affordances of writing. In other cases, differences are the result of applying captioning guidelines too rigidly and not allowing for flexibility in the face of contextual influences.

The Prophetic Power of Dots and Dashes

Savvy caption readers are primed to read ahead, combining fast reading skills with (1) a desire to experience as much of the visual landscape of the film as possible and (2) a willingness to make predictive inferences as they read and experience the film. Because the findings on predictive reading are mixed, this section can only be speculative and exploratory in applying these findings to captioning. Even so, this section begins to lay the groundwork for a fuller account of how readers engage temporally and cognitively with captions and how captions provide unique opportunities for readers to make predictions about where the action and dialogue are headed. Future research can build on this groundwork by testing it on a range of viewers.

No research studies have been done on whether caption readers engage in predictive reading strategies, but a number of studies have explored whether and, if so, how readers make predictions as they read traditional kinds of texts. According to David Allbritton (2004, 309), "Predictive inferences (inferences about what will happen next) allow the reader to go beyond the text and presumably construct a richer understanding of a narrative." The findings are "mixed" (310): "Some studies have found no evidence that predictive inferences are generated during reading" (Calvo and Castillo 1998, 249), while others have explored the constraints that shape inference making. Context plays a large role, not surprisingly. Contextual factors influencing inference production include "reading goals or purposes," "individual differences in the salience and personal relevance of the information," "individual differences in working memory capacity," "whether the necessary information is foregrounded in the text," and "time course," which refers to the amount of delay between reading a sentence or text and being able to name the target inference (Allbritton 2004, 310). These factors are applicable to closed captioning. Clearly, individual differences in reading goals, personal motivation, and memory capacity affect how readers make sense of closed captions and what inferences they draw from the text. Caption readers need to be familiar with narrative forms and major genres of storytelling—"the forces of plot and genre," in Colebrook's (2004, 14) words. Ironic awareness grows out of readers' advance knowledge of the major stories or "myths" of their time and place. Readers also need significant experience with closed captioning conventions. They need to be motivated and engaged with the text too. Finally, they need time to read and process the text. The title of a research article sums it up: "Predictive inferences take time to develop" (Calvo and Castillo 1998). Yet time is scarce on the caption layer, since each caption is usually on the screen for just a few seconds. Caption readers, unlike readers of traditional texts, can't easily or always go back and reread what they missed, at least not without a DVR, DVD, or other random-access technology that makes it easy to go back and rewatch. Caption studies can begin to explore whether caption readers make inferences as they read and, if so, the conditions under which inference making is possible. In this section, I offer a number of paths we might follow—with the help of data drawn from readers—to explore the nature of predictive inferences in closed captioning.

Caption readers can work with the smallest of clues. For example, even something as seemingly inconsequential as a comma (or even no punctuation at all) at the end of a caption will tell readers that the current sentence or utterance continues into the future and that the same speaker will continue to hold the floor beyond the current caption. If readers can read faster than speakers say their lines, they will get to that comma at the end of a caption and, if they are experienced caption readers, process its meaning before the caption is replaced on the screen with a new caption (assuming captions are set in pop-on style). A trailing comma (or no punctuation at the end of the caption) doesn't tell readers much about the future, but it does begin to suggest how sound and writing differ. Hearing viewers watching the uncaptioned version of a movie will not necessarily know some basic information that caption readers will know before they do, such as (1) how long a spoken sentence is (i.e., whether a spoken sentence continues over multiple captions) and (2) whether a spoken sentence is a complex one. For example, consider a compound or complex sentence that occupies two captions, with one clause pressed into the first caption. A trailing comma (or even no punctuation at all) will signal the continuation of that sentence into the next caption. This advance knowledge can be used in very limited ways to predict turn-taking behavior (e.g., how long speakers will control the floor). Figure 5.2 is a compilation of screenshots from *A Serious Man* (2009), a dark comedy from the Coen Brothers set in 1967. Take note of how ending punctuation (i.e., a comma), or no ending punctuation at all, signals the continuation of the current sentence into the future.

Other types of punctuation can facilitate this practice of glimpsing the future, such as preceding hyphens when used to distinguish multiple speakers in the same caption. Depending on the length of each line—that is, how long each party's spoken contribution is—caption readers may be able to finish reading well ahead of the characters speaking those lines. In a simple call and response, for example, one in which both the call and the response occupy the same caption, caption readers may know the response to the call before it has been spoken. To see how this plays out in real time, consider a clip from *The Order*, a 2003 horror film starring Heath Ledger. (*A Serious Man* uses screen placement,

5.2 **A compilation of six frames from *A Serious Man* with captions that display ending commas or no ending punctuation.**

Each example alerts readers that the current speaker will continue to hold the floor into the next caption. Frames 1 and 2 (reading across) have no ending punctuation. Frames 3, 4, 5, and 6 make use of ending commas. These frames were taken over a timespan in the film of about eight minutes (approximately 0:15:00 to 0:23:00). Focus Features, 2009. Blu-Ray. http://ReadingSounds.net/chapter5/#figure2.

discussed below, instead of hyphens, to distinguish speakers.) The clip presents a simple call and response in a single caption. Mara (Shannyn Sossamon) asks Ledger's character about two small, presumably lost or parentless children huddling together on the floor: "What are they?" Ledger replies, "Orphans." These four words remain on the screen for 3.4 seconds, which translates to a very slow reading speed of seventy-one words per minute (wpm), or just slightly faster than one word per second. This speed is much slower than the average reading speed of 141 wpm for television captions (Jensema, McCann, and Ramsey 1996), giving readers ample time to read the response before it is spoken. The long pause between the spoken question and Ledger's answer (approximately two seconds) enables caption viewers to know the answer to the question before hearing viewers who are listening with captions turned off. To put it differently, two seconds (the length of the pause) is the time it would take, at a reading speed of 141 wpm, to read 4.7 words, and 3.4 seconds (the total length of time the caption remains on the screen) is the time it would take to read eight words at the same reading speed. Given the long pause between question and answer, it might have been more effective to split the call and response into two captions to replicate more closely how the spoken lines are delivered. I will have much more to say about reading speed in a minute. **Media:** http://ReadingSounds.net/chapter5/#the-order.

While preceding hyphens may allow caption readers to read the second speaker's line before the first speaker finishes speaking, they can also act as spoilers. Hyphens are less effective at distinguishing speakers' utterances. Screen placement is a more effective technique because each utterance is spatially positioned underneath each speaker, a technique called "identification placement" (WGBH 2002) or positioning. For example, in a standard two-shot, one speaker's captions are positioned on the left side of the screen, the other speaker's captions on the right. In this way, "positioning carries meaning" (Clark 2002, 85). According to the Described and Captioned Media Program's *Captioning Key* (DCMP 2011d), arguably the public authority on captioning style in the United States, hyphens and "other speaker identification techniques" should never be used to distinguish multiple speakers in the same caption. Screen positioning—"place the captions underneath the speakers"—is preferred over hyphens for multispeaker captions for the simple reason that it is sometimes hard to tell who is speaking when two captioned utterances are centered on the screen and only distinguished by hyphens. Caption readers need to be able to associate captioned utterances with speakers at a glance, especially when two speakers are speaking in the same caption.

Unfortunately, robust screen placement capabilities are rare on the Web. Centered alignment is the dominant mode (although Netflix streaming makes use of identification placement when the DVD source captions do so). Stripped-down interfaces are the norm. Captioned experiences and video interfaces on the web are impoverished, to say the least. Do-it-yourself (DIY) captioning services such as Amara.org trade quality for speed, allowing novices to caption easily and quickly but not do much else besides transcribe sounds. Captions are automatically formatted into a two-line space at the bottom-center of the screen in DIY captioning and autocaptioning (YouTube). While the Federal Communications Commission has proposed "placement" as one of the

5.3   A caption covers onscreen text.
In this frame from an episode of *Arrested Development* ("Indian Takers"), Tobias Fünke
(David Cross) wears a tuxedo in a mid-shot while looking to his right. The yellow onscreen
text, placed bottom center in two lines, reads: "Tobias Fünke / Lindsay's acting husband."
This phrase is more than a basic character description. It holds a double meaning because
Tobias is an actor and he continually makes remarks that undercut his own presumed
heterosexuality. A bottom-center caption covers the onscreen text in this frame: "*who
should have taken more.*" Making matters worse: both text and caption are set in the
same yellow font color. Netflix customers can customize the typeface, size, and color of
the captions but not change their placement. Netflix, 2013. http://ReadingSounds.net/
chapter5/#figure3.

key criteria for ensuring that Internet captions will be "at least the same
quality as the television captions for the programming" (FCC 2012),
no specific requirements for quality television captioning have been
adopted by the FCC at the time of this writing (FCC 2013, Schacter Lintz
2013). Identification placement continues to be an important issue for
advocates of television and web captioning, because bottom-center captioning routinely covers important information such as credits (see figure 5.3). When screen placement options are not available beyond the
bottom-center default setting, the *Captioning Key* directs captioners to
"caption each speaker at different timecodes" instead of using hyphens
(DCMP 2011d). Now that captioning on the web is required by US law
for TV content, and the FCC is moving in the direction of offering
specific quality standards, I am hopeful that captioned interfaces and
tools will become more robust on the Web, starting with increased control over caption placement.

When readers noticeably react to well-timed captions before they are spoken, we have strong evidence that they are reading ahead. While I don't have any user data on this phenomenon, I've experienced it firsthand in my own viewing/reading practices and witnessed it in the reactions of my family members. If you've ever laughed at a captioned punch line before it's been fully delivered in speech, even when that caption has been perfectly timed to appear as the first word of the punch line is uttered, then you have experienced one of the productive tensions between listening and reading. Both of my children—one is deaf, the other hearing—developed at an early age the ability to read ahead that rivaled my own and my spouse's, which suggests to me that this phenomenon is not limited to adults, highly literate readers, or hearing viewers. It is also not limited to poorly timed early captions. When captions start on time or even a beat too late, readers can still read ahead—not all the time, of course, but when conditions warrant and when readers possess relevant genre and situational knowledge. Caption studies can begin to explore more deeply these conditions and what makes them fertile for time travel.

Beyond anecdotal evidence, we can turn to reading speed preferences to provide indirect evidence of predictive reading. Carl Jensema (1998) studied the preferences of 578 participants—a diverse groups of hearing, hard-of-hearing, and deaf viewers aged eight to eighty—by measuring their reactions to thirty-second video segments on three topics that were captioned at different speeds, from ninety-six words per minute up to 200 words per minute (wpm). Each video was captioned at eight different speeds for a total of twenty-four videos. Viewers rated each segment on a five-point scale of "Too slow," "Slow," "OK," "Fast," and "Too fast" (320). The average speed rated as most comfortable (the "OK" speed) by all participants (145 wpm) was "very close" to the average speed of all televisions programs (141 wpm), the latter number being derived from an earlier study (Jensema, McCann, and Ramsey 1996). These numbers are also very close to the speed of spontaneous speech. According to Arthur Wingfield et al. (2006, 488), "While speech in thoughtful conversation may be as 'slow' as 90 words per minute (wpm), average speech rates in ordinary conversation vary between 140 and 180 wpm, and a radio or television newsreader working from a prepared script can easily exceed 210 wpm (Stine et al, 1990)." Contrast the speed of reading print, which varies greatly but averages about 250 wpm (Pickett 1986, 263). As speeds picked up for participants in Jensema's (1998) captioning study, they were still able to keep up, adjusting their reading accordingly. Most participants began to experience "significant difficulty" when speeds hit or exceeded 170 wpm. Hearing viewers in particular—those who "had less experience watching captions" (321)—were found to prefer slightly slower captions. This difference between hearing and deaf/hard-of-hearing groups was found to be statistically significant. At the same time, Jensema's (1998) study found "no relationship between age and comfortable caption speed" (322) or between educational level and perceived comfort levels (323), which suggests that differences can be made up quickly between those who watch captions regularly and those who don't. "People apparently adjust to caption reading quickly, with further practice making little difference" (324). This claim is further supported by the deaf and hard-of-hearing teenagers in Jensema's study, who were "most comfortable at approximately the same caption speeds as the overall viewing population" (323).

The limitations of this 1998 study are worth noting. It was conducted at a time when mandatory captioning of all new television content was not yet required by law. The 1996 Telecommunications Act had passed, requiring closed captioning on all new television programming, but the full rollout of this law would still take another decade. Moreover, some participants did not have technical access to captioned television because they "came from poor, inner-city homes with pre-July 1993 televisions, which lack built-in caption decoders" (323). It would be surprising to find this problem today. Despite the access challenges that we continue to face, decoder chip access is not one of them. Television content is universally captioned today in the United States as required by law (with few exceptions, including TV commercials). Hearing people are more accepting and aware of captioning and the benefits of universal design. We would most likely take issue with Jensema's claim that "Deaf and hard of hearing people tend to watch captioned television daily; hearing people seldom watch [captioned TV]" (321). Finally, Jensema's study was highly artificial. The video clips were not selected from examples of actual television programming but were created by moving a video camera around a poster related to each of three topics "to give the illusion of a moving picture" (319). The resulting videos did not have audio (319, 324), and the video images were designed "to distract the viewer without duplicating information given in the captions" (319). Both of these conditions (no audio, no duplication of writing and image) created an unnatural viewing experience. Without access to audio, hard-of-hearing and hearing viewers could not supplement what they read with what they heard (and vice versa). Without access to images that reinforced what they read, viewers could not read lips or verify the content of what they were reading or their level of comprehension. By "control[ling] for audio information" (319), Jensema essentially created a reading speed test, not a study of actual captioning speed preferences. Despite these limitations, Jensema's (1998) study established a baseline of reading speeds—145 wpm was considered "comfortable," 170 wpm or greater was considered "too fast"—that persist in captioning style guide recommendations and that we can apply to an analysis of predictive reading. In other words, speeds that were rated as "too slow" by Jensema's participants (under 126 wpm) could provide extra time for viewers of real programming to examine the images on the screen, read characters' lips, read faster than speakers deliver their lines, and make predictions from small clues in the captions. This extra time away from the demands of reading should be factored into discussions of reading speed. In addition to reading the captions, "viewers also need to look at the image" (Szarkowska et al. 2011, 376), which has prompted Pablo Romero-Fresco to argue for a concept of "viewing speed" over "reading speed," because viewing speed "takes into account the time necessary both to read the captions and to watch the image. [Romero-Fresco's] preliminary findings show that, for instance, a movie captioned with a reading speed of 150 wpm/12 [characters per second] will give viewers about 50 percent of time to read the caption text and another 50 percent to devote to watching the action on the screen" (quoted in Szarkowska et al. 2011, 376).

Let me now turn to specific examples of how captioning speed contributes to predictive reading. First, consider that end-of-caption em dashes, ellipses, and other marks of punctuation are sometimes used to cut utterances off midsentence or even midword. In such cases, readers are presented with a slightly larger glimpse of the future than provided by a comma at the end of a caption, though without necessarily having any greater level of certainty about what is going to happen. That glimpse is a mixed blessing for viewers, because knowing the future, even vaguely, also mutes its intensity. A surprise is a bit less surprising if we know precisely when it is coming, even if we don't know what the surprise is. When viewers use captions to venture a guess at the future, they change their relationship with the text. All viewers, regardless of whether they are watching with closed captions or not, continually make guesses and revise them on the fly. But caption viewers have an additional layer of clues that tells them *when* an interruption will happen and *what* it might mean. For example, in *A Serious Man* (2009), the ellipsis serves a number of functions when it is appended to the end of a caption, the most popular of which is to signal a midutterance interruption of one speaker by another (see table 5.1). Close to half of the

Table 5.1 Eight pairs of captions from *A Serious Man* showing examples of interruptive ellipses.

171 Sy has come into my life, and I . . .	185 Look, Sy feels that we should . . .
172 Come into your . . . What does that mean? You barely know him.	186 Esther is barely cold!
285 Thanks, Sy, but I'm not . . .	409 The kids aren't . . .
286 Listen, I insist, Larry. There's no cause for discomfort.	410 I'm saying "we." I'm not pointing fingers.
492 He tried to tell me about this thing he's working on, this . . .	541 I was hoping that Rabbi Nachtner could . . .
493 The Mentaculus.	542 That he would . . . He would . . . Yes?
615 Although she is planning to marry Sy Ableman, but they . . .	641 The associates, me, for instance, bill at . . .
616 Sy Ableman!	642 Call for Mr. Gopnik. Danny at home.

Note: The first caption of each pair ends with an interruption, which is denoted by an ellipsis. The second caption of each pair is the second speaker's interruption. Focus Features, 2009. http://ReadingSounds.net/chapter5/#table1.

106 end-of-caption ellipses in *A Serious Man* are interruptive (forty-six of 106). Speakers are interrupted by other speakers as well as by the environment (e.g., a police siren, an approaching tornado). The remaining sixty ellipses can be divided pretty equally between pauses (thirty-one) and falters or stammers (twenty-nine). The ellipsis-as-pause is used to signal a single speaker's pause in the middle of his or her utterance, as in this three-caption sequence: "He just . . . /(INHALING)/ . . . congratulates the *bar mitzvah* boy every week." The ellipse-as-pause is also closely associated in this film with discourse markers (e.g., So . . . , Uh . . . , Hmm . . . , look . . . , I mean . . . , um . . . , Well . . . , you know . . .) and, occasionally, with speech that trails off. The line between a pause and a falter is not always clear. I coded an end-of-caption ellipsis as a falter

if a speaker repeated a word or phrase that bridged two captions ("denigrating you and urging us not to . . . /Not to grant you tenure"), or if a speaker started one sentence only to stop in the middle and begin a new sentence in a second caption ("No, I . . . / Well, yes, okay."). These three functions—interruption, pause, falter—accounted for all of the end-of-caption ellipses in *A Serious Man*.

Punctuation is veiled prophecy. While we can never be certain of what's to come, we can draw upon our situated sense of how punctuation functions—both conventionally and in specific films—to make predictions about the future. Closed captions allow us to see what is hidden or veiled in speech—to *see* punctuation—and to do so ahead of the speaker's natural pauses. The ellipsis, like the em dash (see below), resonates with advance meaning that is not available to hearing viewers who are only listening with captions turned off. When it occurs as ending punctuation in the closed captions of *A Serious Man*, the ellipsis serves three primary functions. Even if we don't know upfront how the ellipsis will function in any particular instance, we do know that its range of meanings is limited, and we can draw upon our situated understanding of characters and scenes to provide additional clues to aid our interpretation of what the ellipsis will signify. For example, the tense (but also hilarious) phone call between Larry Gopnik (Michael Stuhlbarg) and the Columbia Record guy (voiced by Warren Keith) is full of interruptive ellipses (as well as one end-of-caption *faltering ellipsis* and two end-of-caption *pausing ellipses*). As Larry's frustration and confusion mount, he interrupts the caller repeatedly in an almost predicable pattern of ellipses. Or, coming at this from another angle, viewers know that when the ending punctuation is a comma an interruption will *not* occur and the same speaker will continue to hold the floor at least into the next caption. Nevertheless, because the ellipsis is made to serve a number of different meanings in this film, its predictive power for readers is limited. **Media:** http://ReadingSounds.net/chapter5/#serious-man.

When the ellipsis is associated exclusively and by convention with a single type of interruption that is punctuated consistently and repeated throughout a scene or movie, predictions can be made more confidently. *Saturday Night Live* (SNL), the long-running live comedy sketch program, is famous for hanging entire skits on single jokes that are repeated numerous times. The one-joke premise can quickly wear out its welcome, but I want to put aside the question of what makes some skits funnier than others and consider the role of punctuation in facilitating predictions about what's going to happen or, in the case of the one-joke premise, the precise moment when the joke will be delivered again. (*Saturday Night Live* is live captioned in scroll-up style, but the clip I will discuss here has been recaptioned in pop-on style for replay on VH1.) In "Quick Zoom Theater" (2006), a doctor (Steve Martin) and his patient (Fred Armisen) are discussing the patient's test results and other mundane topics. The scene is done in mock classic soap-opera style complete with expectant pauses in dialogue, close-up quick zoom shots of speakers, faces turned directly into the camera on cue, and dramatic horns before news is delivered. The form leads us, at least initially, to expect bombshells to drop in typical soap-opera style. But the patient is fine. Indeed, the entire visit turns out to be routine (with one exception), which conflicts with the hyperexaggerated dramatic form in which the routine news is delivered:

I'VE LOOKED OVER YOUR
TEST RESULTS AND . . .

[DRAMATIC MUSIC PLAYS]

EVERYTHING'S FINE.

The humor is intended to come from the mismatch of form and content: the expectation of a revelation or bombshell is repeatedly thwarted. An ellipsis at the end of a caption signals the syntactic location of the dramatic pause that will initiate the joke sequence. The scene includes eleven end-of-caption ellipses. In other words, caption readers can visually see when the joke sequence will be repeated. They can also see which lines will not be accompanied by the joke sequence (because they do not include an ending ellipsis), as in "NO. BY THE WAY, HOW WAS THAT COOKOUT?" and "WELL, THAT SOUNDS NICE. DID YOU ASK VERONICA?" When a real bombshell is finally dropped at the end of the scene ("AFTER ALL, SHE IS CARRYING YOUR BABY."), the joke sequence is not activated, a fact which caption readers anticipate clearly because there is no ellipsis at the end of the caption. The biggest surprise comes when the musical flourish and quick-zoom technique are not accompanied by an ellipsis but simply follow the end of a sentence:

WELL, GREAT.
THANKS A LOT.

[DRAMATIC MUSIC PLAYS]

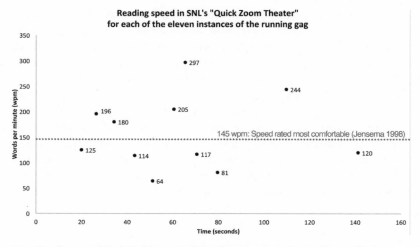

**Reading speed in SNL's "Quick Zoom Theater"
for each of the eleven instances of the running gag**

5.4   Reading speed is plotted for each of the eleven captions that end in an ellipsis on
Saturday Night Live's "Quick Zoom Theater."
The slowest caption is 64 words per minute (wpm) and the fastest is 297 wpm. NBC, 2006.
http://ReadingSounds.net/chapter5/#figure4.

Otherwise, the skit's running gag is always initiated with a caption-ending ellipsis. The captions both help readers make predictions and slightly mute the intensity of the joke by telling us when it will be repeated. The average reading speed of the eleven ellipsis/gag captions is 158 wpm, well within the typical range for caption speed on television. But this single number obscures the wide range of reading speeds, from a low of sixty-four wpm to a high of 297 wpm (see figure 5.4). The slower the reading speed, the more time readers have to process the caption's full meaning and, especially, the ending ellipsis as a primary signal for the running gag.

The most telling examples of prophetic punctuation lean heaviest on repeated contexts (such as a running gag) and on the expectations of savvy viewers who are deeply familiar with basic Hollywood storylines. In such cases, viewers already think they have a pretty good idea of what's going to happen and the ending punctuation seals the deal, so to speak. The best example I know of—in fact, the example that originally set me on the path I'm pursuing in this chapter—is from *Taken* (2008), in which a simple em dash speaks volumes to viewers about *when* the final tense confrontation will be resolved. At the end of the movie, Liam Neeson's character finally tracks down his kidnapped daughter. The kidnapper (Nabil Massad) tries to negotiate but before he can plead for his life ("We can nego—"), Neeson shoots him dead. The movie poster has promised nothing less from Neeson's character: "I don't know who you are but if you don't let my daughter go, I will find you [and] I will kill you." So it shouldn't come as a surprise that, just as Neeson has done with the previous bad guys, this final bad guy will suffer a similar fate. Every other bad guy has been neatly dispatched by Neeson. Why should this last one be any different, particularly when (1) he, unlike the others, is holding Neeson's daughter (whom we haven't seen since she was taken at the beginning of the film), (2) Neeson is pointing a gun at him, and (3) the standard Hollywood script demands a happy reunion of father and daughter? Caption viewers have an extra bit of information in the em dash that allows them to know precisely *when* the bad guy will be shot by Neeson. **Media:** http://ReadingSounds.net/chapter5/#taken.

We surmise ahead of time that the reason the bad guy can't finish saying the word "negotiate" is because Neeson shoots him. It's not hard to guess what's going to happen when we note Neeson's take-no-prisoners style (negotiation is not an option!) and the audience's expectation of a happy ending (i.e., his daughter unharmed and returned to her family; the bad guys pay the ultimate price). The em dash on "nego—" allows us to predict *when* that price will be paid by the final bad guy. (In chapter 4, I shared another example from *Taken* that also includes an ending em dash to signify an interruption or break midsentence: "Attention, travelers. You are not required—") The "nego—" caption is poorly timed too, giving viewers ample time not only to read the line before it is spoken but to make an educated guess about the bad guy's fate. In the featured clip, the caption starts at 00:15.09, the line of dialogue at 00:16.32, the gunshot at 00:16.77. Thus, the difference in time between when the three-word caption appears on the screen and when Neeson fires the gun is 00:01.68, which translates to 107 wpm (3/1.68=107.14/60). In other words, if you can read three words in 1.68 seconds (or, really, two words and an incomplete third word), you can read 107 words in sixty seconds. 107 wpm is clearly on the very slow end of the spectrum. In Jensema's (1998, 324) study, "only about 1 percent" of the participants "would consider 141 WPM too fast," which, even in light of the limitations of his study, is at least suggestive about the potential of readers to have extra time to process what the em dash might mean. If readers interpret the em dash to signify an interruption, they can leverage the slow required reading speed in this scene to venture an informed guess at what (deadly) form the interruption will take.

Another example in this vein is from *The Book of Eli* (2010), a post-apocalyptic Western starring Denzel Washington as Eli. During one tense confrontation in a dusty bar, Eli tells an outlaw named Martz (Evan

Jones), "I don't want any trouble." Martz responds, "Well, that's too bad because you got—" Before Martz can finish his line, Eli slams his head into the bar. We see it coming in the dash at the end of Martz's caption. A dash (or, in this case, an old-style double hyphen) at the end of Martz's speech caption may not tell us precisely what Denzel Washington will do—or more generally, what will happen next—but it does at least alert us to a radical shift in focus. Heading into that dash, viewers are already primed for violence. Fictionalized saloon confrontations, which rarely end with a hug, are further strained in a postapocalyptic landscape in which resources are scarce, lawlessness is the norm, and only the strongest survive. Violence is assumed to be inevitable in such a wasteland. It's just a matter of time. The dash may provide the clue audiences need to recognize when that time has arrived. **Media:** http://ReadingSounds .net/chapter5/#book-of-eli.

There are approximately 2.8 seconds between the appearance of Martz's line on the screen and Eli jumping up to smash Martz's head into the bar, which translates into a reading speed for Martz's seven words of 150 wpm (seven words over 2.8 seconds=150 words over sixty seconds), still easily within the realm of Jensema's (1998) "comfortable" or even "slow" category for some readers. Indeed, anything under 150 wpm has been characterized as "low" by Agnieszka Szarkowska et al. (2011, 364), which is to say that edited captions—captions that have been intentionally slowed down for readers through techniques of reduction and summary—are defined as topping out at 150 wpm. Verbatim captions, by contrast, are "usually displayed at high reading speeds (180 wpm or more)" (Szarkowska et al. 2011, 364). If adult readers are regularly confronted with "180 wpm or more"—and Americans in particular have come to expect and demand nothing less than verbatim captions—then 150 wpm should feel like child's play to some seasoned caption readers. Put another way, 150 wpm may be considered low in terms of reading speed but just right in terms of viewing speed, since the viewer's goal is not simply to read captions but to experience the program (see Neves 2008, 136). The *Captioning Key*, which places a speed limit of "150–160 wpm" on "[s]pecial-interest media for adults" (DCMP 2011a), smartly appeals to an escape clause in its discussion of reading speed: "The presentation rate can be increased if heavy editing radically changes the original meaning, content, or language structure" (DCMP 2011a). Instead of drawing a line in the sand between edited and verbatim captioning, we need to approach captioning rhetorically by starting and ending with an escape clause that places content and context at the fore of any captioning decisions involving reading speed.

When *interruptive punctuation* is combined with a viewer's genre knowledge, the prophetic potential of punctuation is multiplied but may also be constrained by the demands of high reading speed levels. Consider a scene from *Apollo 18* (2011), in which Astronaut Nate Walker (Lloyd Owen), one of two astronauts on the moon's surface, is suddenly pulled into a dark moon crater by some unknown force or creature. Before he can finish his line, he's dragged down on the short tail of an ellipsis mark: "[Nate] You can't help me, but you can save your . . ." Every viewer can probably sense that something bad is going to happen because (1) that scary crater was featured earlier in the film as a foreboding and dangerous spot, (2) Nate has been acting funny and now teeters on the edge of the crater, (3) the dialogue calls up our prior experiences of other movie scenes in which characters go down screaming "Save yourself!" and (4) this is a horror movie (where scary things happen suddenly). *But caption viewers have an extra bit of information in the ellipsis*—if they can read fast enough and have enough prior experience with captions—that allows them to have advance knowledge of the moment *when* something dramatic will happen to prevent Nate from finishing his sentence. **Media:** http://ReadingSounds.net/chapter5/#apollo18.

When I first watched this movie on DVD, I surmised correctly, before Nate was pulled into the crater, that the ellipsis signaled the end of the conversation between the astronauts and the beginning of a sudden shift in focus. It is always just a guess, and it must take place in a split second. I knew (or thought I knew) the scene wouldn't end well, and the caption gave me a clue as to the timing of the bad event. Sometimes we guess wrong. Experienced viewers of closed captioning are continually making educated guesses as they read ahead. At the very least, the ellipsis tells us that something dramatic will happen to prevent the speaker from finishing his sentence. We may not know exactly what will happen, but we glimpse the future nonetheless. Nevertheless, at 236 wpm (ten words in 2.54 seconds), we have very little time to do much more than read and react.

As reading speed increases, readers' ability to make predictions decreases. Specific film contexts and readers' genre knowledge are crucial variables. The more that readers understand what's going on, and the more knowledge they have of basic storylines, and the more experience they have with captioning conventions, and the closer that captioners adhere to consistent and meaningful style guidelines, the greater the likelihood that savvy readers will leverage interruptive punctuation to make correct predictions. Predictive reading is thus a function of readers' genre knowledge, the demands of specific film contexts, and reading speed. *Taken* creates an ideal environment for making predictions by combining standard, feel-good Hollywood fare, a formulaic climactic scene, and an extremely low reading speed. Graphing these variables, we could put reading speed on the vertical axis (from low to high) and adherence to formula on the horizontal axis (from generic to particular). A *low-generic* data point would provide ideal read-ahead conditions, whereas a *high-particular* data point would not. Confounding our calculations would be readers' level of familiarity with movie genres, their level of experience with captions, their understanding of specific scenes in which predictive punctuation is relevant, and captioners' adherence to consistent and standardized style guidelines.

The sudden eruption of violence provides an excellent demonstration of the power of prophetic punctuation. The three examples from *Taken, The Book of Eli,* and *Apollo 18* suggest how punctuation prefigures a major break in the action. Violence is not exactly a surprise in these movies—viewers are primed for it here and in other similar films—but the interruptive punctuation provides an additional clue as we try to make sense of where the action is headed. While captions should never give away the plot, we must also keep in mind the inherent potential of writing over sound to show us the future. That's one of the affordances of pop-on style captioning. Reading the future is facilitated when captions are set in pop-on style, a superior style for prerecorded captioning in which each caption is displayed on the screen all at once (see chapter 2). Nearly all of the video clips in this book—indeed, all DVDs distributed in the United States—make use of pop-on captioning. The other main style of captioning for prerecorded content—scroll-up styling—does not provide the same opportunities for predictive reading. In scroll-up styling, readers only see one new line at a time, making it more difficult to practice the kinds of predictive reading and sense-making strategies that pop-on styling facilitates.

Captioners must continually strive to make the reading process as straightforward and fast as possible. Readers need to grasp the meaning of each caption quickly before it is replaced with a new caption. Style conventions matter because readers learn them and then come to expect and depend on them. Even at the level of punctuation, captioning style should be consistent across captioning companies and responsive to proper usage guidelines. The examples in this section use punctuation inconsistently even as they offer clues about what's going to happen. The em dash and the ellipsis appear to be interchangeable. Whereas both *Taken* and *The Book of Eli* make use of the em dash (including the old-style double hyphen), *Apollo 18* uses an ellipsis, even though all

Table 5.2 Additional examples of ending punctuation featuring em dashes and ellipses.

Source	Caption	Type
30 Rock, Untitled season 2, episode 10, 2008	STUDIO 6H. THIS IS KE--	Interruption (dash)
Psych, "Shawn 2.0" season 5, episode 8, 2010	♪EVERYBODY WANTS TO RULE THE--♪	Pause/trails off (dash)
Futurama, Bender's Big Score 2007	We're not unemploy . . .	Interruption (ellipsis)
Futurama, "Free Will Hunting" season 7, episode 9, 2012	Are you suggesting that any verdict I reach is simply the output of a pre-programmed . . .	Interruption (ellipsis)
Family Guy, "Lois Kills Stewie" season 6, episode 5, 2007	Well, at least it didn't end like *The Sopranos*, where it just cut to black in mid . . .	Interruption (ellipsis)
Castle, "Hell Hath No Fury" season 1, episode 4, 2009	EGOTISTICAL AND COMPLETELY--	Interruption (dash)

Note: http://ReadingSounds.net/chapter5/#table2.

three examples perform the same kind of interruptive work. Moreover, as we've seen, the ellipsis sometimes assumes a heavy burden as a catch-all punctuation mark, serving pauses and interruptions equally. I have collected additional examples in table 5.2 that further support (1) the prophetic potential of ending punctuation to allow readers to make predictions and (2) the claim that em dashes and ellipses are used inconsistently in closed captioning. The *Family Guy* excerpt from "Lois Kills Stewie" (2007) is a compelling example of inconsistency because the ellipsis does double duty as an interruption mark ("where it just cut to black in mid . . .") and, a moment earlier in the scene, as a pause indicator ("only to find out that none of it really happened . . .").

This lack of consistency can slow down readers' efforts to make quick sense of what they're reading, particularly if these same readers were taught in school to distinguish the em dash from the ellipsis. Grammar and style guides agree: use the em dash for interruptions or breaks and the ellipsis for pauses or faltering speech. According to *The Chicago Manual of Style* (16th ed., 2010), the em dash "may indicate a sudden break in thought or sentence structure or an interruption in dialogue" (section 6.84), while the ellipsis "may be used to suggest faltering or fragmented speech accompanied by confusion or insecurity" (section 13.39). That is good advice for captioners. The clip from *The Book of Eli* properly distinguishes between an ellipsis and an em dash in the same clip, the former denoting a pause in speech following a discourse marker

("Uh. . . .") and the latter denoting an interruption in speech ("Well, that's too bad because you got—"). This distinction could have been usefully employed in *A Serious Man* as an alternative to the catch-all approach of using the ellipsis for everything. An em dash is never used in the DVD captions for *A Serious Man*, despite the large numbers of both interruptions and pauses in the film that have, in other films, conventionally required different punctuation marks. By distinguishing stylistically between interruptions and pauses at the ends of captions, captioners can facilitate the kinds of rapid reading and prediction strategies that readers exploit in fast-paced reading environments.

In this section, I pursued two main aims. I wanted to suggest how something as seemingly inconsequential as a punctuation mark at the end of a caption (or even no punctuation at all) can actually provide important clues to readers about what's to come. I also wanted to show how captioners apply stylistic guidelines inconsistently, thus hindering the very strategies that readers employ to make predictions. More broadly, I wanted to deepen and question our current approaches to talking about timing in pop-on style captioning. While discussions of reading and viewing speed can be supple, they are not generally concerned with how readers consume captions. The style guidelines from the Media Access Group at WGBH (2002), for example, give captioners a number of options: captions can be "timed to appear and disappear precisely when the words are spoken," or timed to appear with shot changes "for readability and aesthetic purposes," or "timed to begin before the corresponding audio begins" in order to "increase reading time." But discussions of timing do not usually go beyond a concern for the captioned text to the readers who consume it or the meanings that timing decisions make possible. Even if captions are precisely timed to appear when the first syllable is spoken, all bets are off for when (or even whether) viewers will finish reading the remaining words in the caption. We need a more robust understanding of the ways in which timing intersects with reading speed in pop-on style captioning to produce the kinds of read-ahead strategies I have discussed here.

Ironic Identification

When speakers are named in the captions as part of an effort by captioners to provide parenthetical information about who is speaking, readers gain access to information that noncaption viewers may not have. Sometimes, as we will see, captioners mistakenly give away information too soon. At other times, however, the knowledge conveyed through speaker or character identification is an artifact of writing technology, a consequence of the affordances of writing, which is to say that when captioners write down what was meant to be heard, they change the viewer's relationship with the text. The translation from sound to writing is never without remainder or tension, even when the goal is "equal access." By "ironic identifications," I refer to the ways in which the captioned (reading) experience can sit in an ironic relationship with the uncaptioned (listening) experience, similar to how dramatic irony invokes the audience's superior knowledge over what the characters possess.

How and when captioners should identify characters' names are rarely as simple as the style guides suggest. The style guidelines from WGBH (2002), for example, direct captioners to add a speaker identifier "[w]hen it is not possible to use placement to indicate the speaker (i.e., if the speaker is offscreen)." A speaker ID is usually set-off stylistically from the spoken words so readers can more easily distinguish it from speech. The most common method, assuming the speech is set in sentence case, is to set the speaker ID in uppercase followed by a colon. According to WGBH, "[p]arentheses or brackets may also be considered" as means to distinguish speaker IDs from speech. The DCMP's *Captioning Key* (2011c) starts from the same position as WGBH that speaker identification is necessary when screen placement is not possible. But the *Captioning Key* directs captioners to put parentheses around the names of speakers when those names aren't spoken but provide information. No reason is given for this firm advice. The *Captioning Key* also goes into more detail about identification practices, from a guideline for identifying speakers with no names (e.g., "male narrator"), to a guideline for combining placement with identification, to a warning about "not identify[ing] the speaker by name until the speaker is introduced in the audio or by an onscreen graphic."

These guidelines aren't always followed, particularly the instruction to wait to identify speakers by name until we have been introduced to them, as we will see below. But even when the rules are followed, when captioners are simply identifying off-screen speakers who are known, the resulting identifications may still give an edge to caption readers in viewers' efforts to identify who is speaking offscreen. The reason has to do with the nature of identification itself in different modes (reading vs. listening). Whereas a speaker identifier in writing is immediately known to readers, the same offscreen identification through listening may be delayed while listeners process the character's voice. If a speaker has a less distinctive voice or is not a regular on the show, listeners (but not caption readers) may need more time to process what they are hearing. I know I do. For example, consider the common film technique of starting a new scene with a panning or tracking shot while an offscreen character begins speaking. The camera pans to where the speaker is located or tracks an object of interest while the speaker, who is not yet visible, begins talking. If the speaker is a recurring character, located offscreen, then of course his or her name needs to be identified in the captions. But in naming known speakers, captioners give readers a jump on information about who is speaking. On HBO's *Game of Thrones*, the track or pan is a common method of transitioning to new scenes when speakers, identified immediately in the captions, begin speaking offscreen. Table 5.3 gives some examples from season 2, episode 1 ("The North Remembers," 2012) of offscreen speaker identification at the start of new scenes for different shot types.

With major or recurring characters, listeners may be able to identify who is speaking offscreen just as quickly as caption readers who can read the names of speakers. But I would always put my money on caption readers knowing ahead of listeners, because identifying a voice through listening alone is never a sure thing (some hearing people anecdotally claim to be better at it than others) and, I would argue, identification through listening always takes more time than identification through reading a speaker's name on the screen. In the case of minor or non-recurring characters on *Game of Thrones* (e.g., Shae, Pycelle, Leadranach), captions not only tell us who is speaking but also clarify the names of speakers with unusual-sounding names. In the preface, I gave an example from one of the *Harry Potter* movies, but I could just as easily have drawn from *Game of Thrones* with its complex mythology and strange names. More importantly, speaker identifiers can give caption readers a jump on what's going on by providing immediate access to the identity of speakers before those speakers become visible. In these cases at least, this advance knowledge can't simply be chalked up to poor captioning but rather to the affordances of writing as a technology for making knowledge visible, durable, and more efficiently communicated.

These captioned identifiers create a different experience of the text, one that is eagerly and unabashedly informative and perhaps less suspenseful but also stripped down to its bare essentials. The four examples from *Game of Thrones* involve speakers vocalizing words in the transition from one scene to the next, but speaker identifiers may also be attached to nonspeech vocalizations (e.g., characters grunting, screaming, sighing, whistling, etc.). For example, in the middle of this episode's council scene (the scene starts with Pycelle's caption in table 5.3), someone is

Table 5.3 Some examples of offscreen speaker identification in *Game of Thrones*, "The North Remembers"

Description	Caption	Context
A **tracking shot** of a raven cage being brought into the council room as Pycelle begins speaking offscreen. The speaker ID appears at the start of the tracking shot.	PYCELLE: A raven arrived from the citadel this morning, Your Grace.	New scene
An **establishing shot** of a soldier outside the castle at Winterfell. Someone identified in the captions as Leadranach, a nonrecurring character, begins speaking over the establishing shot before the scene cuts to the inside of the castle and eventually shows us Leadranach addressing the young lord at Winterfell. The establishing shot lasts 6.35 seconds. The speech and captioned identification of Leadranach begin one second into the establishing shot and approximately eight seconds before we see Leadranach.	LEADRANACH: My lord, may the old gods watch over your brother	New scene
A **panning shot** over the city that settles on Shae standing on a balcony overlooking the city. The caption appears 2.58 seconds into the panning shot and about five seconds before we see Shae.	SHAE: This city stinks.	New scene
An **audio offset** of almost two seconds as the camera lingers on the previous scene while the audio from the next scene begins. Robb begins speaking as the camera stays with the previous scene, which is fitting given the previous scene's intensity. It's as though the intensity anchors the camera to the previous scene, and the audio of the next scene dislodges the camera.	ROBB: You're Ser Alton Lannister?	New scene

Note: A number of new scenes in this episode open with establishing or panning shots as characters begin speaking offscreen. HBO, 2012. DVD. http://ReadingSounds.net/chapter5/#table3.

heard whistling while entering the council room in session. The caption identifies the whistling sound and the name of the whistler immediately: (TYRION WHISTLING TUNE). But the camera is much more patient as it shows two reaction shots from members of the council before panning to reveal the identity of the whistler. The caption identifies Tyrion 4.38 seconds before the camera does. As a major character on the show, Tyrion Lannister's voice is distinctive. Peter Dinklage received top billing in season 2 and an Emmy award for the role of Tyrion in 2011. But while Tyrion's voice is recognizable to viewers, his whistling is not. Nonspeech sounds do not have the same distinctiveness as speech sounds. That's why members of the council turned to look in the direction of the whistling to discover the identity of the whistler. It might have been different if one character had a reputation for whistling, which is not the case on *Game of Thrones* but could apply to The Governor (Philip Blake) on *The Walking Dead* and his penchant for creepy whistling in multiple episodes from season 3. In other words, context and history always matter.

It's possible that savvy viewers could surmise that the whistler must be Tyrion because Tyrion has just arrived from the North, telling King Joffrey in the previous scene that he can't "stay and celebrate" with the King because "there is work to be done." But that knowledge is independent of the whistling sounds. Indeed, whistling sounds, like many other nonspeech vocalizations such as sneezing sounds, do not typically provide enough clues on their own to identify their makers. Captioners also need to be mindful of how scenes like this one are constructed to create momentary suspense. There's a reason the camera waits patiently to show us the identity of the whistler and why the whistler comes out of the shadows like a mystery being disclosed. Captioners need to follow the lead of the show over the presumed need to identify characters as soon as they are known. In this case, the premature identification of the whistler ruins whatever suspense the scene was intended to instill in viewers. If the scene sets up a momentary connection or sense of identification between the council members and the audience (because both groups do not know the identity of the whistler), then the caption ironically breaks that bond by revealing the whistler's identity right away. That's why the captioner should not have been so eager to name Tyrion. Just because he is offscreen and a known character doesn't mean he needs to be named, despite what the style guides say. Speaker identification guidelines must be interpreted flexibly and contextually. Captioners must pay special attention to how scenes are visually framed to create suspense and foster identifications between characters and the audience. **Media:** http://ReadingSounds.net/chapter5/#game-of-thrones.

Premature speaker identifications are fairly common, perhaps because of the tendency among captioners to identify offscreen speakers by their script name regardless of context. The *Captioning Key*'s guideline about waiting to "identify the speaker by name until the speaker is introduced in the audio or by an onscreen graphic" could offer the needed flexibility that I think some situations demand (DCMP 2011c). In too many cases, captioners jump the gun on the narrative, delivering readers a name from the future and using it before audiences are expected to know it. Let me share a few telling examples of how premature speaker identifications can compromise narratives. *Unknown* (2011) is a thriller starring Liam Neeson, January Jones, and Diane Kruger. The first major plot point upon the couple's arrival at a hotel in Berlin involves a forgotten briefcase and a taxi ride back to the airport to retrieve it. The taxi driver is referred to as "driver" by Neeson and is otherwise assumed to be just another service worker in the city. But as the taxi driver swerves to avoid a refrigerator that has fallen off a truck, the captions name her prematurely: [GINA SCREAMS]. Now caption readers know what others do not: no matter what happens in the moments following the taxi's crash into the river, the taxi driver will return later in the movie. She has been named; she's not unknown. Even though she flees the scene, we now know that she will return at some point. To maintain the integrity of the narrative, the captions should have referred to her as "driver," even if audiences might recognize the driver from other films (Gina is played by Diane Kruger) and even if the captioner needs to refer to the same character as "Gina" when she returns later in the movie. **Media:** http://ReadingSounds.net/chapter5/#unknown.

Paul (2011) is sci-fi comedy starring Simon Pegg and Nick Frost. Audiences will most likely have seen the ads for the film that identify the alien as Paul. But they probably don't know how Paul got his name. The captions nearly give it away in the opening scene, twenty-nine seconds into the movie, when a family dog is identified as "Paul": (PAUL THE DOG BARKS). The movie has given no indication at this early point that the dog is related somehow to Paul, the alien who has been featured in ads for the movie. The girl who lets the dog outside is identified with a speaker ID as YOUNG TARA, which is not only needlessly descriptive (we can see that she is young) but also needlessly implies that she will return later as an "old" version of herself. The cast list indeed refers to the girl as "Young Tara" (IMDb 2011), but that doesn't mean the captioner should copy characters' names wholesale from the script or cast list to the caption track. Captioners need to channel the narrative, not the script or cast list. "Young Tara" is doubly disconcerting because (1) it tells caption readers the name of a character before the movie offers any clue for who she is and (2) it alerts caption readers that this same character will return later as "old/er Tara." Granted, knowing the dog is named Paul almost sixteen seconds before Tara calls him by that name doesn't tell us much. Neither does knowing that the girl will age significantly in the course of the film. But that's not the point. The larger point is that captions must embody the narrative as the narrative discloses information over time. This means adopting naming conventions that reflect what audiences are expected to know at that point. Not "Paul the Dog" but simply "dog." Not "Young Tara" or even "Tara" in the opening scene but simply "Girl." If "girl" needs to become "Tara" when Paul returns much later in the film to the same house after sixty years have passed (and greets a much older Tara), so be it. Listening audiences have to make these kinds of adjustments all the time as they learn more about characters. That's part of what makes watching movies so much fun. **Media:** http://ReadingSounds.net/chapter5/#paul.

Pirates of the Caribbean: Dead Man's Chest (2006) is an action-adventure-fantasy starring Johnny Depp. Early in the film, Jack Sparrow (Johnny Depp) and his crew are held captive on an island populated by cannibals. The listening audience (watching without captions) isn't told explicitly that the natives are cannibals until ten minutes after we see the first native. There are plenty of clues along the way for savvy viewers: a reference to "long pork" (which I discovered later is slang for human flesh), a parrot that says "Don't eat me," Will Turner (Orlando Bloom) tied to a spit like a pig going to a roast, and Jack Sparrow biting one of the toenails on his human foot necklace. In hindsight, the clues are obvious. But it's also easy to miss them and easier to find them when you're looking for them. The movie seems to recognize the need to make the natives' cannibalism explicit for those who might have missed the clues along the way. One of the pirate crew members informs Will Turner (who, like some members of the audience, doesn't know what the natives really are) that the natives "will roast [Sparrow] and eat him." But the captions dampen the anticipation and thwart the process of collecting clues by telling readers the natives are cannibals eight minutes and fourteen seconds before the narrative makes the big reveal to Will Turner. After one non-speech caption identifies the natives appropriately as just natives—(speaks native language)—the remaining descriptions of the natives' language identify them prematurely as cannibals: (speaks cannibals' language), (cannibals' language), (cannibals murmur), (cannibals' language), and (cannibals' language). These five references to cannibals appear in a space of forty-five seconds, with the fifth reference appearing more than seven minutes before the natives' cannibalism is made explicit. Why the captioners abandoned the preferred "native language" for "cannibals' language" is not clear. References to cannibalism in the nonspeech captions prematurely reveal something about the natives that the narrative intends to reveal through a series of more or less oblique clues. In short, the narrative is suggestive in the minutes leading up to the explicit announcement, but the captions are prematurely revealing. **Media:** http://ReadingSounds.net/chapter5/#pirates.

BloodRayne 2: Deliverance (2007) is a straight-to-DVD vampire western set in the late nineteenth century and adapted from the BloodRayne video game series. In the opening scene, a reporter from Chicago arrives looking for a colorful story. In the second scene, an intruder breaks into a family's prairie home, ostensibly murders the parents, and captures the children. The DVD version of the movie identifies the intruder as (cowboy) when he speaks for the first time offscreen. We don't know anything about him yet, except that he's probably the movie's main bad guy and he speaks English with an odd foreign accent, possibly German or Eastern European. The actor who plays the cowboy is Canadian actor Zack Ward, so the accent is an affectation, which is perhaps why it's so difficult to place. The accent is important to identify in the captions, because viewers usually assume that cowboys in Wild West movies are American. At this early point in the film, a generic name for the accent is warranted, perhaps "foreign accent" or even "Eastern European accent." The TV version's captions are too specific, identifying the evil cowboy with the double spoiler: (Billy the Kid, in Transylvanian accent). First, there's no evidence at all at this point that this cowboy is Billy the Kid. Second, calling the accent "Transylvanian" tells viewers only one thing: the cowboy is a vampire. But the audience won't see any evidence that the cowboy is a vampire for at least thirty-five seconds after the speaker identification, when the camera shows a close-up of fresh blood around his mouth. The [hissing] noise the vampire cowboy makes at the end of the clip, which is uncaptioned in the TV version, helps establish the bloody truth we can see. We'll have to wait a bit longer to find out he's also Billy the Kid. **Media:** http://ReadingSounds.net /chapter5/#bloodrayne.

Names humanize characters, making them seem more sympathetic or at least less flat and more palpable. If I had to place a bet on who would be more likely to dive under the freezing water to pull an unconscious Liam Neeson from the sinking taxi in *Unknown*—someone identified as "taxi driver" or someone identified as "Gina"—I'd wager on Gina. Names imbue characters with substance that generic monikers such as "Man 1" and "Female Narrator" do not. A seemingly innocent caption such as [GINA SCREAMS] may unwittingly and prematurely divulge a narrative's secrets. The power of names can be exploited to make otherwise nameless characters seem more human and real. A popular Allstate Insurance commercial from 2012 leverages the power of names, using [Kyle] and [Roger] as speaker IDs for what are otherwise nameless speakers in the uncaptioned version. In other words, caption readers have access to information—in this case, the names of speakers—that listeners do not have. Only Dennis Haysbert is a known entity in the commercial. His voice alone has become distinctive too, thanks in part to this series of Allstate commercials. The other two guys in the commercial are unknown actors who could have just as easily been identified as Man 1 and Man 2. Indeed, the captioning style guides tell captioners to use generic placeholders like "female #1" and "male narrator" for unknown speakers (DCMP 2011c). But there may be good reasons to choose names over generic identifiers, assuming the names are provided to the captioner and not simply made-up. From a marketing perspective, names have more power and greater potential to increase viewers' sense of identification with characters and, as a result, with the product being advertised. Kyle and Roger are more real because they have names. Kyle and Roger are also more easily distinguishable in the captions than Man 1 and Man 2, especially in the fast-paced reading environment of captioned commercials. Captions should not provide commentary, Easter eggs, separate tracks, new names, or open gateways to secret worlds of meaning. While captioners should strive to provide an experience that is as close as possible to what is intended to be experienced through the soundtrack, they should also weigh the potential of specific names to provide an important means, along with screen placement, of helping readers negotiate complex reading tasks. In the Allstate example, readers are required to understand very quickly that Dennis Haysbert speaks through Kyle and Roger: [Kyle with voice of Dennis] and [Roger with voice of Dennis]. Using specific names in this example helps readers understand something that is common in some parts of Europe but unusual on American television: voice replacement or dubbing (Bianchi 2008). **Media:** http://ReadingSounds.net/chapter5/#allstate.

Identities must be revealed in the captions not because they can be discerned by a prescient captioner with access to future events and scripts but because the narrative demands it. Captioners must strive to reflect and channel the intentions of the content creators, even if that means waiting to identify a major character until that character is known specifically. This can be tricky and, within the context of the rule-drive style guides, seem counterintuitive. Captioners seem reluctant to use generic names (e.g., taxi driver, girl, natives, cowboy) for characters that acquire specific identities later on (Gina, Tara, cannibals, Transylvanian Billy the Kid). But audiences are routinely expected to develop knowledge of characters along the way. New characters acquire names and histories over time, sometimes slowly, and captions should follow suit. Even knowledge of which characters are the main characters may be up for grabs initially. Captioners need to *inhabit the narrative*, and captions need to *embody its mysteries*.

Speaker identification doesn't have to be inflexible or premature but can and should evolve with the narrative. In *Man of Steel* (2013), Clark Kent (Henry Cavill) transforms into Superman over the course of the film. This evolution is reflected in the film's spoken dialogue, not surprisingly, but also in the captioned nonspeech information, specifically the speaker identifiers that tell us who is speaking or vocalizing when it isn't visually or contextually obvious. Clark Kent is referred to as CLARK in the first half of the movie:

0:40:51 [CLARK GRUNTS]
0:40:55 CLARK: It's all right, it's all right, it's all right. It's all right.
0:49:15 CLARK: Why am I so different from them?
0:54:25 CLARK: I'm tired of safe.
0:54:33 CLARK: I didn't say that.
0:57:26 CLARK: I let my father die because I trusted him.
1:00:28 CLARK: What?
1:01:11 [CLARK CHUCKLES]

Even after Clark discovers who he is and where he came from ("Why am I so different from them?," he asks his father), and even after Lois Lane (Amy Adams) discovers Clark's secret identity ("I let my father die because I trusted him," he confesses to her), the speaker IDs continue to refer to him as CLARK. It is only after Clark is forced to come out of hiding and reveal his true alien identity to the evil Zod and the US government that the speaker IDs for Clark shift to SUPERMAN. This shift occurs in the captions nearly simultaneously with the naming of the superhero by Lois Lane, with one key difference: Lois is interrupted before she can finish saying the name the audience knows is coming ("How about . . . Super—" at 01:11:40). Two seconds later, Clark/Superman is identified for the first time in the captions as SUPERMAN. The second half of the movie belongs to SUPERMAN:

01:11:43 SUPERMAN: Emil Hamilton.
01:16:59 SUPERMAN: Unh.
01:17:04 SUPERMAN: I . . .
01:17:07 [SUPERMAN COUGHING]
01:17:09 SUPERMAN: Weak.
01:27:56 SUPERMAN: You'll be safe here.
01:29:36 SUPERMAN: Ahh!
01:39:53 SUPERMAN: Mom?
01:40:12 SUPERMAN: I'm so sorry.
01:45:41 SUPERMAN: Yes
01:59:47 SUPERMAN: Argh!

02:07:41 [SUPERMAN GRUNTS AND NECK SNAPS]
02:09:12 SUPERMAN: It's one of your surveillance drones.
02:09:16 SUPERMAN: It was.

The first face-to-face meeting between Superman and Zod coincides with four SUPERMAN speaker IDs in a span of ten seconds (01:16:59—01:17:09). Because they occur in the scene immediately following the interrupted naming scene ("Super—"), these speaker IDs cement Clark's new status as Superman in the minds of caption readers. But no one actually utters the S-word (i.e. Superman) in this scene. In fact, "Superman" is only uttered three times (not including Lois' interrupted S-word) in the entire film. In the span of about eight seconds towards the end of the movie, and after the SUPERMAN speaker ID has already been used nine times, two minor military characters have the following quick exchange about the S-word:

01:45:08 Colonel Hardy's on his way and he's got Superman in tow.
01:45:11 Superman?
01:45:13 The alien, sir.
01:45:14 That's what they're calling him. Superman.

Thus, while the movie plays coy by avoiding a name that is already well known to an American audience, the closed captions use "Superman" liberally, even though Superman is not fully named until very late in the movie. The audience knows the story of Superman, as the interrupted S-word suggests in its teasing way. That may be reason enough to justify its use as a speaker ID before the character is fully named. To an anticipating, knowing audience, the first two-thirds of the name is as good as the whole name. As soon as he is named by Lois, the captions make a marked shift from CLARK to SUPERMAN.

The captioned distinction between CLARK and SUPERMAN pinpoints and honors his transformation. Clark's identity is fluid. He does not leave Clark behind so much as take on a persona ascribed to him. Clark continues to exist. So does Kal-El (or Kal), his birth name, which is uttered twenty-five times in the movie but never used as a speaker identifier. At the end of the movie, when Superman goes back into semi-hiding as Clark, he is identified again as CLARK.

02:11:33 I gotta find a job where I can keep my ear to the ground.
02:11:43 CLARK: Where people won't look twice . . .

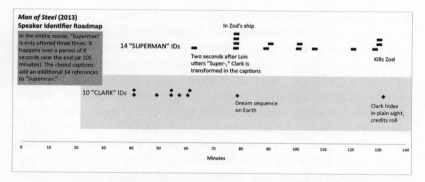

Man of Steel (2013)
Speaker Identifier Roadmap

In the entire movie, "Superman" is only uttered three times. It happens over a period of 8 seconds near the end (at 105 minutes). The closed captions add an additional 14 references to "Superman."

14 "SUPERMAN" IDs

In Zod's ship

Two seconds after Lois utters "Super-," Clark is transformed in the captions

Kills Zod

10 "CLARK" IDs

Dream sequence on Earth

Clark hides in plain sight, credits roll

Minutes

5.5 **A visual representation of the** CLARK **and** SUPERMAN **speaker identifiers in** *Man of Steel.* Warner Bros, 2013. DVD. http://ReadingSounds.net/chapter5/#figure5.

The fluidity and multiplicity of Clark's identity is also suggested in a single CLARK identifier that crosses over the threshold of Clark's transformation into Superman. During his initial encounter on Zod's spaceship as SUPERMAN, Clark passes out and awakes in a dreamlike state on Earth, now dressed sans cape as Clark amidst the backdrop of cornfields and his childhood home. In this dream sequence, Zod greets Clark as Kal before asking, "Or do you prefer Clark?" Within this context, Clark asks a question that is appropriately identified in the captions with CLARK instead of SUPERMAN:

1:18:34 CLARK: How did you find your way to Earth?

Even though at this point in the film CLARK identifiers have given way to SUPERMAN identifiers, this final CLARK identifier (before CLARK returns at the end of the film) makes sense within a dream sequence that features human Clark as opposed to Krypton Kal or Superman. I've mapped these speaker IDs in figure 5.5. In short, *Man of Steel* offers a compelling case study of how speaker identifiers for the same character can be fluid, evolving with the narrative.

This section has focused on speaker identification, but the same set of challenges applies to naming in general. For example, wordplay, puns, and guessing games that rely on listening are difficult to translate into writing without giving away, compromising, or failing to do justice to the joke. Consider a guessing game in which one person hums a television theme song and the other person tries to guess the song. How do you describe the song without naming it—and do so in a way that is accessible and sensitive to the time constraints that caption readers routinely face? There probably isn't a lot of time to describe how the song sounds without naming it. You would probably only be able to use a couple of adjectives or some generic labels for each song in the humming game (dramatic, sweeping, intense, dark, light, comedic, classical, country, etc.). But a short description doesn't really replicate the experience of the game for hearing viewers or allow caption readers to play along. When Stewie and Brian (both voiced by Seth McFarlane) play the theme song game in an episode of the animated Fox television series *Family Guy* ("And Then There Were Fewer," 2010), the television captions prematurely identify each theme song as it's being hummed: (hums theme to Lost in Space), (hums theme to Dick Van Dyke Show), (hums theme to Dynasty). This approach fails to replicate the spirit of the game itself. Stewie's responses to Brian's humming—"Oh, uh, Lost in Space"— simply repeat the captioned identifications. The DVD captions offer a better approach that doesn't identify the theme songs prematurely, opting for (HUMMING TUNE) each time instead. But (HUMMING TUNE) repeated three times doesn't capture the spirit of the guessing game either because it flattens the differences among the tunes and doesn't allow caption readers to play along with listeners. **Media:** http://ReadingSounds.net/chapter5/#family-guy.

Ironic Juxtaposition

Media texts are conglomerations of different elements with the potential to create unexpected, incongruous, ironic, and sometimes intentional meanings when these elements are juxtaposed. Juxtapositions are typically defined as spatial, temporal, or some combination (Goya Martinez 2012, 2). An example of spatial juxtaposition: a Facebook ad placed next to (or nestled among) the status updates in your Facebook timeline. An example of temporal juxtaposition: a TV commercial that precedes or follows other commercials and TV shows. On the web, advertisers continually search for new ways to insinuate themselves into our reading and viewing experiences, using our browsing patterns, profiles, emails, and locations to offer targeted ads. In the latest iteration of Facebook's mobile app, for example, ads are positioned inside of, and visually resemble, the other (noncommercial) status updates in users' vertically scrolling timelines. The goal is not simply to offer targeted ads based on what your friends are discussing but to rhetorically construct the advertiser as part of the stream of your daily life: a diaper ad that shows up in your Facebook timeline two months after *your husband* announces that you are pregnant (Delo 2012). In Facebook's controversial "sponsored stories" program, your status updates, profile data, and pictures can be sold to advertisers and repackaged as Facebook ads (Kravets 2012). But incongruous, even jarring, juxtapositions are still possible. Dove, a beauty products brand, recently faced backlash on Twitter after one of its "self esteem" ads appeared on a Facebook page condoning violence against women (Leber 2013; Slattery 2013). In the context of Dove's "real women" campaigns to raise women's self-esteem and challenge stereotypical views of beauty (Dove n.d.), the juxtaposition of Dove advertising alongside gender-based hate content was ironic to say the least. What made matters worse was Dove's reluctance to join a boycott of fifteen companies that pulled advertising from Facebook (Leber 2013).

Like Facebook, Google's methods of targeting ads to users are systematic, drawing on information from "your Google search queries on the Web, the sites you visit, Google Profile, +1's and other Google Account information to show you more relevant ads in Gmail" (Google 2013). This process is "fully automated, and no humans read your email or Google Account information." Google is also sensitive to potentially troubling juxtapositions and "may block certain ads from running next to an email about catastrophic news" (Google 2013). Yet neither Google nor any other advertiser or distributor can fully control the meanings that are borne out of juxtaposition. Some juxtapositions seem to be obviously ironic, such as a Google ad for police careers ("Become a Police Officer") superimposed on a YouTube video purportedly showing a police deputy kicking a teen girl (Robopanda 2010). Others are more subtle, such as two web ads placed side by side, one advertising a "$430,000 Mortgage for Under $1,299/Month!" and the other asking "What's your IQ?" (Ellis 2007). The unintended irony, which is generated at the intersection of these two ads, is that if your IQ is high enough, you should know that you can't secure a $430,000 mortgage, even with excellent credit and a low interest rate, for anywhere close to $1,299 per month. Juxtaposition becomes ironic when the contrast between elements creates the conditions for criticism or reflective tension. It isn't dependent on advertising; it's merely the type of captioned contrast I turn to below.

Textual irony can generate ironic meanings intentionally or unintentionally. *Always Sunny in Philadelphia*, a television sitcom currently in its ninth season at the time of this writing (2013), regularly and intentionally uses the title of each episode to establish an ironic relationship between a knowing audience and the show's unknowing characters. The episode's title sequence and opening credits follow an opening set-up scene in which one or more main characters make some announcement (e.g., Sweet Dee [Kaitlin Olson] tells the gang that she "beat the system"), followed by an onscreen episode title that offers an ironic correction to this announcement: "Sweet Dee Gets Audited" (season 7, episode 4, 2011). In *Always Sunny*, episode titles often serve as ironic punch lines. At the beginning of another episode, Sweet Dee complains of pain and passes out, the gang shows a lack of concern ("God, what is her problem?" "I don't know"), and the title conveys the truth that the characters don't know yet: "Sweet Dee Has a Heart Attack" (season 4, episode 10, 2008). In a final example, Dennis (Glenn Howerton) insists that he and Dee will "be just fine," which is followed by the ironic title and punch line: "Dennis and Dee Go on Welfare" (season 2, episode 3, 2006). In the course of the episode, Dennis and Dee become addicted to crack cocaine, which is pretty much the ironic opposite of "just fine." The ironic titles are not closed captioned but rather available to all viewers as onscreen text. A similar example comes from *The Big Year* (2011), in which Steve Martin and Jack Black, playing serious bird enthusiasts, deny being involved in a bird-watching contest known as a "Big Year." The onscreen text tells another story as their current totals for the year are displayed for all viewers to see: "Harris: 202" and "Preissler: 232." **Media:** http://ReadingSounds.net/chapter5/#textual-irony.

Textual irony can be unintentional too. Television commercials generate ironic tensions unintentionally when they bump up against each other or against the TV programming they're sponsoring. For example, a Kentucky Fried Chicken (KFC) commercial received a lot of buzz on Twitter in November 2012 after their ad immediately followed a scene from *The Walking Dead* ("When the Dead Come Knocking") featuring a hoard of zombies feasting on the entrails of an unfortunate victim. As the bloody zombie frenzy cuts to a logo for *The Walking Dead*, an announcer says, "Brought to you by KFC." The logo is quickly replaced with an image of a bucket of KFC chicken as the announcer urges viewers to "Come in today and taste why fresh is better" (SpecialK1417 2012). Viewers' reactions on Twitter were mixed, including:

@headgeek666 Scene: Zombies feasting on guts. Cut to: The Walking Dead sponsored by KFC . . . Multiple folks grabbing for meat out of a tub. IRONY IS ALIVE!

@azalben Walking Dead just cut from a guy being eviscerated directly to the message "Brought to you by KFC." #FingerLickingGood

@LAM1086 The KFC 'fresh is better' promo airing RIGHT after 'the walking dead' #teamprison fed that fresh dude to the walkers was perfection

@Stacieieie "The Walking Dead is brought to you by KFC" Seriously? Right after we see some guy get eaten? Gross. #thewalkingdead

KFC was reportedly displeased by the "unfortunate" ad placement (Ives 2012), but it is hard to believe they did not foresee something like this happening on a show known for the consumption of so much "fresh meat" each week. While some viewers were reportedly grossed out by the juxtaposition, just as many (on Twitter at least) seemed to find humor in it (even "genius" and "perfection"). A number of blogs and news/entertainment websites fueled the buzz too, making the entire incident appear more like a win for the advertiser than a crisis in need of a formal apology. Regardless, the juxtaposition was striking and ironic. These ironies have the potential to linger with viewers for an indefinite period of time. For example, when the zombies are captioned as (walkers munching) in an episode entitled "Sick" from October 2012, I don't know whether to be terrified or burst out laughing at the ironic reference to eating. Granted, the "Sick" episode precedes the KFC juxtaposition in "When the Dead Come Knocking" by a month, but day-long marathons and syndication deals ensure that timelines quickly become moot on any television show with staying power. These ironic descriptors are not limited to zombie programs. For example, it's hard not to chuckle when one of the Terminator's rupturing hydrogen fuel cells—essentially a small nuclear device—is described as [SIZZLING] in *Terminator 3: Rise of the Machines* (2003). I tend to associate the sound of sizzle with meat being fried or barbecued, especially bacon. Given that the nuclear device is rupturing (or rather *sizzling*) inside the partly fleshy cyborg body of the Terminator, it makes sense to refer to the sound as sizzling. Even so, [SIZZLING] immediately calls to mind bacon and other sizzling meats. **Media:** http://ReadingSounds.net/chapter5/#unintentional-irony.

Despite the large number of examples discussed online of juxtaposition in popular culture, closed captions are rarely mentioned in such discussions. I found only one example of captioned juxtaposition discussed online, a screenshot posted to Reddit in which a closed caption from *The Tudors* (a Showtime series rebroadcast on Hulu Plus) lingered on the screen during a commercial for IKEA (jeredhead 2013). The commercial featured a small boy on a four-wheeled baby scooter. Lingering on the screen was the inappropriately humorous caption: [Sexual moaning]. While this post to the /r/funny subreddit was not intended to elicit a discussion of ironic juxtaposition, it nonetheless begins to suggest how closed captioning can participate in such discussions. Captions routinely cross boundaries they shouldn't, as when a caption stays on the screen during a commercial break or when one commercial is replaced by another commercial. (Recall that commercials in the United States—indeed, all television programming less than five minutes in length—are not required by law to be closed captioned [FCC 2010].) Television captioners are supposed to insert "clear" or "offtime" signals in the caption file (e.g., CLEAR CC signal at the beginning, CC OFF at the end) to remove previous captions at the start of their own content and to prevent captions from lingering on the screen when their content ends. Even so, the problem of the lingering caption is pretty common. CaptionMax, a major captioning company in the United States, blames the TV networks for "start[ing] the commercial before the 'clear captions' signal gets transmitted" (CaptionMax 2013). But this reason alone doesn't seem to explain the sheer number of final captions that persist beyond their boundaries. A captioner at VTA, a video production company, describes being "amazed at the number of shows and commercials that obviously didn't have this [CC OFF] command. When it is missing, the final captioning stays on the screen until the next captioned material begins playing" (Turner 2010). The VTA captioner, David Turner (2010), puts the problem this way: "If that next commercial is yours, and it isn't captioned, the captioning from the previous piece stays visible over a third of YOUR spot for about 20 seconds, distracting viewers from YOUR message and possibly covering up valuable information on YOUR spot."

Unlike more conventional types of juxtaposition in which elements are placed next to each other in time or space, captioned juxtaposition involves superimposing elements on top of each other. The superimposed or lingering caption is like a foreign body invading a new host. For example, when an episode of *The Walking Dead* ("This Sorrowful Life," 2013) resumed after a commercial break, a Celebrex commercial's final caption lingered on the screen: "For a body in motion." Celebrex is a pain reliever for arthritis. This final Celebrex caption remained on the screen as a small number of zombies wandered aimlessly along a roadside. The unintended irony of such a juxtaposition is located in the degree to which viewers recognize the zombies as pure body too, as mindless but constantly in motion. The Celebrex motto has the potential to be ironic if we think of zombies as having a similar tendency to stay in motion: "A BODY AT REST TENDS TO STAY AT REST . . . WHILE A BODY IN MOTION TENDS TO STAY IN MOTION." Does this ironic link between medicated bodies and zombie bodies subtly encourage us to perform a *zombie rereading* in which pharmaceutical ads seem to be populated with mindless, docile, and sedated bodies in motion? Probably not. But even if we don't go out on this interpretative limb, we may still find that this lingering caption creates an unintended and ironic link between very different "bodies in motion." **Media:** http://ReadingSounds.net/chapter5/#celebrex.

Another example comes from CNN, a cable news channel that, like other live programming channels, makes use of scroll-up style captioning. Live captions on television are typically delayed five to seven seconds, which is the time it takes the stenocaptioner to transcribe what she hears, the transcribed captions to be relayed to the station via Internet connection, and the captions to be delivered to our homes. Some live stenocaptioners work out of their homes, their stenography machines connected directly to a computer, which is connected to the Internet. In this example, the captions are delayed approximately six seconds. It was recorded on May 8, 2013, during a live broadcast outside the Cleveland home of Ariel Castro, who was accused (and subsequently convicted) of abducting and holding three young women hostage in his home for roughly ten years. The women were freed two days prior to this broadcast in a dramatic escape and rescue. As the news coverage live on the scene takes a commercial break, the delayed live captions continue to scroll over the start of a Subaru car commercial featuring an active family with two young daughters. The sisters climb into the trunk of the Subaru at the beginning of the commercial while the CNN captions continue to scroll in two-line style:

INSIGHT FROM KATIE BEARD,

KIDNAPPED AT THE AGE OF NINE,

HELD IN A DUNGEON FOR MORE THAN

17 DAYS.

The CNN captions linger at the start of the Subaru commercial for 6.94 seconds, approximately the length of the live delay. This is common and necessary in live programming that takes a commercial break while the live captions play catch up over the first commercial. The first Subaru caption (a music note) appears seven seconds into the commercial. Seven seconds is a long time for an advertiser in a thirty-second spot (it's 23 percent of the commercial's total length), particularly when those seven seconds are devoted to reporting horrific news. The lingering CNN caption is positioned, albeit unintentionally, for maximum irony. A pause of 2.32 seconds in the middle of the third line ("HELD IN A") gives the two young sisters time to run across the lawn and open the trunk just as the line resumes with the word "DUNGEON." For a split

5.6 **Two sisters get ready to jump into an automobile's trunk in this frame from a Subaru TV commercial.**
At the top of the screen is a lingering CNN caption: "KIDNAPPED AT THE AGE OF NINE, HELD IN A DUNGEON." Source: CNN, May 8, 2013. http://ReadingSounds.net/chapter5/#figure6.

second, the Subaru's dark trunk resembles a dungeon and the young girls become captives (see figure 5.6).

Lingering captions are fairly common in a television environment in which only some commercials are captioned. If advertisers knew that other advertisers' closed captions were infringing on their expensive and limited ad space, they might be more eager to caption their own commercials, if only to have greater control over what appears on the screen. They might also demand from networks greater oversight of "clear screen" signals. I haven't shared examples of captions that cross the boundary from one commercial to another, but they exist as readily as other examples involving the transition from programming to commercial (or vice versa). The final caption in a Priceline commercial, for example, persists on the screen as the next commercial for Chili's restaurant chain begins: "Your accent needs a little work." From the perspective of the Priceline commercial, this line is spoken by a father (William Shatner) to his daughter (Kaley Cuoco). But when the caption loiters on the screen at the start of the Chili's commercial, it has the potential to become something else. An ironic commentary on the accent of the Chili's announcer, perhaps? When captions float free and drift into new contexts, the potential for new associations and new meanings grows. **Media:** http://ReadingSounds.net/chapter5/#chilis.

Misunderstanding caused by the juxtaposition of word and image has led to defamation lawsuits in other contexts. Grimes and Drechsel (1996, 169) describe how "libel suits have arisen where juxtaposition of video and voice-over has allegedly linked passers-by with venereal disease, an innocent neighborhood resident with prostitution, a property owner with slum-tike conditions, an airline with CIA activity, a dairy store with price-fixing, and innocent third parties with accused criminals or criminal activity." (No wonder news segments on the dangers of obesity and smoking avoid showing people's faces in footage recorded on city streets.) One successful defamation suit was brought by a Detroit woman who believed she was being characterized on the news as a prostitute through juxtaposition:

Detroit resident, Ruby Clark, sued ABC News alleging that a story on prostitution left viewers with the false impression that she was a prostitute rather than an innocent resident of a neighborhood that had become the locus of serious prostitution problems. She argued—and an appellate court agreed—that video showing her walking down the street of her neighborhood immediately after prostitutes were mentioned in voiceover could have caused viewers to think she was a prostitute, even though the audio track accompanying the video identified her as a neighborhood resident (Grimes and Drechsel 1996, 170–71).

I do not mean to suggest that advertisers have grounds for bringing defamation lawsuits against networks and other advertisers because of loitering captions and their (un)fortunate and unintended associations. On the contrary, I simply mean to point out that juxtaposition, even when unintended, can have serious, even legal and financial, consequences. Advertisers and programmers should pay closer attention to the boundaries that mark the edges of their content in an effort to shore up the gaps that allow unwanted migrant captions to encroach into their spaces.

Conclusion

Captioned irony is a new term intended to draw our attention to the differences between reading a movie and listening to it. In this chapter, I have been especially concerned with what captions add to the viewing experience, what caption readers know that noncaption listeners do not know and may (or may not) discover in the course of the narrative.

In some cases, writing itself provides inherent affordances that allow readers to get a jump on listeners (e.g., through read-ahead strategies). In other cases, speaker identifications provide clues to readers about the future. When it comes to identifying speakers, captioners need to attend more closely to the narrative, not allowing the script, cast list, or even a style guide to take the lead on identifying characters within an evolving narrative. Like dramatic irony, captioned irony calls attention to what audiences know that characters may not, but unlike dramatic irony, captioned irony is concerned with what captions add to the viewing experience that, for better or worse, is not available to audiences who have the captions turned off. This work is just beginning in caption studies. I imagine discussions of captioned irony contributing to larger discussions of universal design. Arguments from universal design tend to focus on the broad benefits of captioning in areas of literacy development, increased comprehension, and noise canceling. But there are other benefits and effects that are grounded in a concern with how captions interact with audiences, sounds, and narratives. Close readings of captioned texts can offer a new perspective on film interpretation and multimodal composition while suggesting that the benefits of such studies are not limited to those who can't hear or to scholars in disability studies. Or, to put it differently, scholars in disability studies and captioning advocates have much to teach us about the relationships between sound and writing.

SIX

Captioned Silences and Ambient Sounds

As counterintuitive as it may sound, silence sometimes needs to be closed captioned. Captioners not only inscribe sounds in writing but must also account for our assumptions about the nature, production, and reception of sounds. One of our most basic assumptions is that sounds are either discrete (with a clear beginning and end) or sustained (continuous). Sustained sounds, including sounds that are captioned as continuous or repeating (e.g., using the present participle *verb+ing*, as in [phone ringing]) may need to be identified in the captions as stopped or terminated if it's not clear from the visual context. That is, if we can't see the phone being answered or the ring being turned off, the captioner may need to mark the termination of the ringing sound. We also assume as moviegoers that the world is never technically silent. Ambient noise provides context. True silence is rare on the screen. In the real world beyond the screen, the same assumption holds. Sound waves envelop hearing viewers even in "silence." The total absence of sound can only be achieved on Earth artificially in an anechoic chamber, a room designed to block out exterior noise and absorb interior sound waves. Designed to test product noise levels (and not human tolerance levels), the chamber reportedly causes hallucinations and severe disorientation in hearing visitors who spend even a little time in one (Davies 2012):

[silence]

6.1   If a car crashes into a tree but doesn't make a sound, a silence caption may be needed.

In this frame from *The Artist*, smoke billows from the hood of a 1935 Cadillac convertible sedan that has hit a tree. The driver (Bérénice Bejo) is running away from the car and towards the viewer. The camera looks down upon the scene from a second floor vantage point. Caption: [silence]. Right before the car crash, a silent film intertitle goes "BANG!" Because of its placement immediately following a scene in which George (Jean Dujardin) prepares to kill himself by putting a gun in his mouth, we are led to believe that "BANG!" applies to George's suicide attempt when in fact it applies to the car crash. Studio 36, 2011. Blu-Ray. http://ReadingSounds.net/chapter6/#figure1.

Ironically, far from being peaceful, most people find [the anechoic chamber's] perfect quiet upsetting. Being deprived of the usual reassuring ambient sounds can induce fear—it explains why sensory deprivation is a form of torture (Foy 2012).

Yet even in an artificial chamber devoid of sound, the heart audibly beats, the stomach gurgles, and the ears "make a tiny amount of noise" (Davies 2012), which suggests that silence is never truly soundless for hearing people. George Michaelson Foy's (2012) quest for some relief from the noise of city life led him to try out an anechoic chamber, but after hearing "the blood rushing in my veins," "my scalp moving over my skull," and some "strange, metallic scraping noise" that must have been an hallucination, he rightly concludes that "You'd have to be dead for absolute silence." Even as I write these words alone in my house out in the country on a "silent" Sunday, the air conditioner and ceiling fan whir and the dogs make their presence known on hard floors. Silence is contextual.

We also assume as moviegoers that when objects interact with each other in space, they make noise (e.g., a car crashing into a tree, as in the frame from *The Artist* that opens this chapter). Even in outer space, this assumption holds, at least in the movies, even though the vacuum of space is soundless due to the lack of air to carry the sound waves. If this assumption is challenged in some way for artistic or parodic effect, the captioner needs to make it clear to readers that sounds are not being produced in the expected ways. *The Artist* (2011) is done in the style of a silent film and historically set in the years before and after the advent of "talkies." The movie takes itself seriously as a silent film, requiring viewers to read characters' lips and make sense of what they're saying by reading intertitles. Because the silent film style is unexpected today, captions need to make it clear that our assumptions about sound are being thwarted. When lips are moving and objects are interacting in space but no sounds are being produced, silence needs to be inserted into the captions. Captioning should thus account for our expectations about sound itself: how it is produced, disseminated, and sustained. At the same time, the captioner must interpret the soundscape and make judgment calls. Just because ambient noise provides the "key" to a scene, even in moments of relative silence, doesn't mean it will be or needs to be captioned. The iconic hum on the engineering deck of the Enterprise in *Star Trek: The Next Generation*, for example, is never captioned, even though it serves an important function—often just below the listener's conscious awareness—that this sector of the starship is the driving, humming heart. Too often, ambient noise is rendered in the captions as one or two distinct sounds: a dog barking, horn honking, indistinct chattering, and so on. Conveying ambient noise in captioned form is exceedingly challenging without reducing it to single, one-off sounds or stock abstractions that fail to capture its duration and color.

Silence isn't included in our discussions or definitions of closed captioning. Definitions tend to distill captioning down to the act of transcribing, representing, or displaying the "audio portion" of a program, as in Comcast's (2013) definition: "Closed captioning displays the audio portion of a video program—including dialogue, narration and sound effects—as text on your TV screen." More complex definitions make room for nonspeech information (NSI), which provides information, when necessary, *about* sound itself: who is speaking, what language they are speaking, and how to position captions on the screen to convey meaning. For example, consider definitions from the *Captioning Key* and WGBH's captioning FAQ:

Captioning is the process of converting the audio content of a television broadcast, webcast, film, video, CD-ROM, DVD, live event, or other productions into text and displaying the text on a screen or monitor. Captions not only display words as the textual equivalent of spoken dialogue or narration, but they also include speaker identification, sound effects, and music description (DCMP 2011e).

Like subtitles, captions display spoken dialogue as printed words on the television screen. Unlike subtitles, captions are specifically designed for viewers who are deaf and hard of hearing. Captions are carefully placed to identify speakers, on- and offscreen sound effects, music, and laughter (Media Access Group 2002a).

Both of these definitions emphasize the "audio content," as they should. At the same time, they manage to leave a small space for the inclusion of metalevel information about sound. Such metalevel information is usually limited, when it is included in captioning discussions, to speaker IDs and screen placement. But NSI is more than that, and I want to suggest in this chapter that we take a broader view of NSI, one that accounts for the physics and reception of sound itself—the often implicit assumptions we make about the nature and dissemination of sound. Such a broader view would still include the traditional speaker ID and guidelines about screen placement but would now subsume them under a more complex perspective, one that is more attuned to how sounds are produced and experienced.

Silence and rhetorical listening have become increasingly important areas of research in rhetorical studies (e.g., Booth 2004, Brummett 1980, Condit 1993, Glenn 2004, Glenn and Ratcliffe 2011, Palmer-Mehta and Haliliuc 2011, Ratcliffe 2006), as scholars have turned from a view of silence as "simply an absence of text or voice" (Glenn 2004, 2) to silence as "a specific rhetorical art" that is "every bit as important of speech" (4). While research has tended to focus on "the debilitating attributes and effects of silence" (Palmer-Mehta and Haliliuc 2011, 112), silence "as a state to be overcome, an ailment to be remedied, an injustice to be rectified" (111), silence can also be used strategically or productively as a form of resistance to oppression. Speech and silence, according to Cheryl Glenn (2004, 4), are not independent entities but "work together, each shaping and generating the other in a natural dynamism of meaning making." For Glenn, the "stupendous reality is that language itself cannot be understood unless we begin by observing that speech is not only surrounded by silence but consists most of all in silences" (4). For caption studies, Glenn's research suggests a need to rework basic approaches to captioning that are grounded in a positivistic focus on only what can be objectively verified. We should remember that "silence is meaningful, even if it is invisible" (4). When sound cannot be sufficiently described without also accounting for the silence that surrounds it, silence needs to be captioned. Likewise, when silence passes for sound (e.g., mouthed speech), it needs to be captioned to eliminate any confusion about what is or isn't being expressed orally.

An Ontology of Captioned Silences

Silences are captioned on an ad hoc basis. While captioners regularly (but not always) identify silences when necessary, they haven't translated the practice of identifying relevant silences into an ontology of silences that need to be captioned. Such an ontology would help to promote a conversation about standards and best practices. As we will see, silences are captioned inconsistently and, in some cases, not always with the best interests of deaf and hard-of-hearing readers in mind. In this section, I discuss three main types of captioned silence. They are responses to our assumptions about how sound is produced and received.

The Illusion of Audible Speech

When a speaker's lips are moving but no audible or distinct sounds can be heard, the captions should mark the silence with nonspeech information so that deaf and hard-of-hearing viewers don't assume that the speaker is making audible utterances. There is usually little visual difference between mouthing words and speaking them, which is why the former needs to be indicated in the captions to distinguish them from the latter they visually resemble. Silent or partly (in)audible speech can have different causes or explanations. For example, a speaker may be intentionally moving her lips to produce the illusion of speech without the sound—"mouthed" speech. Or, the speech may be mechanically silenced through the imposition of some barrier between speakers. Window glass is a popular silencer, as we will see. This subtle distinction between intentionally mouthed and environmentally silenced speech needs to be maintained in the captions to prevent misunderstanding. It would be technically incorrect to caption mechanically silenced speech as mouthed speech, because the latter is intentional and the former is created by the environment (or by the sound editor, as in a music overlay in which speech is silenced). Sometimes characters' lips are visibly moving but the characters are too far away for audiences to read them, or the mouthed speech cannot be interpreted clearly as distinct words. In such cases, "inaudible" may be an appropriate signifier, since "mouthed" implies that the words being mouthed can be discerned (and therefore captioned verbatim). Before silenced speech can be captioned, captioners need to determine how the speech is functioning in a scene.

Captioners have adopted a number of approaches to identifying silenced words, including (no audio), (inaudible), (mouthing), and (silence). Yet they have not been consistent in applying these options. Captioners need to consider the context and the nature of the silenced words to determine whether the speech should be (or if it can be) identified in the captions. In the examples of silent speech listed in table 6.1, mouthed speech appears in the three clips from *Monk*, the episode from *CSI: NY*, and the episode from *Star Trek: The Next Generation*. The *Star Trek* example is incorrectly captioned as (no audio) even though the speaker clearly mouths "find him" to prevent Captain Picard (Patrick Stewart), who is on the other end of that phone conversation, from hearing her. A better caption in this case would be "(Mouthing) Find him." Even though one of the examples in the *Oblivion* clip is captioned as (Mouthing), the speech is not actually mouthed but rather silenced as part of a flashback scene set to somber, uncaptioned music. A better caption in this case might have been (silent) or "(silent) Hi." In addition to speech that is silenced through sound editing, silenced speech also includes speech muted by window glass, as in the examples from *Nick and Norah's Infinite Playlist* (auto glass), *The Office* (office window glass), and *Twins* (jailhouse glass).

When mouthed or silenced words are not captioned verbatim, readers must usually do two things very rapidly: (1) they must read the caption, and (2) they must direct their gaze away from the caption to the speaker's lips, assuming it's not too late, the character's lips can be clearly seen, and the caption timing allows for it. Or they must perform these actions in reverse: read the speaker's lips (which is neither simple nor always possible) and then read the caption to learn that the words are not being expressed audibly. Because it's easy to miss the meaning of silenced speech, it is good captioning practice, when possible, to identify the silenced words in the same caption that identifies the words

Table 6.1 Examples of mouthed and silenced speech captions.

Source	Caption	Description
Monk, "Mr. Monk and the Red-Headed Stranger" (1.12), 2002	[Mouthing Words]	*Mouthed speech:* "We can't stop."
Monk, "Mr. Monk and the Airplane" (1.13), 2002	[Mouthing Words] [Mouthing "Hammered"]	*Mouthed speech:* The first mouthed speech caption stands in for two utterances. Garry Marshall mouths "Very good" and Bitty Schram mouths "Thank you" in response. The exchange happens very quickly. *Mouthed speech:* The second mouthed speech caption puts the mouthed word(s) in the caption, which is preferred in cases where the speech can be discerned.
CSI: NY, "Unspoken" (9.4), 2012	(mouthing)	*Mouthed speech:* "We need to talk."
Star Trek: The Next Generation, "Starship Mine" (6.18), 1993	(no audio)	*Mouthed speech:* "Find him."
Drive Angry (2011)	(INAUDIBLE)	*Distant speech:* While it's clear that the two men are talking to each other, it's unclear what they are saying. "Inaudible" is appropriate in this case.
Nick and Norah's Infinite Playlist (2008)	[ALL SPEAKING INAUDIBLY]	*Silenced speech:* It is not possible to identify what the multiple characters are saying through the car window.
The Office, "Two Weeks" (5.21), 2009	(INAUDIBLE)	*Silenced speech:* It's unclear what Steve Carell says in the parking lot from the camera's vantage point a couple floors up inside the office.
Twins (1988)	[Silence] [Silence] [Silence] [Silence]	The entire interaction takes place through jailhouse glass: *Silenced speech:* "Excuse me." *Silenced speech:* "Vincent. It's you, Vincent." *Silenced speech:* Intentionally unclear to hearing viewers. Played for laughs. *Silenced speech:* Also intentionally unclear to hearing viewers.
Oblivion (2013)	(MOUTHING) Hi. (INAUDIBLE)	*Silenced speech:* The silent speech is captioned verbatim, which is the preferred approach in cases where the speech can be discerned. But this speech is not technically "mouthed." The entire flashback scene is silent except for the overlay of slow, moving, uncaptioned piano music. The next line of dialogue—"Hello."—does not come with the same "mouthing" or inaudible disclaimer even though it is also silent. *Silenced speech:* It's not clear what the father says to the girl, which is why "inaudible" is appropriate here. But in all three lines of dialogue in this clip, the captions do not make it clear to viewers that the scene is done in silent style with music overlay.

Note: http://ReadingSounds.net/chapter6/#table1.

as silenced. This practice is followed in the examples from *Monk* and *Oblivion* in table 6.1, even though the *Oblivion* speech is not technically mouthed but silenced by music: [Mouthing "Hammered"] and (MOUTHING) Hi.

Captioning mouthed speech in verbatim form offers the most efficient solution to the problem of trying to read captions and lips at the same time, but it may also provide a different kind of access for readers. In some cases, mouthed speech may not be intended to be known clearly or fully. Speakers' gestures may provide context and focus our attention on important keywords and meanings. In short, the general contours of meaning may be more important than the specific details of the mouthed speech. Consider a long sequence of mouthed speech from *We're the Millers* (2013): six consecutive turns lasting roughly twenty seconds as two characters silently plot to infiltrate the tent of another couple whom they don't want to wake. David and Rose (Jason Sudeikis and Jennifer Aniston) take turns ordering the other in mouthed speech to enter the tent and steal the couple's RV keys. The general gist of this silent interaction is obvious even with the closed captions turned off, thanks to some exaggerated gesturing that accompanies the mouthed speech. In his first turn, for example, David puts his index finger to his mouth in a gesture for "shhh" or "quiet," points at Rose, points down at the tent, points at himself, points away from the tent in a "keep lookout" gesture, and then gives a thumbs-up sign. As he gestures, he mouths his instructions, which can be clearly read off his lips as a series of keywords: "Shhh. Okay, you . . . go inside . . . while I keep lookout. Okay? Good." This pantomime continues as Rose, in response, holds up her middle finger and mouths a curse to go with it. He silently accuses her of "yelling" at him. She denies it and he agrees to enter the tent. The scene is more about the experience of reading lips, the silent arguing over who will enter the tent, and the irony of David accusing Rose of "yelling" at him even though she is making no sounds. It is difficult to capture this experience in captions. A word-for-word transcript of the mouthed speech, which is what the captioned version offers, treats the mouthed speech no differently than regular speech, with the exception of a manner of speaking identifier that is used twice, once for each speaker's first turn. For example: "[MOUTHING] *Wait. You go in. I'll keep a lookout, okay? Good.*" In other words, the remaining four turns in this mouthed sequence could be mistaken for regular speech by some caption viewers. What's the solution? At the very least, captioners must vigilantly patrol the boundary between voiced and mouthed speech, making it clear which is which when the distinction is significant. They must also take into account the purpose of a scene and aim to convey that purpose through the captions. If a scene puts a heavier emphasis on exaggerated gestures and mouth movements, the captioner must try to give viewers ample time to experience the visual field beyond the caption space, bearing in mind that deaf caption viewers may spend the majority of their time reading captions. In an eye-tracking study, Jensema, Danturthi, and Burch (2000, 5) found that "subjects gazed at the captions 84 percent of the time, at the video picture 14 percent of the time, and off the video 2 percent of the time." Finally, we must keep in mind that a verbatim transcription of mouthed speech, while often crucial in providing full access, will create a different experience of the text by bringing forward and clarifying meaning that might have been intended to be subtle. **Media:** http://ReadingSounds.net/chapter6/#millers.

Montage sequences, which may include silenced speech, can also prove challenging to caption. In a movie montage, speech and ambient sounds take a back seat to music and the dynamic interplay of images. A number of short shots are juxtaposed into a sequence that signifies the compression of space and time. Perhaps the most famous and parodied example is the sports training montage, made famous in *Rocky* (1976). The typical movie montage is set to music, which fuses the otherwise silent and disparate shots together. Some ambient noise may filter through, but it is more common in a montage sequence to edit out all sounds and set the scene to music. At the end of a 2011 midseason finale episode of *South Park* (season 15, episode 7, "You're Getting Old"), Fleetwood Mac's "Landslide" provides a fitting soundtrack to a montage of life-changing events in Stan's life: his parents' separation, his move to a new home with his mother and sister, and his continued feelings of disillusionment and cynicism in the wake of an existential crisis brought on by his tenth birthday. The montage is uncharacteristically touching and without irony, leading some viewers to speculate that the episode was in fact a commentary on the *South Park* creators' own feelings of weariness and apathy over a show that had, after fifteen seasons, seemingly reached the end of its life. Creators Trey Parker and Matt Stone denied the rumors, and new episodes followed (The Daily Show 2011). (As I write this in late September 2013, on the eve of the premiere of season 17, *South Park* has been renewed through 2016.) The emotional tenor of the montage is due in large part to the musical accompaniment of "Landslide"—the original, haunting version sung by Stevie Nicks instead of a parody or cover version—with poignant lyrics about the passage of time that seem to speak directly to Stan's internal struggle to grow up: "Can I handle the seasons of my life?" "Well, I've been afraid of changing 'cause I've built my life around you." "But time makes you bolder, even children get older, and I'm getting older too." The lyrics, which are closed captioned verbatim, are central to the montage. But because only the lyrics are captioned, viewers may not be aware that Stan's father Randy is not speaking audibly when his lips are moving. The challenge for captioners is to determine whether viewers can be expected, in the absence of explicit information in the captions, to draw upon their knowledge of the context to interpret how sounds function in a scene. In this case, some viewers will recognize the montage structure and assume that Randy is not audibly speaking. But a safer bet would be, time permitting in the caption track, to add (inaudible) to the shot of Randy's moving lips. Another example, also from *South Park*, is a parody sports training montage set to 80s-style music that, like the montage from "You're Getting Old," includes silenced speech (i.e., moving lips) that is not identified in the captions. **Media:** http://ReadingSounds.net/chapter6/#montage.

Mouthed speech has become more popular on the Internet as meme culture spreads and comes into contact with the increasing popularity of web video. Animated images spread virally on user-generated news and discussion sites such as Reddit. An animated image saved in the GIF format, for example, can mimic the movement of video, allowing users to create and share short clips from their favorite television shows and movies. Each animated GIF file comprises a series of images that function like movie frames, so that when the GIF file loads and plays automatically, the effect is a short, often continuously looping, animation. But GIFs do not support sound, which is why animated GIFs made from TV shows and movies will often integrate closed captions (burned onto the surface of the animation for all to read). For example, the top result on a search for "The Office" on Giphy.com, a GIF portal, is a short animation of Jim Halpert (John Krasinski) shown mouthing most of the words to the line "I'm boring myself just talking about this." An open caption provides access to his mouthed speech. The animations are sometimes short and choppy, and not all of the words are always mouthed, which makes the caption an absolutely crucial component of animated GIFs that rely on silenced speech. In another GIF from a search for "The Office" on Giphy.com, Michael Scott (Steve Carell) is shown mouthing only the last three words of the image caption "I don't even consider myself part of society." The caption provides the full meaning that the lipreader of the mouthed speech can only partially reconstruct. Closed captioning advocates should pay attention to the ways in which silenced speech in animated GIFs encourage a burgeoning *lipreading culture* in which (1) captions are absolutely necessary to allow every viewer to make meaning in the absence of sound, and (2) reading lips and facial expressions are part of the process of understanding and enjoying a good animated GIF. I always try to read the lips of characters whose speech has been silenced in an animated GIF. If I don't get the full meaning right away, I let the GIF loop multiple times. Mapping that speech onto the caption involves figuring out which words were animated and which were left out. Even as web video (with sound and, increasingly, captions) becomes increasingly popular and standard on major websites, soundless animations that encourage both lipreading and captioning awareness contribute to viral web culture. **Media:** http://ReadingSounds.net/chapter6/#animated-gifs.

The Termination of Sustained Sounds

Sounds that are not clearly discrete but, rather, sustained or diffuse may need to be captioned to alert viewers when those sounds have terminated. In other words, some sounds need to be captioned not just when they start but when they stop too. Caption readers may require a reminder that a previously captioned sound has not ended but continues. Examples abound of "[sound] continues" reminders, including:

(ROCK MUSIC CONTINUES PLAYING ON HEADPHONES)—*A Serious Man* (2009)

(distant, panicked yelling continues)—*Aliens vs. Predator: Requiem* (2007)

(LAUGHING CONTINUES)—*An Education* (2009)

[CHANTING CONTINUES OUTSIDE]—*Argo* (2012)

(GUNFIRE CONTINUES)—*Avatar* (2009)

[screaming continues]—*Django Unchained* (2012)

(children's screams continue grating)—*BloodRayne 2: Deliverance* (2007)

(CONTINUES TAPPING)—*Inglourious Basterds* (2009)

[SNARLING CONTINUES]—*The Spiderwick Chronicles* (2008)

(TELEGRAPH CONTINUES CLICKING)—*Lincoln* (2012)

(ALARM CONTINUES)—*Oblivion* (2013)

(slapping continues)—*Two and a Half Men*, "Ixnay on the Oggie Day" (2010)

[ANGRY SHOUTS FROM CROWD CONTINUE]—*The 4400*, "Rebirth" (2005)

♪♪ [Continues]—*Lost in Translation* (2003)

(screaming continues)—*The Walking Dead*, "Sick" (2012)

[indistinct chatter continues]—*The Master* (2012)
(continues screaming)—*Futurama*, "Rebirth" (2010)
[silence continues]—*The Artist* (2011)

Continue captions reduce sustained sounds to discrete captions. They are not examples of captioned silences but may be paired with them, as we will see. They are always in response to earlier captions or to visual clues in the same scenes. For example, thirty-six seconds before (distant, panicked yelling continues) in *Aliens vs. Predator: Requiem* (2007), caption readers are alerted to the onset of yelling: (indistinct shouting in distance). In the example from *An Education* (2009), (LAUGHS) is immediately followed four seconds later by (LAUGHING CONTINUES). **Media:** http://ReadingSounds.net/chapter6/#continue.

The Artist (2011) contains the largest number of continue captions in my sample at twenty-five. This best picture Oscar winner is not only set in the period right before and after the advent of the "talkies" (1927–1932) but done in the style of a silent movie too. One caption in *The Artist* even contains two continues, which is unique to my sample:

- [barking continues]
- [ringing continues]

Most of these continue captions (nineteen out of twenty-five) in *The Artist* refer to music, such as [whimsical, ambling melody continues], which is not surprising given that music, and not dialogue, is the dominant and recurring element in *The Artist*. (Dialogue is handled by intertitles in the style of a silent movie.) But the most unusual and remarkable caption in *The Artist* is [silence continues], a caption that has the potential, I would suggest, to raise public awareness and foster a larger conversation about the nature of captioning itself (see figure 6.2). What [silence continues] suggests in dramatic fashion is that captioning is really about meaning, not sound per se. A "silent" movie that relies primarily on music requires closed captions that are attuned to its needs. In this case, the most basic assumption about sound in film—that movies are driven by speech and ambient sounds—is thwarted in *The Artist*.

Continue captions don't need to be followed by stop or end captions when it is visually obvious that a continuing sound terminates. We can see the students stop laughing in *An Education* (2009) and then turn their attention to orchestra practice. What we can't see is that they continue laughing offscreen when the scene cuts to the object of their amusement—hence the need for (LAUGHING CONTINUES). The same is true

6.2   Silence continues in *The Artist*.
Peppy (Bérénice Bejo) embraces George (Jean Dujardin) in a close-up over-the-shoulder shot from *The Artist*. Caption: [silence continues], which appears sixty-six seconds after the previous caption, [silence]. Studio 36, 2011. Blu-Ray. http://ReadingSounds.net /chapter6/#figure2.

of (GUNFIRE CONTINUES) in *Avatar* (2009), a caption that is needed when it isn't visually clear that Jake (Sam Worthington) is still being fired upon after fleeing from the top of the giant bulldozer. The difference is that (GUNFIRE CONTINUES) is not a response to a previous caption in the same way that (LAUGHS) serves as a predecessor for (LAUGHING CONTINUES). Rather, (CUNFIRE CONTINUES) is a response to a shift in focus from visible gunfire from the barrel of a shotgun-style weapon held by a military guy to the mere sounds of gunfire as Jake flees. The same goes for the telegraph machine sending news from the war front in *Lincoln* (2012). The visual depiction of the telegraph is first accompanied by (CLICKING), and then, forty seconds later, the machine is captioned again as (TELEGRAPH CONTINUES CLICKING) when it isn't visually clear (but is still crucial to know) that the telegraph is still sending information. Even continue captions that appear to be only marginally helpful at best and distracting at worst, such as ♪♪ [Continues] at the end of the arcade scene in *Lost in Translation* (2003), are directly linked to prior visuals or captions, in this case ♪♪ [Pop: Japanese]. Thus, when we can't clearly see that a sound continues, a caption may be needed to indicate that its status is significant and sustained. The continue caption addresses in a small and incomplete way one of the more difficult problems in captioning: how to represent sustained sounds that provide ambience throughout a scene.

The same guideline applies to the termination of sounds. If we can't see that a sound has stopped, and the visual and/or captioned context has fostered the expectation in viewers that the sound continues, then it needs to be terminated in the captions. A sound may be simple and singular such as (beeping stops) in *Aliens vs. Predator: Requiem* (2007) or complex and multiple such as (CHATTERING AND SINGING STOP) in *Avatar* (2009). Additional examples include:

(RECORD PLAYER STOPS)—*Moonrise Kingdom* (2012)
(MURMURING STOPS)—*Lincoln* (2012)
(WHIRRING STOPS)—*Oblivion* (2013)
[music stops abruptly]—*The Artist* (2011)
[jazz music ends]—*The Artist* (2011)
[CHATTERING STOPS]—*Inception* (2010)
(screaming stops)—*Futurama*, "Rebirth" (2010)
(engine whirs to a stop)—*Aliens vs. Predator: Requiem* (2007)
(music stops abruptly)—*American Dad*, "Fartbreak Hotel" (2011)
(car turns off)—*Aliens vs. Predator: Requiem* (2007)
(TURNS TAP OFF)—*A Young Doctor's Notebook*, "Episode 1.1" (2012)
(fanfare ends)—*Aliens vs. Predator: Requiem* (2007)
[RATTLES THEN STOPS]—*Argo* (2012)
[screaming halts]—*Django Unchained* (2012)
[slow, melancholy melody ends]—*The Artist* (2011)
(RINGTONE STOPS)—*Knight and Day* (2010)
(slapping stops)—*Two and a Half Men* (2010)
[THUMPING STOPS]—*Unknown* (2011)
(JIM MUTES TV)—*State of Emergency* (2013)
(jingling stops)—*Family Guy*, "Blind Ambition" (2005)
♪♪ [Ends]—*Lost in Translation* (2003)
[TURNS OFF RADIO]—*Inception* (2010)
[CACOPHONY STOPS]—*Man of Steel* (2013)

Each of these termination captions is preceded by a visual context or preceding caption that warrants it. In *Unknown* (2011), for example, the viewer is inside the MRI machine with Martin Harris (Liam Neeson) as the machine starts up off-screen with [MACHINE THUMPING] and then, a short time later, when it shuts down: [THUMPING STOPS]. This pattern is repeated in other texts, as when some bawdy act in the other room is identified as ongoing and then, a short time later, ends: (slapping continues) followed by (slapping stops) in an episode from *Two and a Half Men* (2010). In *Inception* (2010), [PEOPLE CHATTERING] signals to viewers that Cobb's (Leonardo DiCaprio) singular focus inside the dream space is being threatened once again by the intrusion of his projections. When the [CHATTERING STOPS], we conclude that Cobb has regained focus and the projections have retreated. In these examples, sustained sounds are described with the present participle form of the verb (*verb+ing*), which suggests the continuous nature of the sound. Contrast the caption immediately preceding [CHATTERING STOPS] in *Inception* (2010)—[GLASS SHATTERS]—which suggests, even outside of a specific context, that the sound is discrete and nonrecurring. In this case, a single piece of stemware has broken at the bar, initiating the distraction sequence for Cobb. In other examples, a present participle is not required in order to signal the end of the sustained sounds, because music ends, noise stops, ringtones and car engines are turned off, and TVs are muted, all by virtue of the affordances of the English language. When a present participle is not required—for example, [NOISE CACOPHONY] followed by [CACOPHONY STOPS] in *Man of Steel*—nouns take precedence over verbs. **Media:** http://ReadingSounds.net/chapter6/#stop.

These captions raise questions about human agency and intention. The (slapping stops) in *Two and a Half Men* even though Alan (Jon Cryer) is the agent responsible for the sound's termination. We can't see inside Alan's office when he stops spanking the woman (hence the need for the stop caption), but we know from the context that he is the responsible party. The [THUMPING STOPS] in *Unknown* even though a technician, presumably, turns off the MRI machine. Sometimes the agent can't (and shouldn't) be identified, as when the driver of an approaching car is intended to be mystery in this sequence from *Aliens vs. Predator: Requiem*: (car approaching), (tires squealing), and then (car turns off). Similarly, when an extradiegetic song ends, the captioner doesn't need to identify the sound engineer or producer responsible for the music track. At other times, it may be unimportant or left to viewers to infer agency. Does it really matter which technician turns off the MRI machine? Do we really need to be told who stops slapping, when we already know from the context that Alan is responsible? Do we really need to be told that (JIM MUTES TV) in *State of Emergency*? There's only one person in the room—it must be Jim. The absence of agency in a number of these examples should remind captioners that reading speed needs to be balanced against the need to give readers ample time to infer context and agency from captions. To consider agency is to be reminded that captioning is not simply about identifying sounds but also, importantly, about describing actions. When the Young Doctor (Daniel Radcliffe) (TURNS TAP OFF) in *A Young Doctor's Notebook* (2012), the caption focuses on the action rather than the consequences of that action, leaving viewers to infer that the sound of splashing water stops. This single caption also accounts for the sound of the squeaky tap being turned off. If there was something significant about the sound of the tap handle or the water, these sounds might have been captioned independently. But (TURNS TAP OFF) offers an elegant and concise solution that focuses on the agent's action rather than the sound per se.

Because stop captions are initiated by and acquire meaning from prior captions, they need to be evaluated as part of a sequence of linked captions. A caption that seems to be sufficiently meaningful on its own may fail when evaluated as a member of a pair. For example, a 2011 episode of the animated series *American Dad*, "Fartbreak Hotel," opens with a montage sequence in which Francine (voiced by Wendy Schaal) is stuck in an endless loop of cleaning house and cooking meals for her ungrateful family. Mindless vacuuming features prominently. The sequence is set to Electric Light Orchestra's (ELO) "Mr. Blue Sky" (1978), a Beatlesesque number (think "Day in the Life," only peppier) and undoubtedly one of the most uplifting and joyous songs in the history of pop music. The only cloud I can find on this song is that a few bloggers have complained of its overuse in movie soundtracks over the last decade. In 2012 ELO released *Mr. Blue Sky: The Very Best of Electric Light Orchestra* and a new music video for "Mr. Blue Sky." In the opening scene from this episode of *American Dad*, the ironic contrast between song and scene is blatant, making the scene seem even more ridiculous. The beat of the song is timed perfectly with the action and cuts in the montage sequence—for example, Francine sets down breakfast plates and moves the vacuum back and forth to the beat of the music—to add to the viewing pleasure and to reinforce the repetitive monotony of Francine's life. The lyrics (uncaptioned) that play over the montage come from the opening to the song:

Sun is shinin' in the sky
There ain't a cloud in sight
It stopped rainin', everybody's in a play
And don't you know, it's a beautiful new day, hey
Runnin' down the avenue
See how the sun shines brightly
In the city on the streets where once was pity
Mr. Blue Sky is living here today, hey

But neither the lyrics nor the song title are identified for caption readers, leaving deaf and hard-of-hearing viewers out of the rich interplay of music, meaning, and montage. Instead, the music is captioned with musical notes only. Nine captions make up the scene:

(alarm beeping)
♪ ♪
♪ ♪
♪ ♪
(alarm beeping)
♪ ♪
(gunshot)
(music stops abruptly)
(sighs)

The (alarm beeping) sounds coincide with a new day starting again and the same tired cleaning and cooking routine. The (music stops abruptly) when the teenage son Steve (voiced by Scott Grimes) smashes a beautiful monarch butterfly against the glass sliding door that Francine is cleaning. Francine (sighs). While (music stops abruptly) may be an appropriate caption on its own, particularly as it captures the sudden end to any glimmer of happiness Francine sought in the image of the butterfly and that she might hope to find in her domestic drudgery, it doesn't work well as a paired caption in this case. A music note doesn't provide a sufficient set up for the stop caption. A musical note also denies caption readers access to the rich irony of the scene. **Media:** http://ReadingSounds.net/chapter6/#american-dad.

The pairs of captions discussed here tend to assume a sonic universe of binaries in which sounds are either on or off. The TV audio is either on or muted, the chattering continues or stops suddenly, the machine whirs or doesn't, the voice screams or is silent, the slapping continues or stops abruptly, and so on. A good example of an abrupt change of state from on to off is (JIM MUTES TV) in *State of Emergency* (2013). While many of the sounds listed above do in fact operate as binaries (e.g., the TV audio turns off instantly when muted), not every stop caption captures the nature of the sonic changes being described. Sounds may be more complex than a simple stop caption may suggest. For example, the drone in *Oblivion* (2013) doesn't stop abruptly, despite being captioned as (WHIRRING STOPS). Rather, it powers down. The change from on to off is quick but still gradual, taking approximately five seconds. The process is metaphorically closer to turning off a ceiling fan than a car engine. The question, then, is how much complexity captions can support and readers need. It may be good enough, contextually speaking, to describe the drone as stopped, even if a slightly longer or more complex sonic process is taking place. Captioners can easily go too far in trying to caption sounds with greater precision than the context demands. At the same time, captioning advocates should be critical of oversimplified approaches to describing changes of state that paint over sonic complexity with captioned binaries. **Media:** http://ReadingSounds.net/chapter6/#oblivion.

Examples of captioned modulation provide one alternative to the limitations of the captioned binary. Consider, for example, the sonic process inscribed in a single caption from *Argo* (2012): [RATTLES THEN STOPS]. It describes a change of state in the incinerator as the power cuts off in the US embassy in Iran. Instead of a single state, the caption describes a temporal process as the incinerator grinds to a halt. In another example (also using "then"), the sound of a motor boat fades as it is driven away off-screen in *Beasts of the Southern Wild* (2012): [MOTOR REVVING THEN FADING]. Captions can also inscribe two actions inside a process, using "then," as in *Cloud Atlas*'s (2012) [SIGHS THEN GRUNTS]. Other examples rely on a verb such as "fades" or "dies" to suggest a gradual, less abrupt changeover to silence:

[anguished music subsides]—*The Artist* (2011)
[singing fades out]—*The Artist* (2011)
[tempo slows, music fades]—*The Artist* (2011)
(fades out)—*Bones*, "The Wannabe in the Weeds" (2008)
(music fades out)—*American Dad*, "Great Space Roaster" (2010)
(laughter dies down)—*Family Guy*, "Peter's Got Woods" (2005)
(CHATTER DIES DOWN)—*Skyfall* (2012)
[screaming dies out]—*Galaxy Quest* (1999)
(OVERLAPPING PHONE CHATTER FADES)—*Zero Dark Thirty* (2012)
(ENGINE POWERING DOWN)—*Zero Dark Thirty* (2012)
(BLADES SLOWING TO A STOP)—*Zero Dark Thirty* (2012)
[TURNS VOLUME DOWN]—*Moneyball* (2011)

This process also works in the opposite direction, for example in the descriptions of sounds that become louder, as when Abraham Lincoln (Daniel Day-Lewis) opens the window to let in the sounds of church bells. The bells changeover from (CHURCH BELLS CHIMING FAINTLY) to (CHIMING GETS LOUDER). Other examples include (ENGINE POWERING UP) in *Avatar* (2009) and (low electronic whirring, powering up), (crackling intensifies), and (rattle intensifies) in *Aliens vs. Predator: Requiem* (2007). A description of an increase in volume is nested inside a speaker ID in this example from *Skyfall* (2012): "CROWD (BUILDING IN VOLUME): Oh . . . !" In these examples, sound modulates up or down inside a single caption. While these are not examples of captioned silence per se, they may imply or end in silence through references to fading, dying or powering down, and subsiding. **Media:** http://ReadingSounds.net/chapter6/#fading.

The most dramatic examples of captioned silence come from *The Artist* (2011), a film that elevates silence to art and encourages us to reflect on our most basic assumptions about film sound. We assume that movie images will not only be accompanied by sound but that sound itself is sustained and ongoing in film. What needs to be marked as terminated or completed in *The Artist* is not a single sound per se, such as a ringtone or scream, but sound itself. When sound doesn't function as expected, which is the case in *The Artist's* silent movie approach, captions may be required. *The Artist* includes twelve nonspeech silence captions: eleven [silence] captions and one [silence continues]. These twelve captions are spaced pretty evenly apart, almost like regular reminders to readers, "tics" in the caption track (cf. Davis 2010a). The screenplay for *The Artist*, which was nominated for an Oscar, is surprisingly detailed for a silent movie, but the word "silence" appears only four times in the official screenplay by Michel Hazanavicius (2013). In other words, silence is not poured into the captions from the script but invented by the captioner. In table 6.2, each nonspeech "silence" caption is listed next to the chapter of the movie in which it appears. In figure 6.3, these same captions are graphed visually. The silence captions are well-spaced reminders throughout the film, occurring on average once every 7.34 minutes, with the shortest wait between reminders at 1.1 minutes and the longest at 17.8 minutes. Because *The Artist* on DVD only contains 208 closed captions, over 5 percent of the captions in the movie are [silence] captions. (By contrast, the average feature-length movie on DVD contains around 1,500 captions.)

The Artist is a remarkable film for caption studies, but it is not alone in exploring silence as a significant aesthetic resource or mode of communication. The loss of the audio track becomes fodder for fourth wall comedy in an episode of *Jon Benjamin Has a Van*, a short-lived series that ran on Comedy Central for ten episodes in the summer of 2011. The series follows a fictional television news crew led by H. Jon Benjamin as they report on quirky, made-up news stories. In episode 4 ("Breakdown"), the

Table 6.2 All twelve nonspeech "silence" captions in *The Artist*, with accompanying caption start times and movie chapter numbers.

Start time	Caption	Chapter
0:05:31	[silence]	1
0:09:14	[silence]	2
0:25:40	[silence]	4
0:30:35	[silence]	5
0:32:35	[silence]	5
0:42:31	[silence]	7
1:00:19	[silence]	9
1:07:35	[silence]	10
1:19:16	[silence]	11
1:26:08	[silence]	12
1:32:29	[silence]	13
1:33:35	[silence continues]	13

Source: Studio 36, 2011. DVD. http://ReadingSounds.net/chapter6/#table2.

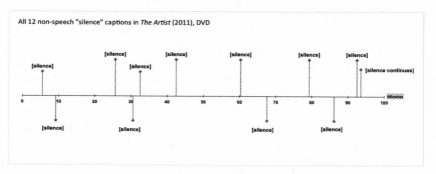

6.3   **The twelve nonspeech "silence" captions in *The Artist* serve as evenly spaced reminders to audiences that film sound is not functioning as expected.**
This graph plots the data in table 6.2. Studio 36, 2011. DVD. http://ReadingSounds.net/chapter6/#figure3.

TV crew is stranded in the desert until a trucker stops and gives Nathan (Nathan Fielder), the crew's sound guy, a ride into town. As Nathan rides off, boom mike in hand, the crew's ability to record audio goes with him, not simply for the show the fictional crew is recording but also for the TV show that we are watching. In this moment, the so-called fourth wall is broken; the means of TV production, which are typically hidden, become visible in an extended joke. This transformation is captioned as [audio crackling] before turning over to [silence]. Jon Benjamin and crew become "deaf" without their sound guy. Likewise, the hearing audience becomes "deaf" when they are forced to follow the narrative by reading characters' lips and handwritten signs exchanged with a deaf man. (This scene could benefit from captioned reminders that no speech sounds are being produced when characters' lips are moving.) It's challenging for hearing viewers to read the handwritten notes while trying to read lips at the same time, and it's in this moment of silent communication—assuming we can put aside the characterization of the deaf man as a country bumpkin—that hearing viewers can potentially identify with deaf and hard-of-hearing viewers. The same potential exists in those fleeting moments when hearing viewers must read the lips of characters who silently mouth words. **Media:** http://ReadingSounds .net/chapter6/#jonbenjamin.

Silence is contextual. The [silence] of *The Artist* is not sonically comparable to the [silence] in *Jon Benjamin*. The former is a deletion of the audio track in the spirit of a silent film. The latter is a fairly loud static buzz. *Jon Benjamin* resorts to buzzing static instead of no audio because the transition to silence, like the breaking of the fourth wall, is so unexpected that it needs to be marked in some way. Total silence could lead hearing viewers to wonder whether something is wrong with their television or the distributor, unlike the *expected* silence of *The Artist*. While the [SILENCE] in *South Park's* "Trapper Keeper" (2000) episode is closer to the static silence in the *Jon Benjamin* episode, it could not serve as a compelling substitute for it. As Kyle (voiced by Matt Stone) approaches the central processor room, we are surrounded by a clash of colors, shapes, and biological and mechanical sounds (heartbeat, flowing liquid, metallic breathing). By contrast, the processor room is color coordinated, geometric, and governed by a soothing electronic, oscillating humming sound akin to the sound of a large computer room. The processor room is not technically silent by any means but seems silent in contrast to the room Kyle just came from. The caption captures that contrast and reflects the contextual nature of captioning. The scene pays homage to the famous scene in *2001: A Space Odyssey* (1968) when Dave (Keir Dullea) enters the processor room of HAL 9000, the spaceship's sentient computer, to deactivate it. When Kyle attempts to deactivate the computer, Cartman (as HAL) tells Kyle (as Dave), "I'm afraid I can't let you do that, Kyle." The original line from *2001* is HAL's response to Dave's order to open the pod bay doors: "I'm sorry, Dave. I'm afraid I can't do that." [SILENCE] doesn't exactly capture the pleasing electronic ambience of the processor room, but it does begin to suggest the wide range of sounds that pass for captioned silence. **Media:** http://ReadingSounds .net/chapter6/#comparison.

The Insertion of Phantom Words

This third type of captioned silence calls attention to differences between the speech track and the caption track. Typically, these differences have been discussed in terms of overly edited captions that summarize speech rather than render it verbatim (Earley 1978). Discussions of edited captions focus on what's lacking in the captions or ,rather, on the surplus of speech sounds that the captions fail to capture. But we might also consider the rare occasions when words are added to the captioned speech that are not spoken. For example, a captioner or subtitler may clean up the speech of a speaker whose first language is not English, adding words to the caption track that don't exist in the speech but helpfully clarify the speech for caption readers. Other examples include curse words that are excised from the soundtrack and replaced in the caption track with [BLEEP], (deleted), or some punctuation like an ellipsis [. . .]. In order to prepare the soundtrack of a movie for cable broadcast on American television, the soundtrack must be sanitized. A cheap and fast way to do this is to delete the offending speech, leaving silence where the curse word(s) used to be. The caption track then marks the silence in some way. For example, in the 2007 movie *The Mad*, which I recorded on the basic cable channel Chiller, the curse words are deleted from the soundtrack, leaving quick bursts of silence behind. On the caption track, (deleted) is used to indicate that the offending audio was removed/silenced: "Shut the (deleted) up!" Another example is from a 2004 episode of *South Park* ("Douche or Turd"), in which curse words are silenced and replaced on the soundtrack with [BLEEP]. But the [BLEEP] caption does not stand in for a "bleep" sound. While the original broadcaster, Comedy Central, uses a bleep sound to block curse words, the syndicator of this episode, WGN, excised the curse words, leaving silent gaps in their wake but keeping the original [BLEEP] captions. Put simply, these placeholders are printed words on the caption track and silences on the soundtrack. The captions are phantoms because they don't correspond to any sounds. **Media:** http://ReadingSounds.net/chapter6/#cursewords.

Another type of phantom caption includes the various bylines and sponsorship messages inserted at the beginning or end of TV programs that do not have any sonic equivalent. For example, no one speaks the sponsorship message that always accompanies the end of the opening theme song sequence in *American Dad*: "Captioning sponsored by

20th CENTURY FOX TELEVISION." In other episodes of this show, I've seen an additional caption added: "and TOYOTA. Moving Forward." This silent information is only available to caption readers. In order to squeeze it into the opening sequence, the final line of the theme song, sung by the chorus as "Good morning, USA!," is sacrificed. The sponsor's name takes precedence, and caption readers are denied access to the final line of the song. Granted, this final line is a repeat of the previous line in the song, sung by Roger the alien (voiced by Seth MacFarlane). But that doesn't justify leaving it out just to make room for an ad. This example suggests how sacrifices may be needed to fit captioning companies' bylines and sponsorship messages into the flow of captions. **Media:** http://ReadingSounds.net/chapter6/#americandad-sponsor.

These sponsorship messages may take up quite a lot of time and space too. For example, the sponsor's message at the start of *CBS News Sunday Morning* (recorded December 1, 2013) takes up 11.36 seconds of the 26-second opening theme: "Captioning made possible / by Johnson & Johnson, / where quality products for / the American family have been / a tradition for generations[.]" These words are phantoms without a voice. They are neither spoken nor available to listeners who have the closed captions turned off. The captions on *CBS News Sunday Morning* are produced in three-line scroll-up style, and the sponsor's message takes up five lines total. The theme music is instrumental, which perhaps justifies filling the screen with an ad for Johnson & Johnson. But the theme music is as iconic on the long-running show as the diversity of sun images that fill the screen between segments. The music is "Abblasen," a fanfare attributed to Gottfried Reiche (CBS News 2012). The origin story for this piece of music reads like something out of a movie: in a 1726 portrait of Reiche by E. G. Haussmann, Reiche is shown holding a coiled trumpet in one hand and, in the other, a small piece of paper featuring a two-line melody in musical notation. This original melody has only survived inside the painting. Whether Reiche composed this melody is unknown. Regardless, the trumpet fanfare has become iconic. A quick search of YouTube will turn up countless amateur and professional versions of the "Abblasen" trumpet fanfare. *CBS News Sunday Morning* currently uses a version of the "Abblasen" performed on piccolo trumpet by Wynton Marsalis (CBS News 2012). The captions make no reference to the theme music, however, even though there is more than enough time and space to do so and despite the fanfare's historical significance and popularity. It would not compromise the sponsor's message whatsoever to precede it with something like "(Wynton Marsalis performs Reiche's 'Abblasen' on piccolo trumpet)." Indeed, theme music should always take precedence in the captions over phantom or voiceless sponsors' ads. Captioning advocates should be critical of any attempt to colonize the caption space with advertisements. The needs of readers to access the program's content must supersede the needs of advertisers to buy up precious caption space. **Media:** http://ReadingSounds.net /chapter6/#cbs-sunday-morning.

To make room for a sponsorship message in the closed captions, the captioner may need to speed up one or more captions, thus compromising the integrity of the dialogue as well as readers' ability to keep up. For example, the sponsorship message at the end of an episode of *Family Guy* ("Into Fat Air," 2012) is on the screen for 9.12 seconds and takes up three consecutive captions, each of which is three lines long:

Captioning sponsored by
20th CENTURY FOX TELEVISION,
FOX BROADCASTING COMPANY

Brought to you by the
VOLKSWAGON Sign Then Drive
event,

where you can drive away in any
new 2012 VW for practically
just your signature.

To make room at the end of the episode, the captioner speeds up the final five captions of the program's dialogue. These five captions span four different speakers. The captions must free up 7.59 seconds before the credits roll to make room for the three-caption ad. While the captions use screen placement to help viewers identify who is speaking, screen placement only adds to the confusion, because the captions no longer line up underneath the speakers. For example, when the scene cuts to Pam Fishman (voiced by Elizabeth Banks) and Ross Fishman (voiced by Martin Spanjers) in a two-shot, two captions are placed beneath Ross but only one is associated with his speech. The second caption is an early caption for the following speaker, Peter Griffin (voiced by Seth MacFarlane). Both captions look like Ross's speech due to screen placement underneath Ross, but only the former is. By the time the camera cuts to Peter speaking, the advertisement for Volkswagen has already begun. Peter's lips are moving while two VW captions are displayed, and in the confusion readers may mistake the VW ad for Peter's

Table 6.3 One example of how captions may be hurried up to make room for a sponsor's message.

Caption	Speech rate	Caption rate	How early?
There are more important things than a petty rivalry.	191 wpm	302 wpm	0.43 seconds
We're just happy that you're okay.	196 wpm	308 wpm	1.65 seconds
PAM: Well, thank you both. You saved our lives.	188 wpm	309 wpm	2.25 seconds
And Peter, I'll never forget what you did for us. You're a good man.	231 wpm	336 wpm	3.4 seconds
All right, you rest up, get better, we ate your son. Bring it up!	253 wpm	410 wpm	4.7 seconds

Note: The presentation rate for speech is compared with the presentation rate for each accompanying caption in the final five speech captions at the end of an episode of Fox's Family Guy ("Into Fat Air," 2012). The units of measurement are words per minute (wpm) and seconds. The final column measures how prematurely each caption appears on the screen by subtracting each caption's start time from the accompanying speech's start time. http://ReadingSounds.net/chapter6/#table3.

speech. Table 6.3 compares each character's speaking speed (in words per minute, or wpm) with the reading speed for each of the five captions. The final column of the table calculates how prematurely each caption appears on the screen by subtracting the caption's start time from the accompanying speech's start time.

To free up the needed time on the caption track, the captions in this clip must move at warp speed, presenting readers with an additional challenge. The clip's speech averages 211 wpm, which is already fast. If the captions were presented at this speed, it's likely that a portion of the audience would have trouble keeping up (see chapter 5). But the captioning rate for this clip is off the charts, with an average speed of 333 wpm, including an astronomical rate of 410 wpm for the final caption (i.e., 14 words in 2.05 seconds = 410 words in 60 seconds). By contrast, the Captioning Key style guide tops out its captioning rate guidelines at 235 wpm for theatrical productions for adults (see DCMP 2011a). In short, it's clear that the needs of readers took a back seat to the drive to make room for the sponsor's advertisement. Not only were the captions displayed up to 4.7 seconds prematurely, they were presented at an exceptionally fast rate. Making room for a silent ad should never come at the expense of providing full access to a program.

There's one more point to be made about these silent bylines and sponsorship messages at the end of programs: because they are typically inserted immediately following a program's final speech or nonspeech caption, they tell viewers when a program is over. They don't exactly give away the ending so much as subtly compromise the integrity of the program. While it is often fairly obvious to viewers when a television show is concluding, the insertion of a captioning company's byline into the narrative before the credits roll removes all doubt. For example, the captioning company's byline at the end of *The Matrix* (1999), rebroadcast on the SyFy channel, appears too early, right after Neo (Keanu Reeves) ends the phone call with the Machines that control the Matrix but before he dons the sunglasses and flies up into the sky. As the pounding rock music begins, the byline caption appears: "CAPTIONED BY THE NATIONAL / CAPTIONING INSTITUTE / —www.ncicap.org—." At this point, regular caption viewers know there will be no more speech or nonspeech captions in this movie, even if they haven't seen this movie before and even though the movie continues for another thirty-five seconds before the credits roll. Granted, there's not much room or need for more dialogue after Neo hangs up the phone—except perhaps right after he puts on his sunglasses—but that's not the point. The byline removes all doubt about which line of dialogue will in fact be the last one. Any remaining mystery or tension is removed prematurely by a caption that announces—in this case, thirty-five seconds early—that the movie's dialogue is done. **Media:** http://ReadingSounds.net/chapter6/#matrix.

Through an analysis of captioned silences, we put needed pressure on our definitions and assumptions. Captioning involves taking stock of the audience's expectations about sound itself. These expectations may be implicit, which is why it is not enough for captioners only to know how to type fast and write well.

Ambient Sounds and the Backchannel

Keynote sounds provide the immersive context, ambience, and mood for a scene. According to R. Murray Schafer (1977, 9), a keynote sound is an "anchor" around which "everything else takes on special meaning" (9). The keynote is the "ground" for a scene, which "exists only to give the figure its outline and mass" (9). The figure is a "signal" or "soundmark," such as dialogue between characters in the foreground of a scene, whereas the ground refers "to the ambient sounds around it" (152). Keynotes "may not always be heard consciously" and yet "cannot be overlooked" (9), hence the challenge for captioners who are tasked with identifying sounds that may not be consciously attended to and yet give body and context to the sounds in the foreground. How do you make caption readers aware of a scene's ambience without unnaturally elevating it to a foreground sound? Keynote sounds have the potential to make a "deep and pervasive influence on our behavior and moods" (9) when they "have imprinted themselves so deeply on the people hearing them that life without them would be sensed as a distinct impoverishment" (10). Schafer offers historical and geographical examples of keynote sounds, including the "sounds of a landscape created by its geography and climate: water, wind, forests, plains, birds, insects and animals" (9–10); the "keynotes of a farm" that "set up rhythms" for people (48); the "sound of the sea for a maritime community or the sound of the internal combustion engine in the modern city" (272); the "preponderance of stone" surfaces in "older European cities" (59); and "the clatter of horses' hooves, everywhere evident over cobblestone streets" (62). While Schafer studied the natural ambience of landscapes, including industrial landscapes, the concept of the keynote can also be applied to the mediated ambience of film and TV sound. On TV shows and in the movies, ambience creates an immersive sonic "ground" for listeners through background music, stock soundscapes, and foley sounds. The subtle keynote sounds of the subway in an episode from the TV show *Castle*, for example, combine with mood music to create atmosphere and dramatic tension. On live TV news, ambience can be created by the sonic properties of spaces themselves, such as the ways in which speech sounds echo off the hard marble surfaces of the US Capitol during satellite interviews with members of Congress. **Media:** http://ReadingSounds.net/chapter6/#keynote.

Ambient background music is especially popular on television commercials. During a commercial break from a Saturday morning news program on MSNBC (recorded October 12, 2013), three of the five commercials were closed captioned. All three relied on instrumental music behind voiceover tracks. A commercial for Liberty Mutual Insurance, for example, uses an instrumental version of The Human League's huge 1986 hit "Human," captioned as [HUMAN LEAGUE PLAYS "I'M ONLY HUMAN"]. At the end of the commercial, three words from the song are sung but not captioned—"I'm only human"—the same three words I've been hearing in my head since the start of the commercial. These three words summarize the point of the commercial: Accidents happen. We're only human. You're covered with this company's insurance policy. A captioned song title is thus necessary, and not simply because the song is well known. In the next commercial during the same program break, a Visa credit card sponsored by United Airlines uses upbeat instrumental music with a strong drum beat and a simple but catchy electronic melody. The music provides a nice accompaniment to the foregrounded customer testimonials about earning rewards points and airline miles. The music feels like a vacation, a happy celebration, a perfect complement to images of the Golden Gate Bridge in San Francisco, the Coliseum in Rome, and the skyline of Rio from the vantage of a cable car. The music is captioned at the very end with two musical notes: ♪ ♪. In the third captioned commercial during this program break, Enbrel, a plaque psoriasis medication, is narrated to the accompaniment of somber classical music, a fitting complement to the expressed feeling of "frustration" about "covering up" but also to the commercial's images of refined living (e.g., dressing up in formal wear to eat fancy dinners and attend classy parties). In this commercial, the background music, while providing context and atmosphere for the narration, is not captioned. **Media:** http://ReadingSounds.net/chapter6/#commercials.

Background music in advertising can have powerful effects. Captioners would do well to familiarize themselves with the body of research on music in advertising. Music is a standard component of television advertising. In a content analysis of 3,456 television commercials collected over a week of primetime network programming, David Allan (2008, 412) found that 94 percent of the total advertisements "contained some type of music": "Of the unique music advertisements, 14 percent contained popular music, 81 percent had needledrop, and 5 percent utilized jingles." (Needledrop is "an occupational term common to advertising agencies and the music industry. It refers to music that is prefabricated, multipurpose, and highly conventional" [Scott 1990, 22].) While there may be "no definitive explanation of *how* music works" in advertising (Allan 2008, 404), a number of studies have suggested that music mediates attention and memory (Allan 2006), complements or "fits" the brand (Macinnis and Park 1991), stimulates moods (Bruner 1990), and impacts audiences' affective and cognitive involvement with the brand (Park and Young 1986). Music has been called a "peripheral persuasion cue" for its ability to increase the audience's involvement with persuasive texts (Park and Young 1986, 11). Not surprisingly, background music has been found to have a powerful effect on audiences' connection with the brand. But that effect is mediated in complex ways that are not yet fully understood. Linda M. Scott (1990, 225) has critiqued the tendency in research on music in advertising to "isolat[e] the music from the message," to assume that the "effect of the music . . . occur[s] independently of meaning or context." Scott (1990, 226) makes a case for the "interdependence of the executional elements themselves—the interaction of visual, verbal, and music elements." She offers a holistic approach to interpreting the meaning of music in advertising, because, as she argues, "music should not be isolated from the complex inter-relationship of verbal and visual symbols that always accompany it in a specific message" (228). In Scott's (1990, 228) rhetorical approach, the "meaning of the music . . . does not inhere in the tones per se." We might say the same thing about sound, that the meaning is not in the sounds themselves but in how they function in specific contexts of sonic and visual interaction (see chapter 3). In deciding whether and/or how to caption music in advertising, captioners should consider the function of music as an interdependent element in a persuasive message. If music is intended to increase subtly the audience's involvement with persuasive texts, then musical note captions by themselves threaten to diminish the integrity and influence of a commercial's musical score.

In the narration-heavy, fast-paced, and highly visual contexts of television commercials, there's usually little time to add a verbose description of ambient music. Speech sounds, especially voiceover narrations, take precedence, with the music often supporting the visual and verbal message. The questions for caption studies are: What gets left out when ambient sounds are not captioned? If ambient sounds serve as "anchors" for meaning and persuasion, do images take their place as the "ground" for meaning in cases where only speech is captioned? If ambient sounds are not attended to consciously and yet have the potential to make a "deep and pervasive influence on our behavior and moods" (Schafer 1977, 9), how do/can captioners become not only consciously aware of ambient sounds but also caption them in ways that support the sonic "figures" of a scene instead of compete with them?

Sonic ambience adds depth, richness, and mood to a scene. To imbue the soundscape with a sense of dimensionality and fullness, captioners sometimes appeal to sound volume and location/distance in non-speech captions:

(eerie, raspy grunting echoing from distance)—*Aliens vs. Predator: Requiem* (2007)
[GUNSHOTS IN DISTANCE]—*Argo* (2012)
(HELICOPTER PASSING IN DISTANCE)—*Zero Dark Thirty* (2012)
JUSTIN: (IN DISTANCE)—*Zero Dark Thirty* (2012)
[thunder rolls in distance]—*The Master* (2012)
[RAPID GUNFIRE AND MEN SHOUTING IN DISTANCE]—*District 9* (2009)
[DOG BARKING IN DISTANCE]—*Extract* (2009)

(truck engine revving, faint digging)—*The Walking Dead*, "Killer Within" (2012)
♪ ♪ [Karaoke Continues, Singer Faint, Indistinct]—*Lost in Translation* (2003)
[Loud Video Game Noises]—*Lost in Translation* (2003)
(WAILING LOUDLY)—*A Serious Man* (2009)
(WHISPERING IN NA'VI)—*Avatar* (2009)
(WHISPERING INDISTINCTLY)—*Moonrise Kingdom* (2012)
[Whispering, Indistinct]—*Lost in Translation* (2003)
[inaudible whispering]—*Silver Linings Playbook* (2012)

In the absence of location indicators such as "in the distance," the captioned soundtrack has a tendency to feel one dimensional. Without location indicators, there's only foreground. Location indicators are intended to create dimensionality by positioning sounds near or far away from the camera/viewer. For example, there are twenty-seven nonspeech references to "distance" in *Aliens vs. Predatory: Requiem* (2007) and fifteen in *Zero Dark Thirty* (2012). Sounds can also be described as loud or quiet with respect to some surrounding context or norm—"whispering" being the most popular option. Without volume modulators, there's only monotone. Such references to location and volume help, albeit in limited ways, to fill out the soundscape with dimensionality and depth. **Media:** http://ReadingSounds.net/chapter6/#distance-volume.

Speech that is described as "indistinct" may serve a similar role, particularly in set-up or establishing shots. As the camera pans the crowd, cityscape, or suburban home before settling on or cutting to the main characters, an "indistinct speech" caption may signal to readers, with the assistance of visual cues, that this speech is separate from the foreground speech, even in the absence of location markers like "in the distance." Perhaps the most overused and generic means of representing background crowd noise is (indistinct chatter):

[CHILDREN CHATTERING INDISTINCTLY]—*Beasts of the Southern Wild* (2012)
[SPEAKING INDISTINCTLY]—*Argo* (2012)
[CROWD SHOUTING INDISTINCTLY]—*Argo* (2012)
(indistinct conversation)—*Aliens vs. Predator: Requiem* (2007)
(muffled, indistinct chatter)—*CSI: NY*, "Unspoken" (2012)
[PEOPLE SPEAKING INDISTINCTLY]—*District 9* (2009)
[INDISTINCT BORG CHATTER]—*Star Trek: The Next Generation*, "Descent, Part II" (1993)
(MEN SCREAMING INDISTINCTLY)—*Life of Pi* (2012)
(LOW, INDISTINCT CONVERSATION)—*Zero Dark Thirty* (2012)
- [announcer chatters indistinctly]—*Silver Linings Playbook* (2012)
(INDISTINCT TALKING)—*Lincoln* (2012)

- [indistinct chattering]—*The Artist* (2011)
[indistinct chatter]—*The Master* (2012)
[indistinct chatter]—*Django Unchained* (2012)
(indistinct chatter)—*Witches of East End*, "Marilyn Fenwick, R.I.P." (2013)

Regardless of their quality and variety, all instances of background speech captioned as "indistinct" end up "sounding" the same to readers when piped through a generic caption. Captioners need to be mindful of simplifying the sonic background with boilerplate captions that flatten meaning. Caption readers are used to seeing the same descriptors again and again across all formats and genres, such as (indistinct chatter), (laughs), (groans), (grunts), (applause), and so on. It's easy for captioners to fall back on a default placeholder for ambient sounds. It's more challenging (and potentially more rewarding for readers) to capture the subtle nuances of ambient sound within the context of the scene or narrative in which it occurs. **Media:** http://ReadingSounds.net/chapter6/#indistinct.

Generic captions create intertextual relationships across different texts. These relationships, which I have referred to as captioned thematics (see chapter 3), suggest patterns or themes within a single movie or TV episode, across episodes in the same TV or film series (see chapter 7 on series awareness), or between two or more different films or shows. These patterns are manifested in the caption track and may be latent or even invisible to (hearing) viewers when the caption track is turned off. Put simply, two or more films become linked when they share the same generic placeholder such as [indistinct chatter]. For example, *Django Unchained*, *The Artist*, and *The Master* are connected in this way despite their significant differences, encouraging us to explore all three chatter captions for instances of overlap and disconnect. Each caption can be interpreted against the backdrop of the others. A reference to (indistinct chatter) in a show I watched the night before writing this—from the television series *Witches of East End*—reminded me of some of the other references to the same caption I have seen recently in other TV shows and films. Within the same program, a repeating caption may also suggest themes or alert readers to forthcoming action (see chapter 5 on captioned irony). For example, in *The Order* (2003), starring Heath Ledger, three of the four references to [Voices Chattering] in the first seventeen minutes and sixteen seconds are visually associated with two creepy kids called "orphans." When the same caption is repeated a fifth time at 26 minutes and 58 seconds, we are primed to expect the reappearance of the orphans, who fulfill our expectations by reappearing moments later in the new context of the cemetery. The caption becomes intertexually linked to the prior appearances of the same caption, its meaning dependent not only on the context of the scene in which it appears but on the context of its prior appearances as well. Themes that become manifest through the caption track may also be available to hearing readers who are listening with the captions turned off. The difference between reading and listening has to do with the way these meanings rise to the surface when captioned. The chattering voices in *The Order* sound like subtle whispers but read like loud announcements. Captions can create a new way of mapping relationships among texts. Scenes that would appear to be unrelated are yoked together on the caption layer because they share the same generic nonspeech captions. We can visualize these relationships when we search a database of closed caption files and gather every example of "indistinct" or "chatter" in the corpus. **Media:** http://ReadingSounds.net/chapter6/#thematics.

The effort to give body and depth to the soundscape is always tested by the degree to which captions tend to equalize and distill sounds. All sounds become equally "loud" on the caption track, despite captioners' attempts to modulate volume and distance. In other words, ambient sounds that are unconsciously or barely attended to, and yet influence the mood of a scene, rise to the surface of awareness when captioned and subsequently compete with foreground sounds. All captioned sounds threaten to become "figures." In chapter 1, I shared two frames from *Extract* (2009) of a stock nighttime suburban soundscape complete with [DOG BARKING IN DISTANCE] and [CRICKETS CHIRPING]. These stock ambient sounds are not meant to be on equal footing with the foreground speech in the scene, but that's what happens when the crickets intrude on the reader's consciousness. In this example, like many others, ambience is problematically reduced to a single sound, a one-off caption. An ambient sound may be sustained throughout a scene—for example, (pulsating tone) or (chattering)—and yet only captioned once when the scene opens. Because captions linearize the soundtrack, ambient sounds, when transformed into captions, do not occur simultaneously with speech but now, when read, precede or follow the speech captions. Sustained or diffuse sounds become discrete captions. For another example of an ambient, sustained sound that is transformed into a one-off caption, consider the [TRAIN RUMBLING OVER TRACKS] in the 2014 television remake of *Flowers in the Attic*. The scene opens with a single description of the ambient engine sound but is dominated by the speech captions of the family talking inside one of the train cars. The scene inside the train lasts about forty seconds. While the engine rumbles throughout the scene to provide its key, the captions only alert us to the sound once at the beginning of the scene and again at the end of the scene as the engine leaves the family behind at the station to the sound of [ENGINE CHUGGING]. The sustained quality of the rumbling is reduced to a discrete identifier. Perhaps a more dramatic example is found in *Skyline* (2010), a low-budget alien invasion thriller that takes place almost entirely inside an apartment building (hence the low budget). Ambient noise is crucial during interior shots to remind us of the massive destruction being wrought by the merciless, giant octopus-like aliens across the city of Los Angeles and especially right outside their high rise windows. As Terry (Donald Faison) walks across the hall to check on his older neighbor Walt (Robin Gammell), the [rumbling outside] serves as a subtle reminder of the alien's constant presence. The quiet rumbling sound resembles the sounds of waves crashing punctuated by ominous thunder or metallic growling. Because we can't see the aliens at first—indeed, Terry looks out Walt's windows and only sees clear skies—the quiet ambient noise is intended to operate as a constant reminder of their fearsome presence. When one of the alien ships buzzes by the windows, the sound is captured as [mechanical whirring] and then as [electrical humming], initiating a sequence in which the ambient presence of the aliens comes to the foreground. But neither the transformation of ambience from keynote to signal nor the sustained and alternating quality of the signal are neatly captured by the three captions. By one-off, then, I mean the transformation of sustained sounds into discrete captions and the challenges of capturing keynote sounds in writing. While I am not suggesting that captioners should overwhelm the caption track with multiple references to the ambient rumblings of trains and aliens, captioners should keep in mind the tendency of writing to distill, reduce, and equalize sounds by transforming grounds into figures. **Media:** http://ReadingSounds.net/chapter6/#one-off.

The soundscape tends to flatten out and lose depth when (1) captioned figures squeeze out captioned background sounds on the caption track (as we saw in the brief analyses of TV commercials), (2) ambient sounds are captioned using default placeholders or boilerplate such as [indistinct chatter], and (3) ambient sounds compete visually with foreground sounds on the caption track. For a detailed example of how captions can blur the distinction between background and foreground sounds, consider *The Happening* (2008), a thriller directed by M. Night Shyamalan about a mysterious airborne illness. The movie contains a number of scenes of crowds of various sizes fretting and murmuring in the background. When the crowd noise is captioned as indistinct (e.g., "chattering"), it remains separate from the verbatim speech in the foreground, albeit equally "loud." But when the crowd's backchannel ambience is captioned verbatim, it comes forward into conscious awareness and becomes as clear as the main dialogue, which is only possible through the clarifying assistance of the captions. Examples of unclear speech brought forward and clarified on the caption track occur throughout the movie:

[Man] We need an extension over here.

I've never seen anything like it.

[Man] Go back to the laughing please.

Sir, we'll need to
check your suitcase.

[Man] Please have your tickets out and ready.
I know, I know you don't want to.

- Do you have a phone?
- No, no. I'm sorry. I don't.

They're not telling us anything.

[Woman] Drive. Just roll up the window.

- We have no communication.
- We don't have a cell phone.

Has anyone seen any other people on the roads?

[Man] I just walked down a quarter mile. It was clean.

[Woman] I'm here. Right here. Keep going.

These background speech sounds remain indistinct to hearing viewers when the closed captions are turned off. They are intended to provide ambience, to reinforce the concerns and questions of the main characters (and the audience) without spelling out these concerns in explicit detail. The crowd chatter is supposed to embody the mystery at the heart of the movie and create unease in the audience. The concerns of the crowd are left in the background as mostly unclear speech sounds. With the captions enabled, the distinction between loud and quiet sounds is downplayed or removed altogether. Because captions make speech and other sounds accessible that are otherwise hard for hearing viewers to understand, not loud enough, spoken too quickly, and so on, the flip side of equalization is increased clarity for viewers who now have full access to sounds that were intended to provide ambience only. **Media:** http://ReadingSounds.net/chapter6/#background.

The clarity that captions give to ambient sounds in certain circumstances is a double-edged sword. On the one hand, captions create access to sounds that are difficult if not impossible to discern otherwise as verbatim speech, regardless of hearing ability. On the other hand, the integrity of the text may be compromised when indistinct ambient sounds are captioned as distinct speech. Put simply, some speech sounds were intended to be unclear for a reason. Consider the much-discussed whisper at the end of *Lost in Translation*, Sofia Coppola's critically acclaimed 2003 film about two American strangers who develop a friendship during lonely stays at a Tokyo hotel. In the final scene, Bob [Bill Murray] whispers something into the ear of Charlotte [Scarlett Johansson] as he prepares to leave her and fly back to the United States. The whisper is captioned as [Whispering, Indistinct], and, despite the best efforts of fans who want to know what he says, it has remained a mystery. In his review of the film, Roger Ebert (2003) says that Bob and Charlotte "deserve their privacy":

I loved the moment near the end when Bob runs after Charlotte and says something in her ear, and we're not allowed to hear it.

We shouldn't be allowed to hear it. It's between them, and by this point in the movie, they've become real enough to deserve their privacy. Maybe he gave her his phone number. Or said he loved her. Or said she was a good person. Or thanked her. Or whispered, "Had we but world enough, and time . . ." and left her to look up the rest of it.

Through its exploration of alienation, loss, intimacy, and mystery, the film suggests in its gentle way that it might be better, really, if we didn't know. The whisper was not meant for us, not meant to be captioned. Its mystery is encoded in a private language that the two participants, alone, can access and decode. Yet fans have continued to speculate. One popular analysis on YouTube (DaeOh 2008), with close to 800,000 views, offers this translation as a result of applying a couple different audio filters to the clip: "When John is waiting on the next business trip . . . go up to the man, and tell him the truth. Okay?" John (Giovanni Ribisi) is Charlotte's husband and Bob is presumably urging Charlotte to tell John the truth about their affair. Despite its popularity, this analysis has only added to the whisper's mystery as it competes with dozens of other interpretations. Perhaps that feeling of being lost in translation themselves is too much for some fans to bear. They yearn to unlock the mystery of that embrace—to be able to hear and know clearly—believing that the whispered message foretells the future of the relationship between the main characters, thus offering to audiences a sought-after resolution to the drama. To offer a verbatim transcription of the whisper, assuming this were even possible, would run counter to the spirit of the film itself. **Media:** http://ReadingSounds.net/chapter6/#lost-in-translation.

Conclusion

This chapter has sought to redraw two main boundaries around captioning. First is the boundary that defines captioning in terms of objectively verifiable sounds. Every definition of closed captioning tends towards positivism by treating captioning as an exercise in translating audio content. But not everything that needs to be captioned can be empirically verified outside of specific visual and narrative contexts. Whether a sustained sound needs to be explicitly marked as stopped in the captions cannot be known by analyzing the sound on its own, just as we can't know which sounds are significant outside of a specific context. Second is the boundary that equates nonspeech information (NSI) with speaker identification and screen placement. A broader view of NSI would scoop up current approaches to providing information *about* sound and place them inside a larger model of how sound functions rhetorically (in specific contexts and for specific purposes) and ideologically (according to the rules and assumptions of how sound works in film, assumptions that don't necessarily hold in the real world).

An analysis of ambience in captioning calls attention to the challenges of recreating sonic dimensionality, simultaneity, and hierarchy in writing. Captions almost inevitably flatten the soundscape as sound is pressed into a narrow caption space and linear writing mode, as speech sounds almost always beat out ambient sounds for room on the caption track, and as captions equalize sounds by blurring the distinction between foreground and background sounds.

SEVEN

Cultural Literacy, Sonic Allusions, and Series Awareness

Captioners need to be able to identify sounds from the past, not simply describe sounds phonetically. A low-brow television show like *Family Guy* actually demands a high level of cultural literacy from captioners (and viewers), because it regularly alludes to sounds from the past. Songs, lyrics, and other television show themes may also need to be identified in the captions. Yet in naming these sounds, captioners convey new information to viewers who may not have been previously aware of the names or the significance of the sounds they're hearing or reading. In this way, the captioned experience always contains the potential to teach viewers something new (see chapter 5). Cultural literacy can also be applied at the global level of the television series: recurring sounds over multiple episodes can take on a special resonance. Captioners need to know which *series sounds* are significant and how they have been captioned in previous episodes.

For the most part, cultural literacy has not been included in our public discussions of quality captioning or what captioners need to know. Accuracy and completeness—two key criteria for measuring quality—tend to be defined technically and simply. But professional captioners, at least those I have interviewed (see chapter 2), recognize the importance of having a wide knowledge base (or "good ear"). That recognition, however, is not always reflected in

7.1   Sounds from the past continue to resonate in the present.
In this frame from *Close Encounters of the Third Kind*, Claude Lacombe (François Truffaut) makes a hand gesture by holding his hand chest high, horizontal with the ground, and fingers held together. In the Kodály method of hand signs, Lacombe is making the *mi* sign, the second of four signs associated with the movie's famous five-note motif. He looks towards the viewer's right. He is joined by three other men, one of whom plays the five-note motif on an offscreen electronic keyboard. All four are looking in the same direction towards the offscreen alien ships on the viewer's right. All but Lacombe are wearing headphones. The five tones are captioned as [NOTES PLAYING ON KEYBOARD] and have, since the movie's release, been invoked in countless other movies and TV shows. Columbia Pictures, 1977. Blu-Ray. http://ReadingSounds.net/chapter7/#figure1.

practice. Drawing on a range of examples, this chapter explores how captioning is more than an empty skill or decontextualized practice. Captioners need to draw on a deep well of knowledge—how much and what kinds of knowledge are always dependent on context—in order to identify significant sonic allusions, including allusions that have been remixed, (re)covered, and/or recontextualized.

Defining Cultural Literacy

Closed captioning is not an empty or mechanical skill. While captioners need to be excellent writers, good listeners, and careful editors, they also need to be *experienced in the world* because meaning is not simply inscribed on the surface of texts. Captioners need to recognize implicit meanings—the inferences in the text—and make them explicit when they are significant. Being a good writer or listener is not enough when it comes to identifying and naming sonic allusions. Cultural literacy for captioners is the ability to be an *ideal reader* of the text, to recognize the intertextual associations that every text creates with prior texts. No text is an island. In the case of captioning popular movies and television shows, cultural literacy demands that captioners possess broad but not necessarily deep knowledge of pop culture past and present: movies, songs, television shows, historical references, general scientific knowledge, important people/places/things/dates, slogans and clichés, popular opinion (*doxa*), and so on. In some cases, information can simply be looked up on the Internet. ("I don't know the name of this song. Let me Google the lyrics.") But expecting an inexperienced captioner to look up every reference assumes that she will recognize every significant sonic allusion in the first place. Allusions can be subtle and easy to miss (see the five-note motif discussion below) when you don't have the cultural literacy to recognize them.

For a rough analogy of what's at stake for captioners, consider an episode from *Family Guy*: "Bango Was His Name-O" (episode 4.29, 2006). The episode ends by poking fun at the show's own fondness for reaching way back into the pop culture archives for its cutaway gags—and, in the process, asking viewers to dig deep into their storehouse of cultural literacy. In this case, the episode spoofs *Soap*, a sitcom from the 1970s that was itself a spoof of the daytime soap opera genre. The *Soap* spoof kicks in at the moment the absurd truth is revealed, in typical soap opera fashion, about the stranger who looks like a grown-up Stewie (voiced by Seth MacFarlane): "I'm not your father, I'm you," the stranger tells Stewie. The narrator intrudes into the scene, as was custom at the end of every episode of *Soap*, to ask "a series of life-or-death questions in a deliberately deadpan style" (Wikipedia 2015). The narrator's final question to the *Family Guy* audience has implications for cultural literacy: "Will our younger viewers get this *Soap* reference?" Revised for caption studies, the question we might ask instead is "Will our younger captioners get this cultural reference?" In this example, the reference is explicit. *Soap* is named by the narrator, and the captioner simply copies down what is explicitly referenced by the speaker. But what if the reference is implicit (which was the point of the spoof), not named but intended to be inferred? Cultural literacy accounts for the unstated inferences that are part of the meaning of a text and can't be read off the surface of the words or sounds because they are extra-linguistic. **Media:** http://ReadingSounds.net/chapter7/#soap.

In his 1988 national bestseller *Cultural Literacy: What Every American Needs to Know*, E. D. Hirsch defines cultural literacy as "the network of information that all competent readers possess. It is the background information, stored in their minds, that enables them to take up a newspaper and read it with an adequate level of comprehension, getting the point, grasping the implications, relating what they read to the unstated context which alone gives meaning to what they read" (2). Hirsch argued "that literacy is far more than a skill and that it requires large amounts of specific information" (2)—"facts" and "formulaic knowledge" (28) about the world that allow people to participate in civic and political life at the national level. Hirsch emphasized content over form because, as he argued, possessing decoding skills alone is insufficient. Literacy is not an empty, decontextualized skill. A reader may be "well trained in phonics, word recognition, and other decoding skills" and yet still struggle to understand "the reality being referred to" beneath the surface of the text (13). Hirsch's answer—what he and his book have become notorious for—is a "provisional" list of 5,000 facts, phrases, concepts, names, and dates that every literate American needs to know (146). Hirsch's list emphasized "the most utilized *doxa* [popular opinions and beliefs] in American culture in the interest of giving historically marginalized students better access to the rewards of our society" (Ellwanger and Cook 2009, 473). The sixty-three-page list—"from abolitionism to Zurich," as *Newsweek* put it (Meacham 2007)—contains dates (1066, 1776), popular phrases and literary allusions ("All the world's a stage"), proper names (Hank Aaron, Athena), scientific and technical terms (covalent bond, condensation point, anaerobic, Doppler effect), geographical places (Central Park, Pittsburgh, PA), vocabulary terms (connotation, concierge, aria), terms from twelfth-grade civics class (filibuster, First Amendment), and similar.

Commentators have tended to focus on Hirsch's list, while dismissing or ignoring the underlying theoretical apparatus that, as I will argue, holds some value for caption studies. According to Paul G. Cook (2009, 489), these commentators have focused on the "*what* of cultural literacy as opposed to the *how*." Hirsch's reduction of American culture to a list of mere 5,000 terms and his simplistic equation of that list with literacy has been called "wacky; it is voodoo education, offering a quick, cheap fix for massive educational problems, while ignoring the intractable realities that underlie those problems" (Scholes 1988, 331). Hirsch's list has also been linked to a conservative, right-wing agenda to take control of the K-12 curriculum (see Ellwanger and Cook 2009, 472). Even more strongly, the book and the list have been described in racist terms: "What riled so many (despite the nagging sense that Hirsch was onto something) was the implied whiteness of that white-collar heritage" (Lovett-Graff 1995, 101). Less strongly but in the same vein, the list has been criticized for representing "a kind of cultural elitism" (Meacham 2007). Hirsch has been accused of reinforcing the status quo—challenging the dominant critical pedagogies of the day—through his book's proposal of a "standardized content" for education (Ellwanger and Cook 2009, 473). "Hirsch was attacked for the limitations of his list, and for the implication that the 'culture' about which we were supposed to be literate was a narrowly defined one that excluded the experiences of women, minorities and immigrants" (Meacham 2007). The presumed lack of multicultural awareness and diversity is captured nicely in a reporter's question at the time of publication—"Why isn't 'Cinco de Mayo' on the list?"—and Hirsch's acknowledgement that he wasn't familiar with the term (Meacham 2007). But perhaps the most damning response comes in the form of a "dubious honor" bestowed on both Hirsch and Alan Bloom's (1987) *The Closing of the American Mind*:

Since their publication, Bloom and Hirsch have been accorded one of academia's most dubious honors—the right *not* to be read. Their names have come to signify a certain set of assumptions about education, democracy, and the university at large. These ideas are thought to be ones that graduate students and professors can glean intuitively. There is an implicit understanding that these books no longer need to be read to be understood (Ellwanger 2009, 475).

Despite these withering critiques, I find something compelling about the very idea of a storehouse of cultural literacy, world knowledge, or *doxa* that will allow captioners to get the deeper meaning of sonic allusions, to get the *Soap* reference even when it is inferred. This storehouse doesn't have to be narrowly defined and elitist, but it should be responsive to the contexts captioners find themselves in. While Hirsch (1988, 29) was interested in the "stable elements of our national culture"— indeed, "stability, not change, is the chief characteristic of cultural literacy" (29)—caption studies should embrace a more rhetorical, adaptive definition of cultural literacy, one grounded not in a discrete list of de-contextualized facts and terms but in the captioner's *experience in and with the world*, particularly the world of pop culture (music, movies, television shows, celebrity culture). The key to acquiring cultural literacy is "practice": the equivalent of "reading a lot of texts, from 'trash' to 'classics'" (Scholes 1988, 330) is, for captioners, consuming a lot of pop culture, past and present, from movies to music to television shows.

Beyond the *what* of *Cultural Literacy* lies a fertile *how* for caption studies that takes us beyond simple transcription or copying. Two facets of

Hirsch's theory are worth stressing here: First, the *theory of inferences*. Hirsch (1988, 34) draws on an iceberg theory in which "the explicit meanings of a piece of writing are the tip of an iceberg of meaning; the larger part lies below the surface of the text." The reader's mind is "highly active," "not only a decoder of what is written down but also a supplier of much essential information that is not written down" (33). The extralinguistic inferences "are part of meaning from the very beginning" (39) because "we always go beyond a text's literal meanings to supply important implications that were not explicitly stated by the words of the text" (39). That's why two speakers who share background knowledge can be "short and efficient, subtle and complex" in their communications. They "can take a lot for granted" (4) that doesn't have to be made explicit. Readers fill in the missing inferences, a practice that Hirsch equates with comprehension. In *The Knowledge Deficit* (2006), Hirsch defines the "skill of comprehension [as] basically the skill of filling in enough of what has been left unsaid—that is, filling in enough blanks—to make sense of the text" (quoted in Clark 2009, 517). We can't show students how to mine a text for and grasp the meaning of inferences in the text by teaching critical thinking skills only because, Hirsch argues, inferences depend on access to content that "no course in critical thinking skills . . . could ever generate" (29). That same access to content or shared knowledge is denied when we focus on narrow decoding skills training (13). The test for Hirsch is not whether a reader can decode a text but rather whether a reader can "grasp the meaning of any piece of writing addressed to the general reader" (12). Grasping inferences is what making sense of a text is all about.

The second aspect of Hirsch's cultural literacy that holds potential value for caption studies is what I'm calling the *theory of haziness*. Cultural literacy demands that readers possess broad but not necessarily deep cultural knowledge. According to Hirsch, readers aren't required to dig down into any topic in order to access and understand the inferences in writing aimed at a general reader. Hirsch defines the "crucial background knowledge possessed by literate people [as] telegraphic, vague, and limited in extent" (26). He refers to "haziness" as a "key characteristic of literacy and cultural literacy" (14). The knowledge needed for comprehension can be imprecise (14), superficial (15), vague (15), abstract (54), stereotypical (16), and contain "little detail" (54). In short, we don't need to know a lot about the Civil War (54), canaries (58), or any other topic for that matter to comprehend texts addressed to a general readership. Hirsch's extended example of the "canary" schema suggests that readers only need to know the "primary associations" of the schema (canaries are small, yellow, they can sing, they are kept in cages [59]). The "primary associations must be available to us in milliseconds" (59). They can be assumed by writers and thus do not need to be made explicit. If writers always had to make explicit these primary associations by "retreat[ing] to the rudiments of knowledge," "ordinary discourse would be so lengthy and intricate as to obscure its own point" (59). As we travel down the schema to secondary, more remote associations ("canaries have backbones," "canaries are descended from reptiles" [58]), we arrive at knowledge that general readers are not expected to know on the spot and thus can't be assumed or inferred by writers. These assumptions about the distinction between primary and secondary associations are problematic, of course, because they tend to obscure the "variability" in audiences and contexts that rhetorical scholars are deeply interested in (Condit 2013). Nevertheless, the larger point about inferences remains an important one for readers of texts, including captioners: meaning is at least partly implicit in every text.

This theory of haziness resonated with me as I put this chapter together. It became clear as I researched a number of sonic allusions— the five-note motif from *Close Encounters of the Third Kind*, the "Cantina Band" song from *Star Wars*, the sound of a telephone's "busy signal," the "Yakety Sax" song made famous on *The Benny Hill Show*, the theme from *The Rockford Files*, the Doppler effect—that only superficial knowledge is needed to recognize and accurately caption sonic allusions in popular movies and television shows. My research unearthed a wealth of information about these sounds and their histories that, while interesting and fun to learn, isn't ultimately needed for the effective captioning of sonic allusions. For example, the "Cantina Band" song actually goes by another name within the *Star Wars* universe ("Mad about Me"), but the captioner doesn't need to know that in order to make this song accessible to a mainstream audience (i.e., when the song is used in the soundtracks of other films). In fact, captioning the song as "Mad about Me" instead of "Cantina Band" is likely to make the allusion less accessible. The popular name may be more accessible than the technical or true name anyway, which explains why "The Benny Hill Theme" is usually an acceptable substitute for "Yakety Sax," even though "Yakety Sax" is technically the more accurate choice.

Cultural literacy remains an important but underexplored and rarely discussed topic for caption studies. Even though we can't equate the practice of reading sounds with the practice of reading words, important similarities remain between the two practices that allow us to adapt the concept of cultural literacy for caption studies:

- Just as print literacy can't be reduced to an empty skill, the same goes for captioning. To be a culturally literate captioner requires more than an ability to decode sounds or represent sounds phonetically. It is a fallacy to assume that captioners don't need special training (Neves 2008, 135).
- The meaning of a sound is not simply inscribed on its surface. A theory of inferences calls our attention to the underlying world of implicit meanings that captioners must access in order to make them available to readers. Captioners must be more than good writers and careful listeners. They must possess cultural knowledge to make sense of inferred meanings.
- When captioners correctly identify the inferred meanings of sounds, they are making visible the intentions of the author/producer. A theory of cultural literacy is also a theory of authorial intention. "This is what interpretation is: the art of spelling out or inferring authorial intention, or the tacit, unsaid, in some sense invisible yet nonetheless logical implications of ideas, implications upon which one builds a case for the probability or validity (not certainty) of a particular understanding of a particular authorial intention" (Clark 2009, 511). Given that captioners strive to channel the content producers' intentions, with the captioner serving as proxy agent, cultural literacy offers one way for us to talk about quality captioning in terms of what the producers intended. Authorial intention is a troubled concept in literary studies, as English professors long ago declared "the art of interpretation is dead" and "insist[ed] that we cannot know what a book means" (511). But in captioning we need some way of talking about what the content producers intended, even as we recognize the fundamentally subjective and interpretative nature of closed captioning (i.e., Clark's [2009] probable but not certain intentions).
- Cultural literacy is broad but not deep. Captioners need to identify sonic allusions at a superficial or shallow level only. Captioners don't usually need to be experts in any specific allusion, but that doesn't make the task any less demanding on them, because captioners must possess an enormous amount of knowledge over a very wide terrain. This claim applies only to captioning for a general readership (e.g., popular culture texts), which is the focus of this book.

In what follows, I work through these ideas in the context of a number of examples from television shows and movies, paying special attention to missed allusions in the closed captions.

Supplying the Missing Inferences in Captioning

The most careful, most attentive captioner in the world can't hope to write adequate captions without a broad base of knowledge about the world. Spelling out *precisely* what kinds of knowledge are required is difficult if not impossible, but we can speak broadly about (1) the role that inferences play in any text, and (2) the important role captioners play in recognizing what's being implied so they can supply that missing information in the captions. These inferences are, by definition, not explicitly referenced in the text. They may or may not be made explicit in the script. When captioners miss what's being inferred in the text because they lack the experience, knowledge, or cultural literacy, they cannot sufficiently mirror the intentions of the producers in the captions.

Anchoring Musical Descriptions

The types of background knowledge needed are dependent on the nature of the show. For the same reason that it's a good idea to assign a sign language interpreter or real-time captioner (CART) with a background in biology or the hard sciences to provide interpreter services for a biology lecture, it's important to assign a captioner with the requisite background knowledge, who is mindful of the show's sonic potential to incorporate (e.g., cite, remix) prior sounds. (Below, I'll go even further and suggest that captioners need to possess a good working knowledge of the show or movie itself—how previous episodes have been captioned and which sounds have resonated historically on that specific show.) This demand for requisite background knowledge—an ability to recognize and correctly identify significant sonic inferences—places quite a burden on captioners, because allusions to prior sounds can range widely over the landscape of sonic possibilities. A television show like *Family Guy* regularly references pop culture texts/sounds from the last forty years. It would be a mistake to assign this show to a captioner who doesn't already possess a deep knowledge of pop culture. Likewise, *Paul* (2011) pays homage to the sights and sounds of iconic sci-fi movies from the 1970s and 1980s: *Close Encounters*, *Star Wars*, *E.T.*, and others. The best captioner will possess, at a minimum, a working knowledge of these (and other) sci-fi movies.

A careful interpretation of the soundscape is compromised if the captioner doesn't recognize the sonic inferences at play. Consider a captioned Honda commercial from 2011. The Honda commercial depicts a familiar family scene: the car-ride sing along on summer vacation. In this case, one multicultural friend is along for the ride to ensure an acceptably diverse television scene. The commercial opens with one of the kids vocalizing ("bum-bum"). The rest of the gang joins in as the song takes on layers of richness and meaning. Vocalized guitar sounds, shaken crushed ice sounds, and a vocalized melody build quickly to a crescendo as the family collectively belts out one line from the song. Instead of being captioned verbatim, the line is captioned unhelpfully as ♪♪ [All Singing]. For readers who are not familiar either with this commercial or with the specific heavy metal song in question (which I am purposefully not naming yet), it is difficult to discern the song title from the captions alone:

[Woman on Radio, Indistinct] ♪ Bum-Bum ♪

♪ Bum-bum,
bum-bum, bum-bum ♪

- ♪ Ai, ai, ai ♪
- ♪ Bum-bum ♪

- ♪ Bum-bum, bum-bum ♪
- ♪♪ [Ice Rattles
Rhythmically]

♪ Bum-bum,
bum-bum, bum-bum ♪
♪♪ [Imitates Guitar Noise]

♪♪ [Vocalizing Up-tempo
Heavy Metal Song]

♪♪ [Vocalizing
Continues]

♪♪ [All Singing]

The redesigned,
8-passenger Pilot.

Smarter thinking.

From Honda.

It's not clear why the specific song title and artist were not identified in the captions or why the line from the song was not captioned verbatim.

Both are needed to convey the full meaning. The song in question is Ozzy Osbourne's signature hit, 1980's "Crazy Train." Whether caption readers have enough knowledge to infer the song title with only a handful of vague sound descriptions is doubtful. While ♪ Ai, ai, ai ♪ is a pretty big clue for fans of Ozzy, we can't hang viewers' identification of the song on a vague phonetic description of a nonspeech sound. **Media:** http://ReadingSounds.net/chapter7/#honda.

I've shared this commercial with a group of grad students and with an audience of academics at a popular culture conference, and a fraction of both audiences (all hearing, presumably) were able to guess the reference to "Crazy Train" before I played the clip. They told me that "Ai, ai, ai" provided a major clue to the identity of the song. Nevertheless, no matter how descriptive and sensitive the captions appear to be, they will be insufficient for all but a narrow range of viewers if the captions do not cite the inference that unites them, in this case the name of the song: "Ozzy's 'Crazy Train.'" Ozzy's song is being inferred sonically, and it's incumbent upon captioners to make that inference explicit by including in the captions the name of the song and the verbatim lyric ("I'm going off the rails on a crazy train"). Otherwise, the sound descriptions and vocalizations have a tendency to float free. The artist's name and song title provide the anchor for free-floating descriptions.

Crediting the Uncredited Narrator

The unseen or omniscient narrator—that is, the narrator who isn't also a character in the narrative—is usually captioned using a speaker ID such as "Narrator" or a more specific variation, such as "Male Narrator." The *Captioning Key* (DCMP 2009, 19) offers the following guideline: "When a speaker cannot be identified by placement and his/her name is unknown, identify the speaker using the same information a hearing viewer has (e.g., 'female #1,' 'male narrator')." But such a simple guideline can't stand up to scrutiny. A narrator may be "unknown" within the narrative but known through his or her other film or television roles. Moreover, it's not always plainly obvious, despite the plain way in which the guideline is written, what "information a hearing viewer has." Some hearing viewers will recognize the narrator's voice as the voice of a famous actor, while others won't. In such cases, the captioner must side with the viewers who know the narrator's identity, even if— and this is important—the narrator is not listed in the credits. What I'm suggesting is that the captioner should serve as an *ideal reader* of the text.

The Other Guys (2010) is narrated by Ice-T. His voice is unmistakable, or at least unmistakable to those hearing viewers familiar with it, even though his work is uncredited in the movie and in the captions. IMDb (2010) lists Ice-T as "Narrator (voice) (uncredited)," while the DVD captions simply refer to him as NARRATOR for the bookend narrations he provides at the beginning and end of the motion picture. Captioners need to dig into their storehouse of cultural knowledge in order to identify narrators as well-known as Ice-T, whether such roles are credited or not. *The Other Guys* spoofs the cop buddy flick and opens with an action-packed car chase, making Ice-T, with his longtime role on *Law & Order* and earlier brushes with the law (see his Wikipedia entry), an appropriate choice to narrate this cop movie. That choice needs to be shared with caption viewers, even if—and this is where simplistic captioning guidelines fail—that choice is uncredited. More broadly, celebrity voice actors need to be credited in the closed captions for their endorsement work. Celebrities are increasingly lending their voices in television spots: Jon Hamm for Mercedes, Robert Downey, Jr. for Nissan, Tim Allen for Chevrolet, John Krasinski for Esurance, to name a few (Flint 2012). In Krasinski's case, [Male Announcer] defeats the purpose of his celebrity voice: to create product awareness and brand differentiation. If the advertiser paid up to "seven figures" to land a celebrity voice actor for their commercial (Flint 2012), the captioner needs to identify the actor by name because the actor's voice, from the advertiser's perspective at least, was the most significant and expensive element of the ad. Whether celebrity endorsements are worth the money (Daboll 2011) is beside the point. Whether some listeners will even recognize Ice-T's or Krasinski's voice without help from the captions is also beside the point. **Media:** http://ReadingSounds.net/chapter7/#narrator.

Arrested Development offers another example of an uncredited but well-known narrator. Ron Howard, the legendary child actor, *Happy Days* star, director, and Oscar winner, provides uncredited narration for the show's original fifty-three-episode run over three seasons on Fox, as well as the fifteen episodes of season four, which were released as a Netflix original in May 2013. Howard is also executive producer for the show. When Howard is providing narration for the show (which he does multiple times in each episode, including during the opening credits), the captions either identify Howard's voice as [Narrator] or set his speech in italics to denote offscreen speech. But the narrator needs to be named in the captions when the narrator is as well known as Howard and also because the narration (like the show itself) is occasionally self-referential. For example, when Howard takes issue with one of the characters calling George Michael (Michael Cera) "Opie" in "Public Relations" (2004), the joke hinges on viewers knowing that Howard is the narrator *and* that he played Opie as a child on *The Andy Griffith Show*. Knowing only one of these pieces of information is not enough to get the joke. Howard's response to the Opie reference, "[Narrator] JESSIE HAD GONE TOO FAR, AND SHE HAD BEST WATCH HER MOUTH," makes no sense unless you know that Ron Howard is the narrator and that he played Opie. A smart, multilayered joke for a smart show. Captioners need to be culturally smart to recognize that the joke depends on knowing Howard's identity. In this scene, it is less important but still useful for the captioner to know the reason for Ron Howard's objection: "Opie" is the speaker's shorthand way of calling George Michael clueless and naïve. **Media:** http://ReadingSounds.net/chapter7/#opie.

Not Just Any "Five Tones"

Hirsch's (1988) *Cultural Literacy: What Every American Needs to Know* tends to be remembered for its list of five thousand things every American needs to know. While both *Star Wars* and *Star Trek* made it on the list, *Close Encounters of the Third Kind*, released the same year as *Star Wars* (1977), did not. And yet *Close Encounters* has burrowed into the American pop culture psyche just as deeply, in some ways, as *Star Wars*. Case in point: the so-called "five-note motif" that serves as a form of communication between the aliens and the humans inhabiting every character in the movie who comes in contact with the aliens. Countless films and TV programs have paid homage to the same five note sequence.

Composer John Williams tried out "300-plus examples of five notes" before he and Steven Spielberg (the film's director) settled on the five notes chosen in the film (Reed 2011). The tones are pleasing to the hearing ear. The scale they belong to—the A-flat major pentatonic scale—has been called a "happy floaty dream" by one music blogger: "There is little tension. The intervals are widely spaced; there are no semitones" (Los Doggies 2008). At the end of the film, the five tones take center stage when the alien spaceships descend on Devils Tower, Wyoming. As a technician plays the tones on a synthesizer keyboard, a large screen displays each note as a color-coded bar of light. The aliens respond in kind with tones and lights emitting from their ships. A dance of music and light ensues as the humans and the aliens communicate in a rapid flurry of musical notes, colors, and flashing lights. During this communicative dance, the basic five tones "go through a number of variations. These include changing the register, placing some of the notes in different octaves, and varying the tone colour through instrumentation" (Loomis 2010). In the final minutes of the film, the motif expands into a glorious symphony of sound. Throughout the film, the five tones are captioned as [PLAYING FIVE NOTES] or something very similar. **Media:** http://ReadingSounds.net/chapter7/#fivetones.

The five tones make a compelling case study of the challenges and possibilities of describing nonspeech sounds. The tones do not simply add extralinguistic ambience or mood. They're central to the plot, a motif for the movie itself, a message or greeting from outer space, a form of communication, a mathematical set of relationships packed with potential meaning. Spielberg was clear, as Williams recalls, that the five-note motif "should not be a melody. It should be a signal" (Reed 2011). The five notes are potent signs, and *Close Encounters* unpacks their meaning across multiple modes and communication systems. At a basic level, the tones stand for musical notes and a musical sequence. Prior to the scenes at Devils Tower, the notes are *G–A–F–F (one octave lower)–C*. At Devils Tower, the notes are *B-flat–C–A-flat–A-flat (one octave lower)–E-flat* (Loomis 2010). The relationships among the tones is described in the movie this way: "Start with the tone . . . Up a full tone . . . Down a major third . . . Now drop an octave . . . Up a perfect fifth." The tones are also translated into solfège as Re Mi Do Do Sol and then into the Kodály method of hand signs. The scientists tracking the alien encounters send the five tones back into space, receiving a series of coordinates in return: 104, 44, 30, 40, 36, 10. At the end of the film, the musical motif is translated into colors: pinkish-red, orange, purple, yellow, white (Loomis 2010). Finally, the five notes have been compared to a greeting— "hello"—with each note corresponding to one letter in the word. The repetition of the same note one octave apart thus corresponds to the two Ls in "hello." This correspondence is taken even further in the claim that the sounds themselves *sound like* "hello." Finally, on the web, the most popular way of visualizing the motif is to put the notes on a conventional musical staff. But viewers have also offered their own phonetic representations of the sounds:

Bah [bi] bah [bom] baaaaaaa (TV Tropes n.d.)
da [DA] da [da] DAA (Gray 2008)

Effectively translating the motif into words using the resources of the English alphabet and the affordances of type (e.g., superscript) is fraught with difficulty (see "Ai, ai, ai" above). But my point here is that *Close Encounters* places the five-note motif at the heart of a number of communication systems: musical scale, sound, gesture, number, color, word, and light. The movie itself is about the possibility of communicating with intelligent beings from another world in the absence of a shared spoken language.

No online analysis of the five-note motif mentions the closed captions. That's not surprising. Closed captions are rarely discussed outside of accessibility contexts. But *Close Encounters* is a movie about the possibility of communicating across modes and barriers, which is precisely what captions are designed to do. The Kodály system of hand signals used in the film, insofar as it calls to mind a full-fledged signing system (e.g., American Sign Language), opens up a space for us to talk about the larger themes in the film in terms of accessibility. In other words, the problem of communicating across modes might be recast as an accessibility issue.

Close Encounters has had a major impact on sci-fi, despite Hirsch's omission of the film from his (admittedly incomplete) list. The film appears high on every "best of sci-fi" list: 11 (*The Guardian*), 20 (*IGN*), 6 out of 100 (*Popular Mechanics*), 7 out of 50 (*TimeOut London*), 16 (*Wired*'s readers' choice poll), to name just a few. Even though the five tones are rarely alluded to in popular culture today, captioners need to recognize what the notes mean and be ready to cite *Close Encounters* in the captions of other movies and television shows. While this point may seem obvious, it bears repeating: *captioners too often miss sonic allusions that are intentional and relevant.* Let me offer four examples in quick succession of how the five notes have been captioned in other contexts. (This review is not intended to be exhaustive but merely suggestive, the result of four chance close encounters with the five notes in a short span of time.) Of the four examples, the only one that captured the inferred meaning of the sonic allusion was a television episode of *Supernatural*, in which the five notes are captioned as [THEME FROM "CLOSE ENCOUNTERS OF THE THIRD KIND" PLAYS]. The five-note motif is an integral part of a scene in *Supernatural* involving an encampment of UFO enthusiasts. It makes sense that the five notes would be included in an episode about aliens. In contrast, consider the missed inferences in the following three examples, inferences that are plainly obvious to culturally aware listeners and likewise should have been conveyed to readers of captions:

· *South Park* pays homage to the famous Devils Tower scene in *Close Encounters* while spoofing and remixing it at the same time. In "Over-Logging" (2008), the scientists use a synthesizer keyboard to communicate with the Internet in an obvious reference to *Close Encounters*. A giant modem representing "the Internet" is perched atop a raised platform. In *Stargate*-style, a metal ramp leads to the modem. The pianist plays the first three notes of the five-note motif, which are captioned as [plays three high-pitched synthesizer notes]. When the Internet doesn't respond, the pianist [plays again] the first three notes followed, after a pause, by the remaining two notes, captioned as [plays low-pitched notes]. But the final note is off-key, contributing to the spoof. Because the captioner misses the inferred meaning (*Close Encounters*'s five-note motif), she also misses the spoofed, off-key final note. **Media:** http://ReadingSounds.net/chapter7/#ce3k-southpark.

· *Paul* (2011) pays homage to sci-fi movies of the 1970s and 1980s, including *Star Wars*, *E.T.*, and *Close Encounters*. One fan site offers a compilation of all the sci-fi references in the film. It's quite a lengthy list. Not just any captioner will be up to the task of identifying the significant sonic allusions to classic sci-fi in *Paul*. The five-note motif is subtle and would be easy to miss were it not for the visual context that supports it. Indeed, the producers hit us over the head with the obvious reference to *Close Encounters*. The large firework that plays the five tones—captioned unhelpfully as (ELECTRONIC MUSIC PLAYING)—is called "The Five Tones" (see figure 7.2). In response to the firework's name, Clive (Nick Frost) says, "Seems rather fitting," which is hard to read as anything but a metatextual address to the audience. Yes, it does seem rather fitting to include a firework named "The Five Tones" in a movie that pays homage to classic sci-fi. The movie's climactic scene—the scene where the firework is set off—takes place at Devils Tower, the site of all the fuss in *Close Encounters*. But just in case it still isn't obvious, one of the official movie posters asks: "Who's up for a close encounter?" (Wikipedia 2013a).

· *The Watch* (2012) also has fun with the sci-fi genre. In a backyard encounter with an alien lurking in the bushes, Ben Stiller tries to reach out to it by humming and singing some "friendly tones." He sings "hello." He also offers gum with an extended hand and by repeatedly singing the word "gum" to the tune of the five notes. But the inference remains hidden in the DVD captions: (SINGING) ♪ Gum, gum, gum for you ♪ This scene also arguably makes a reference to *Avatar* (2009), a box office juggernaut that's become associated with the catchphrase: "I See you." (See the book length study of *Avatar* entitled *I See You* [Grabiner 2012].) Both James Cameron's (n.d.) movie script draft and the DVD captions for *Avatar* capitalize the word "See" but only when it refers to the deeper forms of Seeing associated with the Na'vi people. (Note the lower case "see" in the directive of one of the nonalien human characters: "You just need to get in the habit of documenting everything—what you see, what you feel—it's all part of the science" [Cameron n.d.]). Capitalizing "See" in this scene from *The Watch* might have been a nice touch, a way of highlighting the allusion, drawing an explicit connection between these two films in the form of a shared capital letter. **Media:** http://ReadingSounds.net/chapter7/#ce3k-watch.

7.2 **It's hard to miss this obvious allusion to *Close Encounters.***
In this frame from *Paul*, a large canister firework called *The Five Tones* is shown in close
up. A small firework to the left of the canister is shaped like a saucer UFO. Caption:
(ELECTRONIC MUSIC PLAYING). Universal Pictures, 2011. Blu-Ray. http://ReadingSounds.net
/chapter7/#figure2.

No matter how carefully captioners try to describe nonspeech sounds,
captioned descriptions will be flat and only minimally useful if captioners don't also identify the allusions. By naming the allusions, captioners
give body to individual descriptions.

Generic Conventions as Literate Knowledge

Cultural literacy isn't simply a list of facts about the world—what the
"five tones" mean, who Ice-T is and what his voice sounds like—but includes genre knowledge as well. Movies rely on, exploit, hybridize, and
flout genre conventions and audience expectations. Genres are alive.
They are both (re)constituted through specific texts and shape the production of those texts. Captioners need to be aware of how individual
films and television shows participate in larger generic frames: the sitcom, the horror flick, the reality show, the cop buddy flick, the sketch
show, and so on. Genres are defined "as dynamic fusions of substantive,
stylistic, and situational elements and as constellations that are strategic responses to the demands of the situation and the purposes of the
rhetor" (Jamieson and Campbell 1982, 146).

One substantive element of the horror genre, for example, is the moment at the end of most horror movies when we think the bad guy (or monster, alien, evil force, etc.) is finally dead, only to see him or it rise once more. That moment is a lie. Captions need to convey that lie to us, in whatever sonic form or style it takes, particularly when that lie is genre defining. For example, at the end of *Halloween 5: The Revenge of Michael Myers* (1989), rebroadcast recently on AMC, young Jamie Lloyd (Danielle Harris) believes the evil Michael Myers (Don Shanks) is dead after his car hits a tree, because the (horn blaring) suggests that his body is resting lifelessly against the car horn. A look of relief crosses her face, until the (horn stops blaring) and the chase resumes again, the movie making good on the horror genre's promise of a premature death and resurrection for the bad guy. Captioners need to be attuned to this genre-defining moment, taking care to caption both the lie and the truth when they are sonically dependent. The horror genre may also appeal to music to convey the lie. Consider what happens at the end of *Alien 3* (1992) on DVD. After the alien has been covered with molten lead and presumed dead, the background music tells us it's time to celebrate the alien's demise. The moments following the molten lead bath are intended to be celebratory. How do we know? The music. There are three functions of ambient music in this scene: the pounding, anxious drum beat preceding the alien's capture (0:00–0:12); the celebratory music of the alien's presumed death (0:34–1:30); and the jarring sounds and rapid beat of the resumed chase (1:15 until the end). The triumphant, celebratory music starts just as the rain of lead is ending and continues until the alien leaps out of the molten lead pool to strike one last time. Accompanying the celebratory music is the reaction of Morse (Danny Webb), who clearly believes the alien is dead: "[Hysterical Laughter] All right, bug!" Morse provides the cue to viewers that it may be okay now to breathe a sigh of relief. The music—with its orchestral majesty, high notes, and glorious melody—is key to drawing viewers into believing the alien must certainly be dead. But the music is not captioned. **Media:** http://ReadingSounds.net/chapter7/#halloween5.

The triumphant music that starts at the end of the lead bath is intended to lull us into a false sense of security about the status of the alien. That's how so many of these horror flicks work: (1) we think the alien or serial killer is finally dead, (2) we breathe a sigh of relief, and then 3) it/he rises again to scare the hell out of us one last time. From this perspective, the music is not merely incidental or ambient but *genre defining*. The celebratory music is a generic cue. The cue doesn't have to take the form of music. Music is simply one way of tricking viewers

7.3 A sound wave visualization of the featured clip from *Alien 3*, annotated according to type of sound.
http://ReadingSounds.net/chapter7/#figure3.

into thinking it's finally safe to relax. The beast/alien/bad guy/serial killer/monster is dead now—oh wait, no! That's why the music should have been captioned. Even though the music is deceiving, it's part of the horror genre. The music is absolutely instrumental to conveying the lie at the end of so many horror films. Figure 7.3 annotates the sonic and musical forms that comprise this scene, with a special focus on the triumphant flourish following the alien's lead bath. **Media:** http://ReadingSounds.net/chapter7/#alien3.

Tricking viewers is also one function of ambient sounds in video games. In *Game Sound* (2008), a book about video game music and sound design, Karen Collins discusses how ambient sounds—"music, or ambient dialogue, or . . . outdoor environmental sounds" (92)—can create a particular mood. Ambient sounds "can be used to prepare a player for a particular situation, or to trick the player into thinking an area may be safe when it is not" (92–93). In movies, ambient sounds can also trick us. That's precisely what the triumphant music at the end of *Alien 3* is intended to do. Because the end-of-flick trick is integral to the horror genre, it must be captioned in such a way that caption viewers are mistakenly led to believe that the area is finally safe.

An Ear for the Remix

As we've seen, sonic allusions are not always simple reproductions or copies. Ozzy's "Crazy Train" may be revocalized by a fictional family singing in a fully equipped Honda Pilot. The five tones may be put into words ("Gum, gum, gum for you") or spoofed with an intentionally wrong final note. While captioners have as much responsibility to identify and cite the unaltered pop song as they do the complex remix, the latter is often more interesting to caption studies because the remix raises important and sometimes difficult questions about the differences between a caption and an annotation. How far should captioners go towards excavating the complex history of a sound or score? At what point does a sonic allusion become transformed and remixed into something new, detaching itself significantly from its source? Are sonic allusions always relevant? When do they vie for space with other significant sonic descriptions? The answers lie in the rhetorical work nonspeech sounds perform in specific contexts and under specific constraints of space and time. What information should be included depends upon the meaning of the sound within specific narrative contexts.

Captioners need to have an ear for the remix. Consider the "Cantina Band" song, which has been remixed, reworked, and covered a number of times on film since it was included in the original *Star Wars* movie (1977). George Lucas's vision for the tune, as conveyed in the liner notes of the *Star Wars* soundtrack, went like this: "Can you imagine several creatures in a future century finding some 1930s Benny Goodman swing band music in a time capsule or under a rock someplace—and how they might attempt to interpret it?" (Tip Top Music 2008). Like the five-note motif from that other major sci-fi film released the same year, the "Cantina Band" music has become a familiar piece of sci-fi history for culturally literate hearing viewers. The "Cantina Band" has been subject to a number of parody treatments, remixes, and cover versions. I will mention just two here, because I'm more interested in exploring the line between annotations and captions than providing a survey. For the first example, consider *Paul* (2011) once more, which is a treasure trove of visual and sonic allusions to sci-fi movies of the past. The bar scene in the movie includes a cover version of "Cantina Band," which is captioned unhelpfully as (BAND PLAYING COUNTRY MUSIC). The cover is a faithful reproduction, albeit in a different genre: both versions share the same tempo and melody, though with less jazz swing and more country twang in the cover. Jazz high hat—"the beacon in jazz timekeeping" and key to producing that "classic swing sound" (Fidyk 2012)—is replaced in the *Paul* cover version with a solid drum beat. The melody is muted in the cover version (compared with the original) but still clearly present, even though it is no longer being carried by the wind instruments. (According to composer John Williams, the original song used "one trumpet, two saxophones, [and] one saxophone who doubled on clarinet" [Tip Top Music 2008].) There should have been no confusion here about how to caption this music in *Paul*. Caption readers need both the movie title and the song title: country version of *Star Wars*'s "Cantina Band." Because *Paul* pays homage to classic 1970s and 1980s sci-fi films, this information should have been delivered in the closed captions. **Media:** http://ReadingSounds.net/chapter7/#cantina-paul.

With access to such information—info that culturally literate hearing readers possess—caption readers can more easily build interpretative and thematic frameworks and participate more fully in making sense of movies like *Paul*. For example, when we know that the country band is playing the "Cantina Band" song from *Star Wars*, we can try out different interpretations that are not available to us otherwise. Is this entire bar scene, not just the music, some commentary on the cantina scene in *Star Wars*? Are the three main characters in *Paul* (two men and a woman) supposed to be some version of Luke, Han, and Leia from *Star Wars*? Should we look for seedy, nefarious characters here who are threatening our heroes? Should we expect a violent cantina-style confrontation, similar to the violent outburst that sends the heroes fleeing the cantina in *Star Wars*? Is that black pickup truck in the parking lot at the beginning of the scene (which we encountered earlier in the movie) a metonym for Darth Vader? We can easily go too far with this line of interpretation, but my point is that such a line is only available to us at all if we have access to the significant sonic allusions.

A second example is from *Team America: World Police* (2004), which includes a scene set in a Middle Eastern bar that is reminiscent, in some ways, of the cantina bar scene in *Star Wars*. The setting includes a cantina-style bar band, some alien-looking characters (including an alien-looking guy in a gas mask), and a strangely familiar song, entitled "Derka Derka (Terrorist Theme)." The song has been called "a blatant (and intentional) knock off of the Cantina Band music from *Star Wars*" (Goldwasser and Brennan 2005). Other sources agree that "Derka Derka" is a parody of "Cantina Band" (e.g., Wikipedia's "Team America" page). One unofficial source claims that "Derka Derka" is the "Cantina Band" played backwards (The Soundtrack Info Project n.d.). While "Derka Derka" may be an intentional spoof of "Cantina Band," it's also a less direct or obvious version than the cover version included in *Paul*. The difference in these two examples is between a cover version and a parody version. "Derka Derka" suggests that captioners need to be mindful of the various styles for delivering sonic allusions (parody, knock off, remix, new genre, backward version). They must not only listen but watch too, using visual, verbal, and sonic cues, because the *Star Wars* parody becomes obvious only through the visual cues. Without the visual cues, "Derka Derka" becomes harder to recognize as a parody. Regardless, the subtle parody is lost on the DVD captioner, who opts for (band playing jaunty music). **Media:** http://ReadingSounds.net/chapter7/#cantina-teamamerica.

If "jaunty" fails to capture the allusion to *Star Wars*, then what other words would be sufficient and how much detail should the captioner provide? Captions are not annotations. The captioner's job is not to comment on the text or provide extradiegetic information. Captions are not information bubbles containing fun facts and other trivia. We are not watching VH1's *Pop-Up Video* or a so-called "enhanced episode" of *Lost* (Lostpedia n.d.). At the same time, captioners need to have a deeper knowledge of what they're captioning so they can determine whether an allusion exists, whether it is significant, and, if so, how to fit a description of the allusion to the narrative. Where do we draw the line between too little and too much? How deep should the captioner go? Do we need to know that "Derka Derka" is almost certainly a parody remix of "Cantina Band"? Absolutely. Do we also need to know that "Derka Derka" is "Cantina Band" played backwards (assuming that's true)? Probably not. Unlike *Paul, Team America* does not pay homage to *Star Wars* and other classic sci-fi films. Indeed, "Derka Derka" has been called "a toss away track [that] doesn't really do anything for the flow" (Goldwasser and Brennan 2005). The need to supply readers with relevant information must be balanced against the needs of the narrative and the constraints of space and time on the caption track. Nonspeech captions must be as efficient as possible. The captioner's goal is not to excavate the history of a sound—charting its various influences, discussing the nature or quality of the remix, and so on—but to capture in words the essence of the sound in context. In this example from *Team America*, the sonic aspects of the bar scene frame the visual aspects to produce a parody of the *Star Wars* cantina scene. A sonically aware caption might thus be: (parody of *Star Wars*'s "Cantina Band" song). If the caption is timed carefully to appear while the band is playing, it doesn't have to state the obvious ("band playing").

No sound is an island. Every sound is a text that indexes other prior sounds, styles, and genres. Some intertextual associations are more explicit and direct (e.g., the country version of "Cantina Band" in *Paul*); others are less direct (the parody/reverse version of "Cantina Band" in *Team America*). Charles Bazerman (2004, 86–88) discusses six levels of intertextuality for written texts, from the most explicit (e.g., directly quoting information from another text) to the least explicit (e.g., idioms and "recognizable kinds of language, phrasing, and genres" [88]). These intertextual levels reflect an intertextual view of language: "We create our texts out of the sea of former texts" (83). Although Bazerman is focused exclusively on written texts, his theory also applies to nonspeech sounds and music. Captioners need to identify the intertextual influences, "citing" the most explicit sources by name and identifying the more generic, less explicit influences. For the most part, captioners have done a fair job of identifying the less explicit levels: the genres, styles, and musical instruments (see "jaunty music" and "country music" above). But they have not consistently been able to identify the more explicit source material: song titles, uncredited but well-known voice actors, musical phrases from famous movies of the past, and so on.

Intercultural and Scientific Literacies

Cultural literacy is an expansive term covering numerous knowledge domains and perspectives. Even when we restrict ourselves to the English language for native speakers or readers of English, we must still attend to a multitude of concerns. In this section, I consider intercultural literacy from the perspective of language variety (e.g., British vs. American English), and scientific literacy from the perspective of the needs of a sitcom audience. Differences between American and British English can pose problems for some readers of closed captions. For example, how would you caption the sound a telephone makes when it fails to connect to another line because the recipient is already on the line with someone else and there's no automatic call-forwarding to voice mail? (I'm trying not to name the sound just yet, which is why my description is unnecessarily longwinded and confusing.) If the sound can be captioned in more than one way, how do you choose the best way? What if the option you prefer depends on the variety of English you speak or write? **Media:** http://ReadingSounds.net/chapter7/#mystery-phone.

Odds are good that American listeners and British listeners will have different names for this sound. Americans have "busy signal"; Britons have "engaged tone." The first term is native to me as an American. Busy signal is my unquestioned reality. That's what the sound *is* to me. The second term, I'm embarrassed to admit, I did not know until recently. Nonspeech sounds need to be considered from the perspective of the language variety typically spoken (or signed) in the DVD region in question. (There are eight DVD Regions across the globe. North America is in Region 1, Great Britain in Region 2. A DVD movie encoded for Region 1 may not play on a standard [region-restricted] DVD player sold in Region 2 and vice versa. Thus, "DVDs sold by Amazon.co.uk are encoded for Region 2 or Region 0. Region 2 DVDs may not work on DVD players in other countries" [Amazon.co.uk n.d.].) While it may not be feasible to localize English captions for each region, captioners should remember that less common British terms may cause confusion for some American DVD caption readers (and vice versa). This point becomes especially important at a time when most large captioning companies in the United States are reportedly outsourcing their captioning work to India and the

Philippines (Pond 2010). Indian captioners educated in or influenced by a British system need to be familiar with the conventional terms used by American viewers, who will purchase Region 1 DVDs to play in their North American DVD players. It's hard to know just how much US captioning work is sent overseas because outsourcing isn't exactly something companies like to publicly brag about.

The Adjustment Bureau (2011), starring Matt Damon, is a quintessentially American movie based on a short story by American sci-fi author Philip K. Dick, set in New York City, and driven initially by the main character's political aspirations for the US Senate. The movie relies heavily on NYC for its visual backdrop. The closing credits give thanks to "The People of New York," "The City of New York," Mayor Bloomberg, and a variety of locations and landmarks in the city. Perhaps that's why, given the extent to which the movie seems to be so thoroughly American (with the notable exception of British costar Emily Blunt), I took special notice when a sound I had only known as "busy signal" was twice captioned on the DVD with a British term ("engaged tone") that I wasn't familiar with. The two (ENGAGED TONE) captions occur at 00:23:17 and 00:45:42. Although the movie script written by American screenwriter George Nolfi (also the movie's director) makes no reference to busy signals in these two scenes, the script does describe a "fast busy" sound in another scene involving a busy signal. Given that Nolfi is American, I'm not surprised he chooses "busy" over "engaged." Why does this minor example matter? It matters, I would argue, because it points to the need to consider language variety or DVD region when captioning nonspeech sounds. Language variety (e.g., American English vs. British English) can be considered in terms of cultural literacy. Hirsch (1988, 75) considers country-specific knowledge to be one "domain" of "our national vocabulary" (along with two other domains: "international" literacy and "literacy in English" regardless of country):

every literate person today has to possess information and vocabulary that is special to his or her own country. A literate Briton has to know more about the game of cricket and the Corn Laws than an American. An American has to know more about baseball and the Bill of Rights than a Briton.

To this perspective I would add that cultural literacy is not simply about different topics (cricket vs. baseball) but also includes different terms for the same topic (engage tone vs. busy signal). **Media:** http://ReadingSounds.net/chapter7/#engagedtone.

Just because a nonspeech sound is captioned using a familiar term doesn't guarantee that it will be automatically accessible to viewers who may have cut their teeth on a set of competing terms. Captioners should at least keep in mind that British and American varieties of English are separated by hundreds of words with different meanings (Wikipedia 2013c), as well as hundreds of British terms that are not widely used in the United States (Wikipedia 2013b), starting with "engaged tone" for the American "busy signal." While savvy caption readers will rely on the surrounding context to make sense of unfamiliar words (just as savvy print readers have always done), true accessibility must account for language variety. Captions written in English may not be fully accessible to native readers of English, even highly educated readers of English, if they contain words specific to another variety of English with which they are unfamiliar. The most accessible captioning solution will account for language variety in such a way that captions can be *localized* for specific audiences.

When it comes to scientific terms and their corresponding sounds, captioners should not simply guess or fall back on phonetic spellings. At the same time, descriptions must be contextually and audience appropriate. Scientific objectivity has no bearing on the needs of audiences for contextually sensitive captions. A description for one audience may not work for another. What makes a good description depends on the venue or genre, the presumed technical or scientific knowledge of the audience, the function of the sound in context, and how much space and time the captioner has available to describe the sound. In an episode from the first season of *The Big Bang Theory* ("The Middle Earth Paradigm," episode 1.6, 2007), Sheldon (Jim Parsons) dresses up for Halloween as the Doppler effect, which essentially involves wearing a black and white striped leotard that's mistaken at one point for a Zebra costume. For nonscientists, the paradigmatic example of the Doppler effect is a vehicle passing a stationary observer, in which the pitch or frequency is higher as the vehicle approaches and then lowers as the vehicle passes by. This phenomenon is explained in terms of the shortening of sound waves as the vehicle approaches (with each wave reaching the observer sooner than the previous wave) and the lengthening of sound waves as the vehicle passes and recedes (with each wave taking longer to reach the observer than the previous wave). The result is a distinctive high-to-low sound. Sheldon mimics this sound four times in three separate scenes during the episode. The sound is repeatedly and consistently captioned as a "wavering hum": (makes wavering hum), (makes a wavering hum), (wavering hum), (makes wavering hum). How well does "wavering hum" capture the meaning of the Doppler effect? On the one hand, "wavering" captures the undulating, swinging, cycling aspect of all waves. On the other hand, "wavering" misses the more significant meaning of the Doppler effect, what Sheldon calls attention to in his pitch change from high to low. "Wavering" suggests a swinging back and forth, a flickering or wobbling, but what's required is a term or phrase that embodies the characteristic one-way movement of a Doppler wave from high to low frequency. One of the fan-created transcripts of this episode offers a phonetic approximation of Sheldon's vocalization: "Neeeeooooowwwww!" (Big Bang Theory Transcripts n.d.). But phonetic, onomatopoetic translations risk cutting caption readers off from inferred meanings. In this case, however, that risk is mitigated by Sheldon's explicit naming of the Doppler effect. Freed from the demand to cite the inference in the caption, the captioner can focus on providing a description of the sound that embodies the meaning of the Doppler effect and conveys Sheldon's vocalization: less emphasis on "wavering hum" and more attention to the change in frequency from high to low. Scientific literacy in pop-culture captioning requires that captioners understand technical or scientific concepts but adapt them to the needs of programs and audiences. The (wavering hum) caption suggests that the captioner did not possess the basic scientific literacy required to understand the rudimentary science behind the Doppler effect. **Media:** http://ReadingSounds.net/chapter7/#doppler.

Toward a Culturally Literate Captioning Workforce

When we talk about job requirements for captioners, we tend to start with writing, typing, and editing skills. We say that captioners need to possess advanced grammar skills. They need to have a good eye for details. They need to be good listeners. They need to have college degrees in English, journalism, communication studies, or a related field. We don't tend to discuss job requirements in terms of possessing what long-time captioning expert Joe Clark (2008) calls "life experience." Clark takes aim at twenty-something college graduates who can't find work in their fields of choice and end up doing offline captioning work to pay the bills (presumably until something better comes along). Clark put the issue this way in a comment on my blog:

If you want to caption competently, you cannot be:
   Someone doing captioning until something better comes along or because it's the only job vaguely related to your B.A. (Hono[u]rs) English degree. Someone with so little

life experience, and experience watching TV of all kinds, that you do not automatically know what people are talking about and know up front nearly every single term they use. Someone who *at any level* thinks you can just look that up when you get a chance.

You have to have enough experience and literacy, cultural and otherwise, to understand what people are saying exactly when you hear it. You then have to know exactly how to render all those words. You cannot sit there and guess. And, as a typical young woman with a B.A. (Hono[u]rs) History would do, just write shit down phonetically because you don't understand it and you've got a quota to meet.

That's why a workforce of young people, mostly female, with commercially-useless liberal-arts degrees are actively harmful to captioning.

Clear? (Zdenek 2013)

From my (admittedly biased) perspective as a liberal arts college graduate three times over, I think Clark is too hard on the liberal arts. His critique of the "typical" history major is an unfair (not to mention sexist) generalization. In captioning, what's true for history majors is also true for majors outside the liberal arts. Would the converse of a liberal arts graduate—someone with a "commercially-useful" degree in finance or engineering, let's say—fare any better when faced with a sonic allusion to a piece of music from the past or a television show theme song from the 1970s? I don't think so. Singling out the liberal arts is unfair. The problem of cultural illiteracy is not the purview of any specific academic discipline. (But I'd still pick the liberal arts—especially English/rhetoric, film studies, political science, journalism, and mass communication—if I had to choose an academic home for captioning, because the liberal arts are intellectually closer to the contexts in which sonic allusions circulate on American television and movies.) The real problem is not the liberal arts but a lack of cultural literacy. Clark's appeal to "life experience" is, I think, an appeal to increased cultural literacy. Neither a phonetic approach (i.e., just spell it like it sounds) nor a Wikipedia approach (i.e., "look that up when you get a chance") will suffice when sonic allusions are involved. Years before we could look it up on Wikipedia, Hirsch (1988, 60) questioned the notion that readers could just look up what they don't know:

It is not enough to say that students can look these facts up . . . When readers constantly lack crucial information, dictionaries and encyclopedias become quite impractical tools. A consistent lack of necessary information can make the reading process so laborious and uncommunicative that it fails to convey meaning.

For Hirsch, the answer to the literacy problem is found not simply in a list of facts about the world but in "specific, quickly available schemata" (60) that provide readers with primary associations for making sense of the implicit meanings in texts. Applied to captioning, we might imagine one schema for *Paul* to be "classic sci-fi films." The ideal captioner would be selected for this film because he or she is familiar with this and presumably other schemata. The culturally literate viewer draws upon the same schemata. The viewer who is deaf or hard-of-hearing relies on the captioner (as an ideal reader or proxy agent) to convey these intertextual associations in the captions.

How much blame should we place on youthful inexperience, which Clark invokes in the image of a "workforce of young people"? Elsewhere, in response to a captioning error in *Daddy Day Care*, Clark (2003) asks, "What 25-year-old female with a poli-sci degree (the typical clueless Canadian captioner) is making this journeyman error?" Should we simply demand, then, that offline captioners be of a certain age? Age is, after all, a pretty good antidote to cluelessness. On the one hand, a workforce of highly experienced (older) captioners is a compelling answer to cultural illiteracy. It's a way of addressing the *Family Guy* question head on: "Will our younger ~~viewers~~ captioners get this reference?" On the other hand, arguing that more experience is better is like arguing that a higher paycheck is better. It doesn't tell us anything we don't already know. It goes without saying in any profession that a more experienced workforce will be better than a less experienced one, but how do we get there without imposing a ridiculous and impractical age requirement on captioners?

Cultural illiteracy is a hard problem to solve. Instead of arguing a lá Hirsch for a list of 5,000 facts every captioner needs to know, I want to conclude with a call to reform our understanding of captioning itself. I appeal to two concepts from rhetorical studies: the *fragment* and the *schema*. We have tended to treat offline captioning as an empty skill, reducible to an ability to read and write well. As such, the typical film or television show has been treated as a bounded, finished object, a complete text in itself to be transcribed into captions. But I want to suggest, drawing on the work of Michael Calvin McGee (2013) in rhetorical studies, that we rethink the work to be captioned not as a whole, finished object but as a "fragment," a "dense reconstruction of all the bits of other discourses from which it was made" (McGee 2013, 231). These bits are themselves fragments and other "apparently finished" texts (231). The concept of the "fragment" directs us outside the text to the influences that shaped it (the source material, cultural influences, allusions, generic conventions, etc.). Just as the rhetorical critic "provid[es] in a formal way the missing fragments of the object of criticism, its influence" (233), the captioner supplies the missing sonic inferences to readers. To do this, captioners must go beyond the finished text, seeing the text as a fragment in a universe of discursive fragments. Captioning in this revised view is not an act of simple transcription because the text is only "apparently finished" when the captioner receives it (231). Put another way, the text to be captioned is not to be treated as "discrete" but "diffuse." "A diffuse text is one with a perimeter that is not so clear, one that is mixed up with other signs" (Brummett 1994, cited in Dickinson, Ott, and Aoki 2013, 340). While films and television shows are, from one perspective, discrete texts with "relatively clear beginnings and endings" (340), they are also, from another perspective, diffuse conglomerations of other texts and prior sounds.

These missing fragments are inferences in the text, unstated or partially stated. When meaning is inferred but not stated explicitly, when that meaning is delivered sonically, and when it is significant in the context of the object being captioned, the captioner must supply the meaning in the captions. Just as the culturally literate reader fills in the unstated inferences, so too must the captioner, drawing on relevant schemata, which are themselves replete with facts and primary associations. I want to suggest that we rethink captioning in terms of schemata and assign captioners to jobs based on their experience with relevant schemata. In the hypothetical *Paul* example above, I suggested that the ideal captioner would already be familiar with classic sci-fi films from the 1970s and 1980s. But we could also consider schemata in terms of genre, assigning captioners based on their experience with similar genres or even with previous episodes of the same show. The captioning department's resident sci-fi expert could oversee or editorially review the captions produced for a sci-fi flick, for example. When captioning a television series, captioners need to look beyond the individual episode, treating each episode as a "fragment" in a larger whole that unfolds over time, to see how recurring sounds across episodes have been captioned (e.g., running gags). Creating continuity across captioned episodes from the same series—what I call *series awareness*—requires that the captioner possess a working knowledge of how previous episodes have been captioned. In chapter 3, I offered an example of a running gag from *Family Guy*—the so-called "hurt knee" gag—that has been captioned in a variety of ways, seemingly with little regard for continuity across recurring examples of the sound. The mantra for series awareness—*caption the series, not the episode*—reminds us that some significant sounds may be missed if the captioner isn't familiar with their importance on the series.

For a good example of why series awareness matters, consider recurring music on a TV show that's identified with a specific character, a kind of leitmotif for that character. The differences between "disco" and "Europe's 'The Final Countdown'" are significant when the show is *Arrested Development* and the character is Gob (Will Arnett). Culturally literate fans associate Europe's 1986 hit "The Final Countdown" with Gob. The song always accompanies Gob's magic act. It has been called "his magic theme" (Arrested Development Wikia, "Storming the Castle"). It first appears in "Storming the Castle" (2004). Searching for the song on the Arrested Development Wikia turns up seven different episodes in which the song can be heard on the show's original fifty-three-episode run. As a recurring sound intimately tied to a major character, the song must be captioned by artist name and title so that the recurring theme can be explicitly linked across episodes. A generic caption such as [Disco] will not suffice to convey the full significance of this song in the series. In an example from "Making a Stand" (2005), Gob's trademark song, captioned as [Disco], plays through a boom box. While [Disco] may work in isolation, it doesn't work from a series perspective or for anyone who is culturally literate in the show. The boom box also plays a snippet from another recurring song in the series—"It ain't easy"—which is undercaptioned as [Country]. Only someone unfamiliar with the show would reduce this recurring song to [Country]. When captions are considered from a series perspective, important themes are more likely to be visible on the caption track. (Postscript: When *Arrested Development* returned for a fourth season in 2013 after being canceled seven years earlier, the intervening time and the show's growing cult status apparently did little to change the importance of "The Final Countdown" in the captions. In episode 7 of season 4, "The Final Countdown" is captioned as (dramatic music playing), even though the song was integrated into advertisements for the fourth season and has arguably become a second theme song for the show itself.) **Media:** http://ReadingSounds.net/chapter7/#finalcountdown.

In short, captioners should be matched with specific types of programs because they possess the experience and schemata to supply the missing inferences, treating the object to be captioned as a fragment in a global exchange of meaning and artistic creativity. Completed jobs regularly go through a review process by a senior captioner or editor (see chapter 2), but missed sonic allusions still routinely get through.

The editorial review process (under the assumption that more eyes are always better) remains an important line of defense against missed sonic allusions. But we also need to continue to challenge the notion that captioning is an empty or generalizable skill. The text to be captioned is a fragment in a context of influences, and the captioner's job is to recognize these sonic influences when they are relevant.

EIGHT

In a Manner of Speaking

Manner of speaking refers to the various nuances of speech and pronunciation. Typically, manner boils down to a speaker's dialect or accent. But manner of speaking also includes any kind of linguistic variation that distinguishes one speaker from another: age, gender, regional differences, pitch, volume, hesitation, intonation, timbre, reverberation, speed, and so on. What happens to these qualities when they are "entextualized" in closed captions? According to Mary Bucholtz (2007, 785), one form of entextualization in discourse studies involves "the process of capturing the fluidity of social interaction on the printed page" for the purposes of linguistic analysis. Essentially, closed captioning is a process of entextualizing speech and nonspeech sounds through the creation of a written text that has been "recontextualized" (798). A caption file might be understood, then, as a soundtrack recontextualized for the purposes of time-based reading and with the needs of deaf and hard-of-hearing viewers in mind. What happens to meaning, and manner of speaking in particular, when they are entextualized in writing and recontextualized as closed captions? One answer that I will offer in this chapter is that closed captions tend to *formalize* speech by mimicking, for the sake of accessibility and uptake speed, conventional written English. For the most part, as we will see, linguistic variations in pronunciation are scrubbed from the written caption file. What's left of pronunciation or accent will typically be handled (if at all) by a nonspeech identifier such as (IN SLOW-MOTION), while the speech that accompanies it bears little trace of the manner in which those words are

(STAMMERING)
There's some mistake.

8.1   Manner of speaking identifiers carry a heavy burden.
In this over the shoulder shot from *A Serious Man*, Larry Gopnik (Michael Stuhlbarg)
talks on the phone at his university office. He wears a dark tweed suit jacket. His desk
appears fairly well organized if a bit cluttered. Immediately visible items include three
books stacked into one pile, one sheet of paper and a manila folder, an intercom speaker,
a small black voltage device, a rotary phone, a protractor, a ruler, a magnifying glass,
and a large tan desk mat. These are items we might have seen on the desk of a 1960s
professor of math and physics. Larry's frustration and confusion with the bill collector on
the other end of the line are expressed in a manner of speaking identifier, (STAMMERING),
that precedes his speech: "There's some mistake." The speech itself sounds like "Ah, ah,
there's some mistake." While the manner identifier tells us how the speech sounds, the
speech captions themselves continue to be rendered in standard English. The manner is
fully carried by the nonspeech identifier. Larry does not have a speech disability, but this
caption raises questions about how speech disfluencies such as stammering and stutter-
ing are represented in closed captioning. Focus Features, 2009. Blu-Ray. http://Reading
Sounds.net/chapter8/#figure1.

pronounced. If the speaker is drunk and slurring his words, the only
clues in the captions will typically come from the manner of speaking
identifier that introduces the drunk speech: (drunken slurring). The captioned speech itself, however, will be perfectly "sober," so to speak—that
is, entextualized as standard written English.

Entextualization has political implications. When speech is conven-
tionalized in the process of being transcribed, and differences in pronunciation are scrubbed, every speaker tends to "sound" the same in
the captions. If the captions don't make use of screen placement to help
readers identify who is speaking, it becomes easier to confuse one speaker
with another in multiturn captions, as speech becomes interchangeable when multiple speakers occupy the same scene. The "process of
discourse transcription is also a process of social ascription," according

to Bucholtz (2007, 801), since decisions about how to represent speech may also imply social status, whether warranted or not. Standard English captions create a world of formal, sober, identical, high-status speakers. Speech becomes flattened out in the captions, disembodied. Likewise, deviations from the norm, which are perfectly acceptable in captioning when warranted, not only stand out against this formal background of standard English but potentially mark such speakers/captions as lower status and, perhaps, less educated. For example, "broken" English is a stigmatized category when attributed to nonnative speakers of English (Lindemann 2005). Could the same be said of "broken" English captions? In describing manner of speech in closed captioning, we need to account for those instances in which a nonspeech identifier such as (HEAVILY ACCENTED) becomes, for whatever reason, insufficient and must be supplemented or replaced with nonstandard alternatives such as phonetic spellings. How do we chart and explain the cracks that sometimes form in the captioned façade of standard English speech?

Drunk Speech but Sober Captions

If a character's manner of speaking is significant, it needs to be described in the closed captions. Captioned speech may be qualified in any number of ways: with more volume (yelling, shouting), with less volume (whispering, quietly, indistinctly), at an unusual or altered pitch (in a deep voice), with emotion (angrily, with sarcasm, sobbing), in an altered state (drunken slurring), with an emphasis on certain words, in a different voice (mocking, baby talk, impersonation), with a thick accent (British accent), in a specific language (in English), with hesitation (stammering), or with a singing voice. Manner applies to the speaker's words but also to the listener's interpretation (if the listener is another character who is experiencing sounds in an altered state). Even a speaker's normal voice may need to be noted in the captions if her normal voice represents a change from a previous state (e.g., a speaker goes from impersonating someone to speaking normally again). Of course, marking speech as "normal" doesn't tell us much about it, except that it has returned to sounding like it used to, whatever that means. The most common way of announcing a speaker's manner is through a nonspeech caption that usually accompanies a character's speech. For the sake of analysis, I distinguish manner of speaking identifiers from simple nonspeech captions because the former are always accompanied by speech that tells readers how the speech is inflected. Manner of speaking captions answer the *how* question: how is this speech pronounced? Thus, because adverbs also answer questions about how actions are performed, manner of speaking captions are often phrased, where the English language allows for it, as adverbs (robotically, distantly, unenthusiastically, gently) or adverbial phrases (in distorted voice, in silly voice). The examples in table 8.1 are intended to reflect a broad range of manner types: drunk/slurred speech, distorted voices, impersonations, normal speech, foreign accents, slowed-down speech, whispering, crying, bored speech, wavering speech, and robotic speech. Certainly we could add other types to this list, as well as additional examples of the types already listed. Whispered speech is perhaps the most popular type as it cuts across genres in a way that robotic speech or drunk speech does not.

Every speech caption could theoretically be accompanied by a manner ID, because "even in the high reaches of academia, everyone speaks with an accent that betrays their geographical origin" (Cavanaugh 2005, 133). But just because everyone speaks with an accent doesn't mean that every accent should be named in the captions. Rather than riddle a caption track with parenthetical references to the manner in which every word is pronounced, the captioner must decide when manner of speaking is significant within the context of the narrative. Typically, the nonspeech manner caption will carry all or nearly all of the burden of informing caption readers of the manner in which words are pronounced. That's because speech captions will most likely follow the conventions of standard written English, with some allowances for informal or colloquial forms that are characteristic of everyday speech, assuming these alternative forms are spelled more or less conventionally (e.g., phonetic spellings [*murda bizness* for "murder business" in a music video], g-dropping [*bringin, tryin*], contractions and reductions [*ain't, gonna*], vocalized pauses and exclamations [*hmm, like, um, ohh*], types of aphaeresis [*'bout, 'cause*]), and subject/verb disagreements. Experienced caption readers can process captions quickly in part because they are intimately familiar with the conventions of formal and informal English orthography, grammar, syntax, and style.

Manner of speaking identifiers are typically *front loaded* onto speech captions. In all of the examples in table 8.1, the nonspeech identifier precedes the speech that follows it. It is rare to find a manner identifier wedged into the middle of a speech caption. On the one hand, it makes sense from an accessibility perspective to separate parenthetical information about how words are pronounced from the words themselves, so as not to disrupt the flow of reading midstream. On the other hand, front-loaded manner identifiers aren't well suited to speech that becomes

Table 8.1 Examples of manner of speaking identifiers along with a sample of the speech that accompanies them.

Source	Manner of speaking identifiers and accompanying speech	
It's Always Sunny in Philadelphia, "Mac's Banging the Waitress" (4.4), 2008	(drunken slurring): It's a little late, isn't it?	
District 9 (2009)	WIKUS [IN DISTORTED VOICE]: Cut some cake.	
Scott Pilgrim vs. the World (2010)	(DISTORTED) Who do you think you are, Pilgrim?	
Aliens vs. Predator: Requiem (2007)	DARCY (electronically distorted): Sam? Sam?	
An Education (2009)	DAVID (IN SILLY VOICE): And I'll tell you what the first thing	
Twilight: New Moon (2009)	(IN BELLA'S VOICE) Gran, I'd like you to meet . . .	
The Office, "Scott's Tots" (6.12), 2009	- [imitating Kevin] THIS IS KEVIN MALONE. IS DAVID THERE? - TELL HIM I'M MAD AT JIM, 'CAUSE HE'S ASKING US TO GIVE MONEY TO PAM	[imitating Stanley] THIS IS STANLEY HUDSON. JIM HALPERT IS A MENACE. [imitating Toby] IT'S TOBY FLENDERSON. LISTEN, THINGS ARE GETTING REALLY BAD DOWN HERE.
Family Guy, Spies Reminiscent of Us" (8.3), 2009	(normal voice): I'm John Wayne	
A Serious Man (2009)	CLIVE (HEAVILY ACCENTED) Dr. Gopnik, I believe the results of the physics midterm were unjust.	
The Office, "Stress Relief," Part 1 (5.14), 2009	(IN A BRITISH ACCENT) A throne for your highness.	
Scott Pilgrim vs. the World (2010)	(IN SLOW-MOTION) Wow!	
Scott Pilgrim vs. the World (2010)	(WHISPERS) Don't go!	
Avatar (2009)	(GENTLY) Shh. Easy. Easy.	
Aliens vs. Predator: Requiem (2007)	(sobbing): Sheriff. (sobbing deeply): She's dead! She's dead . . .	
Avatar (2009)	(DISTANTLY) He's in. Jake, can you hear me?	
Family Guy, "Family Goy" (8.2), 2009	ANNOUNCER 2 (bored): And Meg.	
The Big Year (2011)	(UNENTHUSIASTICALLY) Yah!	
American Dad, "Great Space Roaster" (5.18), 2010	(voice wavering) : Why? Why would you do this?	
Elysium (2013)	[ROBOTICALLY] No, I am okay, thank you. [IN NORMAL VOICE] Understood.	

Note: http://ReadingSounds.net/chapter8/#table1.

inflected in the middle of a caption, such as a speaker who begins to choke up in the second half of a speech caption. A larger concern, however, is that each manner of speaking identifier typically only appears once per scene (or once per inflected speech type), even if the speaker takes up multiple captions or speaking turns. It's hard to demonstrate this claim using only the data in table 8.1 (without tripling the size of the table), but it's clear from these more detailed explanations:

· In "Mac's Banging the Waitress," a 2008 episode of *It's Always Sunny in Philadelphia*, Charlie's (Charlie Day) drunk speech spans eighteen captions and is interspersed with thirteen sober speaking turns from Mac (Rob McElhenney). But only the first drunk speaking turn is marked in the captions: (drunken slurring). Granted, perhaps only one reference to Charlie's drunk speech is needed. After all, we can see clearly that he's drunk. It's not like he sobers up by the time he gets to the end of the scene. Still, caption readers are forced to remember that every one of Charlie's unmarked, seemingly sober speech captions is in fact slurred. There are no reminders beyond the initial manner ID, despite the fact that another speaker (who is not drunk) repeatedly takes turns speaking. This approach—one manner identifier even for long speeches that span multiple turns—is standard in captioning.

· In an example of distorted speech from *Scott Pilgrim vs. the World* (2010), the final battle between Scott (Michael Cera) and Gideon (Jason Schwartzman) is produced in the style of a video game. Gideon's health is drained. His body flickers and changes color. His speech is marked by a (DISTORTED) identifier that appears only once, even though the distortion effect continues for another twelve speech captions that are interrupted only when Gideon (COUGHS) after the eighth caption. If we assume that his speech continues to be distorted throughout this scene, we do so only on the basis of our prior experience and the visual clues (e.g., his video-game body takes a flickering form), not because of any additional help from the captions to tell us that the distortion continues.

· In an episode from *The Office* ("Scott's Tots," 2009), Dwight (Rainn Wilson) impersonates three coworkers during three separate phone calls to the corporate office. Each impersonation is front-loaded with a single manner identifier such as [imitating Kevin]. When Dwight's impersonation of Kevin (Brian Baumgartner) is interrupted by a two-caption response from the secretary speaking on the other end of the line—NO, HE'S IN HIS WEEKLY STAFF MEETING. / CAN I TAKE A MESSAGE?—the impersonation continues but without any reminder that Dwight is still impersonating Kevin. Perhaps it's obvious that the impersonation continues, but we must rely entirely on context and our memory.

· At the end of *Scott Pilgrim vs. the World* (2010), the stylized video-game defeat of the evil Gideon is celebrated in slow motion as video game coins rain down from the sky and Ramona (Mary Elizabeth Winstead) says, "(IN SLOW-MOTION) Wow!" This is followed by Scott replying, also in slowed-down speech, "Yeah, wow." But only Ramona's speech is attached to a manner qualifier. We are left to assume that the same set of conditions holds for Scott's speech too. Even though two different speakers are impacted by the same inflection, only one manner identifier is used.

Manner identifiers often carry nearly the entire burden in the caption track of telling readers what's special about the way words are pronounced. Ramona's "Wow!" isn't slowed down phonetically (e.g., "Wooooowwww!"). The slow delivery of her speech is carried only by the parenthetical reference that precedes it: (IN SLOW-MOTION). Take away the nonspeech manner caption and there's nothing left in the speech caption to indicate slowed-down speech, except for an exclamation point, but an exclamation point might instead suggest excited or loud speech. Similarly, if you remove the manner identifier from "(WHISPERS) Don't go!" (also from *Scott Pilgrim vs. the World*), you're left with a speech caption that, thanks to an exclamation mark, could actually and ironically mean the opposite of whispering. Play this game: remove the nonspeech manner qualifiers and try to guess which speech captions go with which types of speech. You might get some help from an exclamation point (is that sobbing? or perhaps yelling?) or from multiple ellipses close together (which might indicate halting, drunk speech). But for the most part, speech captions do not betray the manner in which the words are uttered. The nonspeech manner identifier alone does that. Tearful speech rarely *looks* tearful when captioned. Slowed speech doesn't look slowed down in the captions. Distorted speech looks exactly like the perfectly formed (undistorted) speech captions that surround it. Charlie's drunk speech doesn't look/read drunk in the example from *Always Sunny*. The nonspeech manner identifier tells us that he is "drunken[ly] slurring" his words, but the words themselves appear to be quite sober. There's a hesitation or two (marked by ellipses) and a repetition of "my" and "we." Otherwise, the speech resembles any other speech captions— drunk or sober—we're likely to encounter, perhaps with just a bit more marked hesitation. I don't see any significant differences between Mac's sober speech captions and Charlie's drunk speech captions. In fact, Mac's speech is marked by the same hesitation: "Well, um . . . I went on a long walk, so . . ." The slurring is carried entirely by the manner identifier. Indeed, most speech captions, regardless of manner, accent, or level of emotion, tend to look the same, because captioning converts speech into standard written English. Put another way, Charlie's drunk speech captions could also be recited sober. There's nothing inherently sloshed about the way his speech is captioned. The drunkenness comes from the manner caption that introduces his speech. What marks speech as drunk (slurred, hesitant, irregular, slowed, incoherent, garbled) is evacuated from the speech captions themselves, reduced to a single description, and carried by that description at the start of the drunk sequence.

This is the standard approach to documenting manner of speaking, then: when speech captions need to be marked with an inflection of some kind, one nonspeech manner identifier is called into service to bear nearly the entire burden. But speech captions do sometimes bear traces of their embodiment. In a scene from *A Serious Man* (2009), for example, the weight of one manner identifier is partly distributed among the speech captions, which is to say that the (HEAVILY ACCENTED) speech of Clive Park (David Kang), a foreign exchange student, is reflected in his stilted English speech. The speech captions provide small clues that the speaker is not a native speaker of English. The formal phrasing ("the failing grade"), misuse of articles, stilted vocabulary ("unjust"), subject/verb disagreement ("it cover"), and other grammatical infelicities ("I was unaware to be examined . . .") suggest, even without the nonspeech manner caption, that the student does not have a strong command of spoken English. Nevertheless, the accented words, when captioned, continue to conform to standard conventions of spelling, which leaves the manner identifier with the responsibility of conveying how those words sound when pronounced. In this example, it might have been more effective, then, to name the origin of Clive Park's accent too—"(Korean accent)." **Media:** http://ReadingSounds.net/chapter8/#seriousman.

When transcribing speech, the linguist faces a challenge similar to that of the captioner describing manner of speech. In a 1991 article in *American Speech*, Ronald Macaulay discusses the inherent limitations of phonetic transcription: "Any transcription, no matter how detailed, is an interpretation of the tape and necessarily selective in what it includes or leaves out" (282). The researcher can strive to offer an "exhaustive" transcript, but such a transcript would almost certainly be less accessible. Because a detailed phonetic transcript "is not easy to read," it would take "longer to comprehend" (282). Moreover, it would contain quite a bit of annoying "code-noise" because it "would provide a massive amount of information that is largely irrelevant" (Labov and Fanshel 1977, quoted in Macaulay 1991, 282). It "may look more authentic, but the value of any transcription depends upon its effectiveness for the reader" (289). Macaulay suggests, among other things, that "the purpose of the transcription" (282) should drive the process of selecting what to include in the transcript. In the case of the southwest Scottish variety of English that he studied in this article, Macaulay says he "wanted the transcription to be as readable as possible while providing some information on how the speakers sounded" (286). A readable transcript of speech, according to Macaulay, "should not slow the reader's eye, unless there is a particular purpose for doing so that is relevant to the transcribed passage" (289). Finally, "[t]he aim of any transcription," according to Macaulay, "is to make the reader's task as simple as possible." Unlike the linguist, the captioner is not trying to produce a phonetic transcription of speech. Like the linguist, the captioner is ideally driven by the *purpose* of the narrative, text, or scene being captioned when making decisions about how or whether to indicate manner of speaking. The captioner is committed to making captions *as readable as possible*. And the captioner is concerned with ensuring that captions do "not slow the reader's eye." No wonder that speech captions, regardless of the manner in which that speech is pronounced, reflect conventions of standard written English, which are simply more accessible to literate readers than a collection of phonetic, unconventional spellings.

Speech captions are readable and accessible to the extent that they reflect standard or conventional forms, especially well-worn linguistic conventions of spelling, punctuation, grammar, syntax, and so forth. Captions that approximate or reflect standard forms will be more accessible and easier to read than captions that endeavor to display phonetically or typographically how that speech sounds. Nonstandard or unconventional spellings are rarely used in closed captioning. But standard written English tends to squeeze out so much of the linguistic variation that distinguishes one speaker from another: age, gender, region, pitch, volume, hesitation, intonation, timbre, reverberation, speed, and so on. In "Sounding Cajun: The Rhetorical Use of Dialect in Speech and Writing," Sylvie Dubois and Barbara H. Horvath (2002, 264) suggest that writing homogenizes speech:

People can often use their conscious or unconscious knowledge of dialectal variation to achieve some rhetorical effect: friendliness, humor, earthiness, honesty, nostalgia, and a host of other possibilities. But in writing, standardization imposes a special problem for using linguistic variation rhetorically. Written languages homogenize much of the linguistic variation that identifies a speaker's background, and if writers want readers to know a narrator's or a character's social and geographic background, they either have to state it explicitly or break the rules—primarily, but certainly not exclusively, the spelling rules.

Violating rigid spelling rules rarely happens in closed-captioned speech, with the possible exception of captioned music videos, which may be only slightly more likely to invoke lyrical spellings (such as *murda biz-ness* for murder business in Iggy Azalea's 2014 video for "Fancy"). Even informal forms of speech (*'cause, gonna*) follow well-known conventions when written down. As a result, writing tends to homogenize speech. Every speaker tends to "sound" the same in standard written English—flattened and disembodied—unless the transcriber "breaks the rules." But breaking the rules can slow readers down, making texts less accessible by introducing unfamiliar spellings or new variants that take longer to process. Within this context, the nonspeech manner identifier makes sense as a solution to the problem of inscribing in writing *how* words are pronounced. To get the most out of the manner identifier, caption readers must rely on visual clues about or around the speaker to embody the flattened, homogenized writing style of standard speech captions.

Informal but Conventional Forms

The title of Ronald Macaulay's (1991) article about entextualizing dialect in writing, "Coz it izny spelt when they say it," would make a poor speech caption. We could improve access to it, while maintaining its informal feel, by replacing *Coz* with *'Cause* and *izny* with *isn't*. We could also add a manner identifier for dialect if we thought our more accessible alternative did not sufficiently capture the quality of the speech. These trade-offs are sometimes necessary because a purely phonetic rendering of speech compromises its accessibility. What's important is not whether the captioned writing is formal but whether it is conventional. Readers can quickly interpret the meaning of informal forms if they are presented in conventionally accepted ways. If someone says something that sounds like *coz* or *cuz* (and they mean the shortened form of *because*), the best option, at least when captioning for American readers, may be to caption that sound as *'cause* (or even as *because*). While *coz* is an accepted variant spelling for the same conjunction, *'cause* is the standard form in the United States. Some examples of informal forms from DVD closed captioning:

[laughs] That's 'cause you knows what I like.—*Django Unchained* (2012)

So how 'bout no bags this time—*Django Unchained* (2012)

'Cause I know I'm not gonna like the guy.—*Kevin Hart: Let Me Explain* (2013)

(IN SQUEAKY VOICE) We're all gonna die!—*Ice Age 3* (2009)

[Pat] How you gonna pay for it?—*Silver Linings Playbook* (2012)

And it's not like you're not tryin'.—*Lost in Translation* (2003)

[clears throat] But this thing's gotta move, all right?—*Killing Them Softly* (2012)
You don't think he sounds kinda goofy?—*Lost in Translation* (2003)
Y'all better learn how to survive now.—*Beasts of the Southern Wild* (2012)
Get outta here.—*The Faculty* (1998)

These colloquial forms need not create comprehension problems for most readers. While *'cause, gonna,* and *tryin'* may not be formal, they are conventional. But they do come with political and cultural baggage. Speech that is presented as colloquial or rustic may also, by implication, reflect poorly on the speaker, who may be seen as less intelligent (see Bucholtz 2007, 801; Dubois and Horvath 2002, 265; Lindemann 2005). These implications may be warranted in the context or by the script. Still, care must be taken with them. Colloquial and informal forms may serve as unwarranted code for a speaker's education, class, or race. **Media:** http://ReadingSounds.net/chapter8/#nonstandard.

When an Associated Press (AP) reporter dropped some of the final g's from his transcript of a 2011 speech by President Obama, the potential racial implications became a subject of debate in the news media. Speaking to the Congressional Black Caucus, Obama told the audience, according to the official White House blog (Lindsay 2011):

Take off your bedroom slippers, put on your marching shoes. Shake it off. Stop complaining, stop grumbling, stop crying. We are going to press on. We've got work to do, CBC.

But AP reporter Mark S. Smith (2011) offered this transcript, made lighter with fewer g's:

"Take off your bedroom slippers. Put on your marching shoes," he said, his voice rising as applause and cheers mounted. "Shake it off. Stop complainin'. Stop grumblin'. Stop cryin'. We are going to press on. We have work to do."

To the charge that Smith (2011) was being racist, using coded language to signal black dialect (see Christopher 2011), conservatives countered that Smith was only aiming for an accurate transcript, one that showed the President adopting a black dialect for the sake of "pandering to a black base" (Stableford 2011). Regardless, no one seemed to deny that g-dropping has racial implications. At issue was whether it was justified to attribute dialect and nonstandard speech to a US president in a news report. The style guides for the AP and the *New York Times* stress the importance of scrubbing dialects from transcripts:

The Associated Press Stylebook says, "Do not use substandard spellings such as *gonna* or *wanna* in attempts to convey regional dialects or informal pronunciations, except to help a desired touch or to convey an emphasis by the speaker." *The New York Times Manual of Style and Usage* is even stronger, saying that the results of using such dialect "are likely to be subjective (since they depend on the home region of speaker and listener) and will strike at least some readers as patronizing. After all, national leaders and corporate executives have been known to say 'gonna' and 'hadda' and perhaps 'y'all' " (Perlman 2011).

Closed captioning is not beholden to the conventions of news reporting. If someone says *gonna* or *'bout* in *Django Unchained* (2012), for example, then it may be warranted (script and context permitting) to caption the speech as accurately or phonetically as possible, so long as any nonstandard spellings are widely accessible or conventional. Indeed, Quentin Tarantino's (n.d.) script for this spaghetti western and slave-era narrative set on the eve of the Civil War is littered with evidence of dialect: *'bout, tryin', gonna, coulda, ain't, y'all, 'cause,* and so on. As it deals explicitly with race and class relations in antebellum America, *Django Unchained,* repeatedly appeared in the results as I searched my collection of official DVD caption files for colloquial speech. But closed captioning isn't just about representing regional or ethnic dialects as accurately as possible. Captioning has an important role to play in all kinds of discourse, including news reporting. How captioners should entextualize a speaker's dialect (including whether to scrub that dialect) may not always be straightforward. Just because someone says *gonna* doesn't automatically mean that's how the word should be captioned. It may be more appropriate to caption *gonna* as *going to* and *'cause* as *because.* The same goes for hesitations and pauses like *um* and *uh,* which are typically removed when the aim is to create a more formal transcript. At the least, captioners should be mindful of the associations some readers may make between nonstandard informal English and a speaker's race, class, and/or intelligence. While the transcript of a president's speech is formalized in major news articles and, presumably, in the official closed captions that accompany the speech, the same formal approach would not necessarily hold for other speakers in less formal occasions. For example, *Kevin Hart: Let Me Explain* (2013), a movie featuring the stand-up comedy of Hart, an African American, is rife with nonstandard variants. Out of 1,409 total captions on the DVD, thirty-three (2 percent) contain *'cause* and eighty-four (6 percent) contain *gonna.* Hart's informal style and the occasion are reflected appropriately in the captions and without any of the controversy about racial code words from news media pundits that accompanied the AP's treatment of President Obama's speech. **Media:** http://ReadingSounds.net/chapter8/#obama-hart.

The alternative to nonstandard but conventional forms is to spell out dialects phonetically, using unconventional spellings or neologisms, which is rare practice in captioning but sometimes deemed necessary. The challenge for captioners is to ensure that every nonstandard spelling or neologism is still readily accessible to readers who are working under time constraints. Readers are always more interested in what's happening away from the caption space than trying to decode neologisms or words that are spelled funny. Madeline Kahn's faux-German accent, played for laughs in *Blazing Saddles* (1974), is immediately accessible to readers when presented as a series of neologisms in which the letter R in select keywords is replaced with the letter W and sometimes V: DWESSING WOOM, FAW OUT, WED WOSE, WOMANTIC, SHEWIFF, REFWESHED, BWIGHT, TWU, and so on. Kahn's rhotacistic speech pattern is significant—indeed, it's the first thing commentators mention about her role in this classic comedy. The captions are simple but effective. Phonetically accurate they are not, however. Kahn's spoken accent continues even in the absence of clues in some of Kahn's speech captions: WON'T YOU EXCUSE ME FOR A MOMENT / WHILE I SLIP INTO SOMETHING A / LITTLE BIT MORE COMFORTABLE? In other words, the captions are contextual rather than narrowly concerned with accuracy. Because readers are given a sufficient number of reminders at key moments, the accent continues to resonate for us. **Media:** http://ReadingSounds.net/chapter8/#blazingsaddles.

Another example of rhotacism (which is defined as mispronouncing Rs or substituting other letters, especially W-sounds, for Rs and Ls) comes from *The Big Bang Theory*. Barry Kripke (John Ross Bowie), a minor but recurring character and foil for Sheldon (Jim Parsons), has a speech impediment. In fact, Barry's rhotacism, like Kahn's, is the first thing that commentators and wikis mention about his character, after his job title and meanness. Barry Kripke can't say his own name without invoking his identity and reducing it to a speech pattern. Number one on John Ross Bowie's (2012) "Answers to Kripke's FAQs" is whether the impediment is real: "1—No, I don't talk that way in real life." Number three is whether the impediment is entextualized in the script: "3—Good question. No, the impediment is not written into the script, I transpose it on my own." In other words, any clues in the captions about Barry's rhotacism must come from the captioners themselves. In "The Beta Test Initiation" (2012), Barry's impediment is absent from his own speech captions, despite being his defining trait on the show. When he demonstrates that the voice recognition on his iPhone ("Siri") doesn't work, his speech is transcribed into standard English captions: "Voice recognition on that thing is terrible. Look. Siri, can you recommend a restaurant?" It is only in Siri's responses that his rhotacism becomes visible in the captions: "I'm sorry, Bawwy. I don't understand 'wecommend a westau-want.'" Barry seems unaware that the problem is his own voice, which might explain why his speech captions don't imitate his impediment—his speech is presented from his point of view. But this explanation is insufficient because it cuts caption readers off from Barry's rhotacistic speech except in the rare instance when it is a subject of conversation or is parroted back to him by others. Barry's captioned rhotacism was brought to my attention by Stephanie Kerschbaum, a disability studies scholar and regular viewer of *The Big Bang Theory* who became aware of Barry's speech impediment in this episode of season 5, because this was the first time she had seen Barry's speech spelled phonetically. Barry's interactions with Siri cannot be explained without resorting to phonetic spellings. The question is whether Barry's earlier interactions on the show, which always play his impediment for laughs, should also have been captioned phonetically. On the one hand, it might quickly become tedious and taxing for viewers if every R-sound and L-sound in Barry's speech were replaced with Ws. On the other hand, Barry's speech pattern should not come as a surprise to caption readers in the fifth year of the show. Right after Barry speaks to Siri for the first time, you can see the awareness on the faces of Sheldon and Raj (Kunal Nayyar). They know and have always known, but they have never made fun of Barry's speech, which is why his pronunciations can surprise longtime viewers of the show who depend on captions. Caption readers should have the same knowledge and awareness as the characters and the listening audience to aid their interpretation of this scene. Barry's distinctive speech patterns shouldn't be sprung on caption readers in the fifth year of a series. Because *rhotacistic* is a technical, less accessible term, it may not work as a manner identifier in *The Big Bang Theory*. The *Blazing Saddles* approach, which relies on replacing some Rs with Ws, could be an effective, phonetic alternative that doesn't overburden readers when limited to short bursts. **Media:** http://ReadingSounds.net/chapter8/#bigbangtheory.

This example from *The Big Bang Theory* suggests that manner of speaking may only become visible in speech captions when it can't be avoided, such as a joke that depends on the audience having access to a speaker's distinctive pronunciation patterns. At all other times, standard English serves as the norm in speech captioning. Apart from this scene, Barry's speech in other episodes looks no different than the speech of every other character on *The Big Bang Theory*. When speech is captioned phonetically, then, we take notice. It stands out. *Star Wars: Episode 1—The Phantom Menace* (1999) is chock full of ethnic characters, racial stereotypes, and accents: the British-sounding Jedi (namely Ewan McGregor as Obi-Wan), the East Asian-inspired Neimoidian of the Trade Federation, a number of droid/robot voices, the Emperor's deep and gravelly voice, the Caribbean-sounding Jar Jar Binks, and the Jewish merchant named Watto. All of these accents (and more) are presented in the first eleven minutes of the film. In a withering critique of the film in *The Village Voice*, J. Hoberman (1999) describes the Neimoidian as "fish faces [that] talk like Fu Manchu," the "Third World" Jar Jar as "a rabbit-eared ambulatory lizard whose pidgin English degenerates from pseudo-Caribbean patois to Teletubby gurgle," and Watto as "the hook-nosed merchant" who is "the most blatant ethnic stereotype." But only Jar Jar's dialect is marked in the closed captions (unless we count Yoda's famously inverted grammar and one or two unusual English words in Watto's captioned speech, such as "thee"). Every other character is represented in the captions in neutral, standard English. By contrast, Jar Jar's speech captions reflect a colorful, ethnic patois through phonetic spellings and nonstandard grammar. Here are just three examples among many in the DVD captions:

Mesa culled Jar Jar Binks.
Mesa your humble servant.

Ex-squeeze-me, but de mostest
safest place would be Gunga City.

My forgotten.
Da bosses would do terrible tings to me.

Granted, Jar Jar's accent is the most blatant and perhaps the easiest to caption phonetically. But for caption readers, Jar Jar's speech is the only one that is specially marked as nonstandard, even though the film strives to be a galactic melting pot in which everyone has a noticeable accent. I'm not suggesting that captioners should identify speech as British or East Asian in a movie that doesn't take place on Earth. That wouldn't make sense. I am simply pointing out that the captioned landscape does not routinely provide access to information about dialects and, when it does, the results may be uneven, privileging the dialect of one character (Jar Jar) or appearing in only one scene (Barry). **Media:** http://ReadingSounds.net/chapter8/#starwars.

When phonetic information is needed but missing in the captions, the oversight can be significant. In the "Margaritaville" episode of *South Park* (2009), Stan (voiced by Trey Parker) attempts to return a fancy drink mixer to Sur La Table (a store specializing in cookware and kitchen products). In the scene's running gag, the store employee not only pronounces the store's name with a French accent as expected (table is pronounced *tah-bluh*), but every English word with an L-sound ending is treated as French too: cable is pronounced *cah-bluh*, doable is *do-ah-bluh*, able is *ah-bluh*, improbable is *improba-bluh*, and so on. But the captions play it straight, in standard English, and as a result cut caption readers off from the only gag in the scene. Another example is from the animated series *Family Guy*. In the episode "Dog Gone" (2009), Lois Griffin (voiced by Alex Borstein) encourages Brian the dog to join PETA, the animal advocacy organization. But in her New York accent, Lois pronounces "PETA" and "Peter" interchangeably (as PEE-tah), leading to some confusion on Peter's part that resembles an Abbott and Costello "Who's on First?" routine. Lois tries to explain: "PETA is an acronym, Peter." To which Peter (voiced by Seth MacFarlane) responds, "No, I'm not. I'm Catholic." But the captions never make it clear that Lois's pronunciation is the same for both PETA and Peter. Some caption readers will be savvy enough to interpret the joke based on the context, but they do so in spite of the captions. The final joke in the PETA/Peter routine—captioned as, "I think Betty White is in PETA"—makes me think of caption readers when Peter responds, "That doesn't even make any sense!" Readers who rely on captions are asked to share in Peter's confusion rather than laugh at Peter. Lois's accent, like Barry's rhotacistic speech in *The Big Bang Theory*, becomes crucially important for only one scene. Captioners need to be attuned to the changing value of a character's accent so that when it becomes significant, it can be captioned with phonetic sensitivity. **Media:** http://ReadingSounds.net/chapter8/#missingphonetics.

The Limits of Onomatopoeia

Phonetic spellings are tricky. What readers gain in knowledge of how sounds are pronounced may come at the expense of understanding what those sounds mean. Onomatopoeia should be used only when the context provides a clear rationale for its use. Without the support of a clear visual or rhetorical context, neologisms like *chkty* are less accessible and have greater potential to be ambiguous and confusing to readers. (I single out neologisms—made-up words—but recognize that onomatopoeia

can be used effectively with conventional words that imitate sounds—
e.g., splat, hum, buzz, roar, etc.) For example, the first sentence of Bruce R.
Smith's (2004, 127) chapter on sounds in seventeenth-century London
includes three onomatopoetic neologisms:

> The *chkty-chkty-chkty* spilling out of someone else's headphones, the *yeow-yeow* clearing the way for ambulances and police cars, the *bllliiiiii* heralding the banalities of a stranger's one-sided conversation on a mobile phone—these serve as keen reminders that most of us live immersed in a world of sound.

While *chkty*, *yeow*, and *bllliiiiii* might work well for scholars of sound
studies and for the percentage of Smith's inscribed audience who can
hear (i.e., "most of us"), they are problematic for the rest of the population who require efficient and clear closed captions. These examples
of onomatopoeia could probably not stand on their own in the closed
captions without some help from more conventional nonspeech descriptions or a familiar visual context. When Superman (Henry Cavill)
screams in *Man of Steel* (2013), his expression vertically fills the screen,
so that even when his vocalization is captioned unconventionally and
onomatopoetically as "Ragh!," the meaning is nonetheless clear from
the visual context. While the *Captioning Key* style guide recommends
that captioners combine onomatopoeia with description "[i]f the presentation rate permits" (DCMP 2011b), citing a study of viewer preferences from Gallaudet University (Harkins et al. 1995, 18), there's not
usually enough caption space or reading time available to implement
this recommendation as a general principle. The potential for confusion is also too high to warrant such an approach. Onomatopoeia can
cut caption readers off from a sound's meaning and purpose. Consider
example 3 on the *Captioning Key's* "sound effects" page (DCMP 2011b),
in which a motorboat sound is captioned as both [engine idling] and
[rrrrr]. The former tethers the sound to a meaningful visual context
while the latter floats free without the former to anchor it. The visual
context may help us determine that [rrrrr] refers to the engine sound,
but we still won't know for sure what [rrrrr] means without the assistance of [engine idling]. Is the engine in distress (like a growl sound) or
merely idling? Another example from this same source combines [ma-
chine gun firing] with "rat-a-tat-tat" into a single caption. But what the
Captioning Key leaves out of its style guide recommendation is that the
Gallaudet study also states clearly that "[o]nomatopoeia should not be
used alone" (Harkins et al. 1995, 18). I would go further and argue that it
should rarely be used in closed captioning intended for adult viewers. In

8.2   **Backchannel particles such as "mmm-mmm" and "mm-hmm" are easy to confuse when captioned and read quickly.**
This frame from *Orange is the New Black* features an over-the-shoulder shot of Alex Vause (Laura Prepon) speaking to Piper Chapman (Taylor Schilling). Both Alex's statement and Piper's response occupy the same caption: "Well, you're not in the SHU." "Mmm-mmm." Netflix, 2013. http://ReadingSounds.net/chapter8/#figure2.

situations that demand quick uptake by readers and require captioners to squeeze meaning into a very cramped space, function should trump imitation. Explaining the meaning or purpose of the sound—[doorbell rings]—is usually more effective and efficient than trying to replicate the sound phonetically—[ding dong]. Granted, [ding dong] is less ambiguous than [rrrrr], but both lack the fuller clarity that comes from a description of the sound's purpose. **Media:** http://ReadingSounds.net/chapter8/#onomatopoeia.

Another example of a potentially ambiguous phonetic transcription is "Mmm-mmm," which appears in the captions for one episode of *Orange is the New Black*, a Netflix original series, as a closed-mouth variant of "uh-uh," which is itself a variant of "no." In "Tall Men with Feelings" (2013), Piper's (Taylor Schilling) response to Alex's (Laura Prepon) statement, "Well, you're not in the SHU," is captioned as "Mmm-mmm" (see figure 8.2). While the context helps us interpret "Mmm-mmm" (because we can see clearly by Piper's presence that she's not in the solitary confinement wing known as the SHU), the caption remains unnecessarily ambiguous, despite being a good reflection of how Piper's closed-mouth vocalization *sounds*. A series of Ms in print is conventionally associated not with something negative but with pleasure ("mmmmm, that's good pie" or the Campbell's Soup motto: "Mmm Mmm Good") or sometimes hesitation ("mmmmm, I'm not sure," which is likely a variation of "hmm"). "Mmm-mmm" does not mean *no* to readers even though it might mean *no* to listeners when vocalized appropriately. A print alternative to "mmm-mmm" is "uh-uh," which has the advantage, unlike "mmm-mmm," of being included in the dictionary (as is "uh-huh" for *yes*). Another particle, "mmm-hmm" for *yes*, is also common and appears in the captions for another episode of *Orange is the New Black* ("Can't Fix Crazy"). Backchannel particles are complex and comprise a number of closely related variants (see Kjellmer 2009). It's especially easy for readers to confuse them when reading captions quickly (e.g., compare "mmm-mmm" with "mmm-hmm"). In "Sheesh Cab, Bob," a 2011 episode of the animated series *Bob's Burgers*, Bob (voiced by H. Jon Benjamin) responds to his daughter Tina (voiced by Dan Mintz) with a variety of *mmm* sounds/captions that have different meanings: mm-hmm (yes), mm (listening or not sure), mmm-mm (no), hmm (listening or not sure). Bob shakes his head to provide a visual clue to accompany "mm" and give it specific meaning, but the interaction happens so quickly that it's easy to get the *mmm* captions mixed up. At any rate, my point is that given the ambiguity of "mmm-mmm" and the lack of agreement about its meaning as a variant for *no*, captioners need to put meaning ahead of vague phonetic imitation. In the example from *Orange is the New Black*, we need to know above all else that Piper vocalizes *no*, even if that means the captioner has to sacrifice some of the phonetic quality of Piper's answer. Captions are for readers. What works for readers may not perfectly capture what listeners hear, but such a trade-off is sometimes necessary. Likewise, what listeners hear may need to be adjusted in the captions to make sure that meaning is conveyed in ways that are conventionally understood.

A final example of onomatopoeia comes to me via email from Stephanie Kerschbaum. On early seasons of *Sons of Anarchy* on DVD, the spoken abbreviation for San Joaquin is captioned as "Sanwa," a phonetic translation that, like [rrrrr], is highly ambiguous, the result of a misguided attempt to convey what the sound *sounds like* while disregarding what the sound means. Stephanie explained to me that she figured out what "Sanwa" means in season 4 when the same abbreviated pronunciation began to be captioned (correctly) as "San Joa." Unless the way a word sounds (e.g., "Sanwa") plays a significant role in a scene (e.g., wordplay), the meaning of the sound—what it is doing in a scene—should always take precedence over an attempt to spell it like it sounds, even if that means cutting viewers off from phonetic translations. When imitation takes precedence over meaning—as in [rrrrr], "Mmm-mmm," and "Sanwa"—captions can confuse readers even as they strive to describe sounds objectively.

Language Ideology

When the sounds to be captioned are not nonsense or ambiguous sounds like "mmm-mmm" or [rrrrr] but foreign-language words, they need to be captioned verbatim, even if they read like nonsense to readers (like me) who may only know English. In other words, English-language closed captions should not be limited to English speech sounds. Captioners should make every effort to convey the *meaning* of the text regardless of the language in which that meaning is expressed, even if readers can't be assumed to understand the foreign-language speech sounds. A line of Latin or a greeting in Spanish ("Hola") should not be captioned only as (READING IN LATIN) or [BOTH SPEAK IN SPANISH]. There may be exceptions to this rule—context always matters—but captioners should be mindful of the differences, which are sometimes significant, between a parenthetical reference to the foreign language being spoken and the specific words spoken in that language.

When there are extended stretches of foreign speech in a movie, open subtitles are used to translate the speech into English for all viewers in English-speaking countries. For example, everyone sees the English subtitles for the German and French speech in *Inglourious Basterds* (2009), a Quentin Tarantino film that presents a fictionalized account of the Nazi occupation of France during World War II. The subtitles can't be turned off because they are burned onto the surface of the film. In a separate track, the closed captions provide information about which foreign language is being spoken: for example, (SPEAKING GERMAN). The closed captioner isn't responsible for translating the foreign-language speech or placing it on the screen. The open subtitles and the closed captions work in concert to provide a full account of the speech sounds. A total of 150 nonspeech captions in *Inglourious Basterds* are devoted to identifying German speech, from the straightforward (SPEAKING GERMAN), which occurs most often at 112 times, to the more colorful and descriptive: (THANKING IN GERMAN), (TRANSLATES IN GERMAN), (ORDERING IN GERMAN AND FRENCH), (ALL TOASTING IN GERMAN), and (USHER CONTINUES ANNOUNCING IN GERMAN). Language identifiers are typically placed at the top of the screen to leave room at the bottom of the screen for the open-subtitled English translations. In one scene from this movie, seven language IDs are used in a twenty-three-second stretch, one for each speaking turn. Language IDs follow speaking turns in *Inglourious Basterds*, not speakers or captions. The same speaker may be associated with more than one identical language ID in the same scene if her speech takes up multiple turns, one ID for each of her turns. If the same turn requires multiple subtitles, only one language ID at the start of the turn is required (unless the speaker starts speaking a different language in the middle of her turn). In marking every turn with a language ID, even short turns of one or two words, captioners need to be mindful of the cognitive burden this places on readers to read two streams at once—the nonspeech language identifier at the top of the screen and the subtitled speech at the bottom of the screen. In the featured clip from the film, three officers order whiskey in German, which is subtitled at the bottom of the screen: "Whiskey." "Two whiskeys." "Three whiskeys." Each turn is accompanied by (SPEAKING GERMAN) at the top of screen (using placement to position each language ID over the speaker) for a total of eleven words over 3.17 seconds (i.e., five subtitled words and six words in the language IDs). That's a fast reading speed of 208 words per minute. But this number doesn't account for the visual gymnastics required of the reader, who must repeatedly look to the top and then to the bottom of the screen in quick succession: top, bottom, top, bottom, top, bottom. Discussions of reading speed (see chapter 5) do not account for situations in which readers are expected to read two visually separated and distinct streams at once, closed captions and subtitles. Of course, the cognitive load is not quite so taxing if we can assume that the reader can see that these officers are German (and will be ordering in German), thus making the language IDs less important in this scene. Indeed, when coupled with verbatim subtitles, language IDs will almost always be subservient to verbatim subtitles in a scene, even if captions set in all caps will seem to scream out their importance over the more subtly and beautifully rendered subtitles. If we concentrate on reading the subtitles at the bottom of the screen, we aren't disadvantaged if we skip the language IDs during the whiskey scene (or assume that the language IDs are identical). When our expectations about language type can't be tied to the visual rhetoric of a scene, the relative importance of the language IDs increases, as when Diane Kruger addresses the bartender in French as "mon chéri" before ordering in German. Her English subtitles are accompanied by the caption (SPEAKING FRENCH AND GERMAN). But while this language ID tells readers that two languages are spoken by the same speaker in the same utterance, it does not tell us which words are French and which are German. **Media:** http://ReadingSounds.net/chapter8/#inglourious-langIDs.

When foreign speech is used in short bursts, or in the absence of robust subtitles, captioners need to ensure that language IDs do not cut caption readers off from the meaning of the foreign speech. Too often, short phrases or single words in a foreign language are identified only with language IDs, even if the foreign speech is common to some English speakers, such as a familiar non-English greeting (e.g., "Bonjour"). But even if the speech is not common to most caption readers, such as an incantation in Latin, it still needs to be presented in the original spoken language. It doesn't need to be translated for readers, because captions aren't annotations. But the speech does need to be presented as it is spoken, even if some readers will not know what the foreign words mean. As with other aspects of captioning such as sonic allusions (chapter 7), the captioner is an ideal reader of the text who is responsible not only for determining which language is being spoken but also transcribing that speech in the original language. In fact, as we will see, the transcription of the foreign language may be more important than a vague reference to "Spanish," "Latin," or "Sanskrit." When captions trade foreign transcriptions such as "mazel tov" for language IDs such as [SPEAKS IN HEBREW], they subtly imply that only English speech matters, a form of linguistic imperialism. They also separate caption readers from meanings that are tied to specific words, regardless of the language spoken or whether readers understand those words. Even if I don't understand what someone is saying in Spanish, I want to have access to the *sound* of that language, to the music, rhythm, and number of words spoken. Sometimes, with romance languages in particular, I can pick out a word or two based on their root meanings in English. A language ID without accompanying verbatim foreign speech prevents caption readers from accessing the specific music and meaning of foreign speech. At worst, language IDs can imply that the speakers in these films and TV shows actually have a working understanding of the foreign language, when the truth may be that the speakers can do little more than parrot a greeting in that language (which may be the entire point of an exchange).

Inglourious Basterds (2009) makes use of robust English-language subtitles for German and French speech . . . but not always. In a number of instances, single foreign words and phrases are described only with language IDs instead of subtitled verbatim. Just because foreign speech is not subtitled doesn't release the captioner from the responsibility of captioning foreign speech verbatim. The examples in table 8.2 rely on

Table 8.2 Examples of unsubtitled foreign speech in *Inglourious Basterds*.

Timestamp	Speech (not subtitled)	Caption
0:10:52	Oui.	(AGREES IN FRENCH)
0:13:26	Merci.	(THANKING IN FRENCH)
0:15:25	Oui.	(AGREES IN FRENCH)
0:46:37	Au revoir.	(SAYING GOODBYE IN FRENCH)
1:54:59	Arrivederci.	(ALL SAYING GOODBYE IN ITALIAN)
2:00:21	Voila.	(speaking french)

Note: This speech is only identified in the closed captions with a language identifier and a description of the speech act. Universal Pictures, 2009. DVD. http://Reading Sounds.net/chapter8/#table2.

common expressions in French and Italian, so common that subtitles were considered unnecessary. While (AGREES IN FRENCH) is roughly equivalent to "Oui," and "Merci" is a way of (THANKING IN FRENCH), a description of the French speech fails to capture the precise nature of either the agreement or the thankfulness. Such descriptions are vaguer than the words spoken. A caption reader could mistakenly assume that (AGREES IN FRENCH) means that the speaker said "I agree" ("d'accord"), which isn't the same as "oui." Wouldn't it have been more efficient and precise to caption the French words in the French original (e.g., "oui")? The producers likely didn't subtitle these common expressions because they expected listeners to have a working knowledge of some basic foreign terms like "voila," "oui," "merci," and "arrivederci." For the same reason, these foreign terms, "merci" in particular, show up occasionally and intentionally in the English subtitles (IMDb 2009a). Captioners need to present foreign terms to readers in the same way they are presented to hearing viewers. If the hearing viewer has access to "oui" (unsubtitled), the caption reader should have access to the word "oui," not some vague description such as (AGREES IN FRENCH). There's a fine line between being eager to explain the soundtrack to readers and being patronizing. If hearing viewers are assumed in *Inglourious Basterds* to be knowledgeable enough to translate "oui" for themselves, then shouldn't the same be assumed of caption readers? A captioned explanation of a common French expression (as opposed to a direct transcription) may be mistakenly taken as a sign that caption readers aren't considered knowledgeable enough to translate the original for themselves but must rely on a paternalistic captioner.

Explanations of foreign terms may also leave out the manner in which these terms are pronounced. What is missing in (ALL SAYING

GOODBYE IN ITALIAN) is the hilarious way in which Lt. Aldo Raine (Brad Pitt), the American who claims to "speak the most Italian" in his group of American soldiers in *Inglourious Basterds*, pronounces "arrivederci" with a thick Southern US accent. Hans Landa (Christoph Waltz), the Nazi officer with the reputation for sniffing out a ruse and a surprising ability to speak Italian fluently, delights in playfully taunting these fake Italians. Raine's awkwardly pronounced "arrivederci"—the final word in this scene—leaves us with the lasting impression that the Americans have been outsmarted. Yet none of this comes through in a generic description such as (ALL SAYING GOODBYE IN ITALIAN), which may in fact reinforce the opposite impression—that these Americans can actually speak Italian fluently. Instead of emphasizing the Italian, the captioner might have emphasized the American accent: "(with Southern accent) Arrivederci." This same approach could have been used at the opening of the scene, when Raine bids Landa "buongiorno" with the same heavy accent. This time, Raine's greeting is subtitled as "Hello" and captioned simultaneously as (SPEAKING ITALIAN). But what's missing from the straightforward declaration that Raine is speaking Italian is the manner in which he pronounces "buongiorno." In other words, there is a huge difference between Landa's fluent and purposefully fast Italian and Raine's stilted, heavily accented, one-word responses in Italian, even though both are captioned identically as (SPEAKING ITALIAN). The challenge for captioners is to incorporate information about manner of speaking, when it's significant, without adding an even heavier burden on readers to keep up with multiple streams of information in subtitled films. We might begin to ease this burden, particularly in cases where English subtitles are not used, by replacing vague and sometimes lengthy descriptions such as (SAYING GOODBYE IN FRENCH) with shorter and more precise verbatim transcriptions such as "Au revoir." When subtitles are used, captioners are limited in how much additional information they can provide in the second track (as in the "Hello" example). Still, we should be mindful of the implications even a straightforward language ID can make. Even though the speaker may be (SPEAKING ITALIAN), it may be more important to know that he is speaking Italian badly, comically, or with a thick foreign accent. A simple declaration that someone is (SPEAKING ITALIAN) may suggest much more about the speaker's facility with that language than is warranted.

We can see how language IDs may create unwarranted implications when we turn from heavily subtitled films like *Inglourious Basterds* to captioned English-language films and TV shows that incorporate non-English terms or expressions, such as "mazel tov" and "hola." Non-English speech should usually be captioned verbatim in English language programming. In an episode of *Witches of East End* (2013) that happens to be playing on the television as I work on this section, the witches' Latin spells are captioned verbatim, as they should be: "Mysticum flamma aperire pictura." Most likely, the Latin spells were spelled out in the script, making the captioners' work straightforward. But whether or not the foreign speech is written down in the captioners' materials, it still needs to be accounted for. Additional examples of foreign speech that is correctly closed captioned verbatim in English-language programming:

- *Family Guy*, "We Love you, Conrad" (2009). The suave and talented boyfriend of Jillian (voiced by Drew Barrymore) orders in French at a French restaurant. His words are captioned verbatim in French in two captions: "Je voudrais le petit chèvre fondue à la provençale / et le gigot d'agneau bourguignon, s'il vous plaît." Meanwhile, Brian the dog (voiced by Seth MacFarlane), Jillian's former boyfriend, orders awkwardly in English, pointing to the menu: "Yeah, can I have, uh, this third thing down please?" English subtitles are neither used nor required in this scene. Even if we don't understand French, we can appreciate, and need access to, the specific French words. That's the point. Not the meaning of the French food order but the perfect new boyfriend's effortlessness at delivering a full line of foreign dialogue.
- A Taco Bell commercial (2013). This thirty-second English-language TV ad—one of a handful of ads in Taco Bell's "Live Mas" campaign—makes use of a Spanish-language cover of Notorious B.I.G.'s "Big Poppa." The lyrics are not subtitled in English, so hearing viewers who have the captions turned off will only hear Spanish lyrics being sung. The producers reportedly ran B.I.G.'s lyrics through Google Translate (Kiefaber 2013), resulting in an imperfect translation that is both grammatically questionable and also too long to fit the melody. While the song title is not shared with caption readers, the Spanish lyrics are captioned verbatim, including the repeated chorus: "♪ Me encanta cuando me llamas grande papi . . . ♪." Some viewers may be familiar with the original English source (hence the importance of the song title): "I love it when you call me Big Pop-pa." But including the verbatim Spanish lyrics in the captions is a good start.
- *Supernatural*, "Rock and a Hard Place" (2013). Dean (Jensen Ackles) meets a former adult film star named Suzy (Susie Abromeit) who had once starred in some Spanish-language titles. Dean recognizes her. They flirt. Mariachi-style music plays in the background, uncaptioned. She calls him a "BAD BOY," and he responds, "WHY DON'T YOU ASK ME THAT IN SPANISH?" She complies, and her translation is captioned in verbatim Spanish: "¿ERES UN CHICO MALO?" I don't speak or read Spanish, but I get the gist of her question. Even if I had no idea what "chico" or "malo" means, I

wouldn't expect the captioner to translate the Spanish speech into English. In the absence of English subtitles, the closed captioner's responsibility starts with a clear representation of the original foreign speech and ends with a consideration of both the manner in which those foreign words are uttered (e.g., quickly, loudly, etc.) and their larger meaning (e.g., artist name and song title).

In these examples—and others we could just as easily include if we had unlimited space—non-English speech is not a barrier to verbatim closed captioning. Just because the spoken words are neither uttered in English nor subtitled in English doesn't exempt the captioner from representing them faithfully in the original language. **Media:** http://ReadingSounds .net/chapter8/#verbatimforeignspeech.

Too often, when short snippets of foreign speech aren't already subtitled in English-language programming, captions can only be counted on to offer a summary of the foreign speech such as (SPEAKING FRENCH). In all of the examples in table 8.3, a language ID takes the place of an uncaptioned and unsubtitled foreign term or expression. Stephanie Kerschbaum brought to my attention the first two examples (from *The Office* and *The Big Bang Theory*) when she shared them on Facebook. I've grouped her examples with others I have been collecting. Every one of these captions misses the point of the foreign speech by describing the language being spoken instead of captioning the verbatim foreign speech itself. In part 1 of *The Office's* two-part "Stress Relief" episode, Michael Scott (Steve Carell) doesn't know Sanskrit when he tries to lead his employees through a relaxation exercise following Stanley's (Leslie David Baker) heart attack. Michael is a mostly clueless buffoon whose only exposure to the Sanskrit mystical syllable "om" (also spelled "aum" in English) came most likely through popular culture and not through some sustained study and practice of meditation. He doesn't know that "om" is Sanskrit and he doesn't practice meditation. By captioning his vulgar appropriation of "om" as (CHANTING IN SANSKRIT), the captioner implies much more about Michael Scott's knowledge than is warranted. Michael Scott becomes a bit less clueless and a bit more refined and cultured when he is associated with a caption that not only implies he knows a classical language and has studied meditation but that he may be more sensitive to other cultures and ways of thinking than we have given him credit for. In other words, a single caption can subtly change a character's reputation or integrity in ways that the producers didn't intend. This problem could have been solved with verbatim speech instead: "Om."

Table 8.3 Examples of foreign speech captioned with language identifiers only.

Source	Speech (not subtitled)	Caption
The Office, "Stress Relief" Part 1 (5.14), 2009	Om. Om. Om. Om.	(CHANTING IN SANSKRIT)
The Big Bang Theory, "The Fish Guts Displacement" (6.10), 2012	Mazel tov.	[SPEAKS IN HEBREW]
Moneyball (2011)	- Hola, senor. - Hola.	[BOTH SPEAK IN SPANISH]
Cloud Atlas (2012)	Ma chère Ma chère?	[CAVENDISH SPEAKS IN FRENCH] [SPEAKS IN FRENCH]
The Office, "Stress Relief" Part 2 (5.15), 2009	"¡Me das una úlcera cada vez que me despierto y tengo que venir a trabajar para ti, para ti!"	(YELLING IN SPANISH)
Cabin in the Woods (2012)	Dolor supervivo caro, dolor supplemus caro, dolor ignio animus.	(STARTS READING) (READING IN LATIN)
Groundhog Day (1993)	La fille que j'aimera/ Sera comme bon vin/ Qui se bonifiera/ Un peu chaque matin.	(SPEAKING FRENCH) (SPEAKING FRENCH) (SPEAKING FRENCH)
The Internship (2013)	Vaya con Dios, mi amor. Porque está bravo en Barcelona y fantástico.	(SPEAKING SPANISH)
Parks and Recreation, "Sweet Sixteen" (4.16), 2012	Sprechen sie. Platz. Steh auf.	- [speaks German] - [speaking German]

Note: The foreign speech is not captioned or subtitled. http://ReadingSounds.net/chapter8/#table3.

Similarly, in *The Big Bang Theory*, Howard's father-in-law, Mike Rostenkowski (Casey Sander), doesn't speak Hebrew. He and his family—including Howard's wife (Melissa Rauch)—are Catholic. Howard (Simon Helberg) is Jewish. When Mike asks Howard, "You know how to shoot craps?," Howard explains that he's "not a stranger to dice games" because he was "the Temple Beth-El Hebrew School Yahtzee champion." Mike's response, "Mazel tov," is dripping with sarcasm. We can see the look of disdain on Mike's face and hear it in his gravelly, tough-guy voice. Through his sarcasm, he suggests to Howard that playing Yahtzee in elementary school does *not* prepare one to shoot casino dice. Opting for [SPEAKS IN HEBREW] over "Mazel tov," the captioner substitutes one complex meaning for another very different one. The point isn't that Mike speaks Hebrew or even understands Hebrew, despite what the caption implies. Rather, the meaning is located in the manner in which Mike replies to Howard and the common expression he uses (but most likely doesn't fully grasp). What matters—and what the caption fails to capture—is how Mike says "mazel tov": the sarcasm in his voice and face, but also the subtle way in which we are reminded that Mike is an outsider to Jewish tradition (and thus also an outsider to Howard's world).

A more effective alternative would have been to combine the manner of speaking with the verbatim foreign speech, while giving up on the need to identify the language: "(sarcastically) Mazel tov." In this context, who cares if "mazel tov" is Hebrew? The language identifier by itself, in the absence of verbatim speech, does nothing to advance the narrative and suggests too much that is not warranted. While a reference to Hebrew could help readers who didn't know that "mazel tov" was an expression associated with Howard's world, the captioner isn't a language teacher. The show assumes we know what Mike is doing without stooping to explain it to us, and the captioner should follow suit.

In the featured clip from *Moneyball* (2011), Billy Beane (Brad Pitt), the general manager of the Oakland Athletics, has just acquired a relief pitcher from the Cleveland Indians, Ricardo Rincón (Miguel Mendoza). They greet each other for the first time just outside the A's locker room. Rincón, who hails from Mexico, greets Beane in Spanish: "Hola, senor." Beane parrots the greeting back to Rincón in an awkward four-syllable stammer that sounds like "eh uh oh la" (or "H- H- Hola"). This interaction, albeit without the stammer, is included in the movie script in the verbatim Spanish (Zaillian and Sorkin n.d.). But the caption for this interaction—[BOTH SPEAK IN SPANISH]—cuts readers off from its specific nature and purpose. We need to know as caption readers that they are not just *speaking* Spanish but *greeting* each other. The vague caption may lead readers to mistakenly assume that they are saying much more than "Hola." We also need to know that Beane is not entirely comfortable with this non-English greeting. Beane's slight awkwardness is subtle but important. In the absence of verbatim foreign speech, the language ID implies the opposite of the truth—that Beane, like Rincón, is a fluent Spanish speaker when he really isn't. Beane's relationship to Rincón is purely as a baseball manager who wants to win games. It's not personal and has nothing to do with Beane's knowledge of Spanish or his intercultural sensitivity. Beane's slight stumble over the common Spanish greeting reminds us of this. The captioner should weigh the need to convey the complex meanings packed into even the smallest interaction with the time available to convey them. Meanings must be prioritized. At the very least, this interaction needed to be conveyed in verbatim Spanish. Knowing that Spanish is the language of the greeting is helpful but unnecessary. Screen placement can be used to locate each "Hola" underneath each speaker. Because the stammer plays a subtle but important role in the narrative, it could be represented for Beane as "Uh, hola" or possibly "H- H- Hola."

In some cases, the use of a language identifier instead of verbatim foreign speech can take up more precious space in the captions than the verbatim speech that's being described. In *Cloud Atlas*, multiple stories are told over six eras. The official movie website puts it this way: *Cloud Atlas* "explores how the actions and consequences of individual lives impact one another throughout the past, the present and the future" (Warner Bros. 2012). One of the storylines involves a vanity book publisher, Timothy Cavendish (Jim Broadbent), who is being extorted by the thuggish associates of one of his jailed authors for some of the profits of the author's book. So Cavendish tries desperately to raise £50,000 in one day, making a number of phone calls to friends and associates to no avail. Each recipient hangs up on him, but their actions are obscured with a neutral caption that is repeated for each phone conversation: [LINE DISCONNECTS]. The line doesn't "disconnect" on its own. A more accurate description would have highlighted the action taken by each recipient to end the call with Cavendish (e.g., "Recipient hangs up"). One call is presumably to a former sweetheart, whom Cavendish addresses twice as "ma chère," once as a greeting at the start of the call and again as a question after she hangs up on him a moment later. We never hear her voice—in fact, the scene never cuts away from Cavendish to show us the people on the other end of these calls—and it's possible she never says anything at all after answering the phone. His initial greeting, "ma chère," is captioned as [CAVENDISH SPEAKS IN FRENCH]. Two words are thus traded for four, which doesn't sound so bad until you put it in terms of number of symbols (eight symbols in "ma chère" versus twenty-eight in the nonspeech caption). Because this phone call begins with Cavendish offscreen, the nonspeech caption must also identify the speaker, which adds to the character count. Readers must also read this caption very quickly because it's only on the screen for 0.95 seconds. That's a reading speed of 252 words per minute (4/0.95 = 252/60). Any caption left on the screen for less than one second is easy to miss (and frustrating for readers). To remedy this, the captioner might have started the caption just under a second later, when the scene cuts to Cavendish and a speaker identifier would not be required (i.e., [In French] would then have been sufficient). Delaying the first caption for this particular phone call for one second would have cut two words from the caption, but that still leaves the more serious problem of the vague language identifier, which denies readers access to the specific content of the French speech. We don't need to know that Cavendish speaks in French. We need to know that he says "ma chère." The name of the language is outside of the context of the narrative, even if it might be helpful to readers who would associate French with the language of love. By naming the language at the end of the brief call as [SPEAKS IN FRENCH], the caption not only leaves out the specific words he speaks but also the manner in which Cavendish repeats "ma chère" with inflection after the woman hangs up on him.

These examples involve one or two foreign words that are common to many English readers: *mazel tov*, *hola*, and *ma chère*. Even if some readers don't know what these foreign terms mean, it's not the captioner's job to translate foreign speech into English or even to tell readers which language is being spoken. Sometimes, the name of the language is relevant to the narrative and needs to be captioned, as in the interplay of captioned language identifiers and subtitles in *Inglourious Basterds*. But the captioner is not a teacher of Hebrew, Spanish, or French. As we move from short, fairly common expressions to longer, more complex foreign phrases and sentences, the same principle holds. Foreign speech needs to be captioned verbatim. Whether anything more than a direct transcription is required depends on the context. I don't speak Spanish, but I still deserve to have access in the captions to what Oscar (Oscar Nuñez) yells at Michael Scott (Steve Carell) during the comedic roast of Michael in part 2 of *The Office's* "Stress Relief" episodes. The caption for Oscar's quickly delivered and animated rant is (YELLING IN SPANISH), which is how I would summarize what Oscar is doing since I don't speak Spanish. But it's not how I would caption Oscar's foreign speech. Only through a complete transcription will readers have access to the rhythm, music, and meaning of the Spanish words: "¡Me das una úlcera cada vez que me despierto y tengo que venir a trabajar para ti, para ti!" Of course, access to the Spanish words through print is not the same as access through listening alone (for English-only hearing viewers anyway). When written down, the Spanish becomes more meaningful to me than when I just listen to it. These trade-offs are sometimes necessary in captioning, given the different affordances of listening and reading. For example, I can now see words I recognize because they look like English or Latin words I already knew: úlcera (ulcer), venir (come), trabajar (work). I wasn't able to hear these words when I simply listened to the speech and tried to make sense of it. I can now surmise through reading (but not through listening alone) that Oscar, who is yelling at Michael, is complaining that Michael gives him an ulcer when he comes to work. That's close enough to the precise English translation (which the captioner is not responsible for providing): "You give me an ulcer every time I wake up and have to come to work for you, for you!" If the choice is between

(YELLING IN SPANISH) and a direct transcription of that yelling, captioners must choose the latter in order to convey more precisely the nature and function of the speech. Yelling is not what he's saying but the manner in which he's saying it. Hearing viewers have access to the specific sounds Oscar is making. At the very least they can hear the repetition at the end ("para ti, para ti") and may know what these words mean in English. Caption readers deserve the same opportunity to hear the content of the yelling and make full, partial, or no sense of it. In revising this caption, captioners should privilege the verbatim Spanish. Manner of speaking (i.e., yelling) is important but also supported through clear visual clues in Oscar's facial expressions.

Spanish is more familiar and accessible to English speakers and readers than Latin, despite Latin's significant (and usually unnoticed) impact on the English language, a fact I came to appreciate during two semesters of grueling college Latin and, later, when I was studying English vocabulary terms and their root meanings in preparation for the graduate school admissions test. Yet few Americans, myself included, can comprehend Latin sentences. Regardless, the captioner's responsibility doesn't change just because the foreign language is obscure or inaccessible. *The Cabin in the Woods* (2012) provides an instructive example. Included on a number of "best of" lists for 2012, *The Cabin in the Woods* revives the horror/slasher genre by "turning the splatter formula on its empty head" (Travers 2012). The well-worn conventions of the genre are brilliantly manipulated, celebrated, and satirized by writers Joss Whedon and Drew Goddard. *The Cabin in the Woods* is not just another slasher flick, despite the seemingly generic cabin in the woods and the five college student stereotypes. One familiar convention is the incantation that unwittingly activates some evil force or monster. In *The Cabin in the Woods*, the incantation is a series of three Latin phrases that are read by one of the college students, Dana Polk (Kristen Connolly), from the diary of an abused girl who used to live in the cabin: "Dolor supervivo caro, dolor supplemus caro, dolor ignio animus" (IMDb 2012). A rough English translation is "Pain outlives the flesh. Pain raises the flesh. Pain ignites the spirit" (IMDb 2012). But as I discovered when I consulted a colleague for her Latin expertise, the Latin in the movie is ungrammatical, most likely the result of the English phrases being run through an automated English-to-Latin translator. Nevertheless, the Latin phrases, grammatical or not, contain clues to English readers about the meaning of the incantation, clues that every viewer deserves the opportunity to make sense of. For example, *dolor* gives us English words like "dolorous" and may suggest suffering and pain. *Caro* may suggest flesh or meat through words like "carnivorous" and "carnage." *Animus* may suggest soul, mind, or even reanimation.

The incantation causes the girl who wrote it in her diary and her "zombie redneck torture family" to rise up from the grave. In a clever twist on the genre, the grave turns out to be an underground facility holding monsters from a large number of previous horror movies. If the cabin dwellers had chosen another artifact in the basement over the diary, another monster would have been activated. In other words, the Latin incantation may be the single most important speech in the movie in terms of its effect on the narrative. But instead of being captioned in verbatim Latin, the Latin is captioned as (STARTS READING) and (READING IN LATIN). These captions are not only insufficient but also redundant. In the case of the first caption, we don't need to be told that Dana is reading from a book because we can see it clearly, and in the case of the second caption, we don't need to be told that she is reading in Latin because Dana has already said as much: "And then there's something in Latin." Every viewer deserves to have access to this crucial speech in its original verbatim form so they can make as much or as little sense of it as they can or want. Self-evident speech acts and vague language identifiers prevent caption readers from satiating their curiosity about the specific words being read from the diary or exploring the complex interplay of meanings contained in the magic Latin spell. The Latin does not need to be translated into English captions because the movie doesn't offer a translation. It just needs to be transcribed in Latin.

Language identifiers tease readers with information that they don't have access to. Consider how French is captioned in *Groundhog Day* (1993), starring Bill Murray and Andie MacDowell, during a scene in which Phil Connors (Murray) is trying to woo Rita (MacDowell), who studied "nineteenth-century French poetry" in college. Over dinner, Murray recites some supposed French poetry to her: "La fille que j'aimera / Sera comme bon vin / Qui se bonifiera / Un peu chaque matin." In English, these lines mean "The girl I will love / is like a fine wine / that gets a little better / every morning" (IMDb 1993). The French is taken directly from a 1957 song by Belgian singer-songwriter Jacques Brel, entitled "La Bourrée Du Célibataire." But closed captioning readers are not privy to the French speech, because the lines in French are only captioned as (SPEAKING FRENCH). The French is not subtitled into English. Roughly after each pause in Phil's recitation, this same caption is repeated, even though no one interrupts Phil. Unlike *Inglourious Basterds*, in which language identifiers follow speaker turns (as they should), Phil's single turn is associated with three identical captions, timed approximately five seconds apart (see 0:36.38, 0:41.13, and 0:46.56 in the clip). Because the French poetry is not captioned verbatim, the captions repeatedly tease us, as each new reference to Phil speaking French only reinforces our curiosity to know what he is saying. When he finishes, Rita says, smilingly, "You speak French," to which Phil responds, "Oui." Unlike the recited poetry, this one-word reply in French is captioned verbatim, which suggests that the barrier to full verbatim, multilingual speech captioning is not non-English speech per se (as it is with the uncaptioned examples of "mazel tov," "hola," and "ma chère") but speech that is not immediately familiar to English-only captioners. Breaking through that barrier is crucial, as all nine examples in table 8.3 suggest, to give caption readers verbatim access to all speech sounds, not just English speech or the low-hanging fruit ("oui") that is obvious to English-only captioners. Whether caption readers can be expected to understand foreign speech is, to some extent, beside the point.

These examples are intended to be suggestive and not definitive. We don't yet know how often closed captions support an English-only ideology at the expense of foreign speech sounds. What's important to note is that shortcuts are sometimes taken to get around the problem of transcribing foreign speech. Almost daily, I add examples to my personal collection. On the night I finished drafting this section of this chapter, for example, I sat down with my family and watched a rented movie that had just been released on DVD (*The Internship*, 2013) and, later, a rerun of *Parks and Recreation* on the WGN channel. Both programs included a bit of foreign speech—in Spanish and German, respectively—and both used language identifiers as shortcuts instead of captioning the foreign speech verbatim. In *The Internship*, starring Vince Vaughn and Owen Wilson, Vaughn's character says something in Spanish to his girlfriend about a dream trip to Barcelona: "Vaya con Dios, mi amor. Porque está bravo en Barcelona y fantástico." His pronunciation is terrible and hilarious. The English translation, which is not subtitled, is: "Go with God, my love. Because Barcelona is brave and fantastic." His line of Spanish is only captioned as (SPEAKING SPANISH). In the "Sweet Sixteen" episode of *Parks and Recreation* (2012), Chris (Rob Lowe) teaches Andy's (Chris Platt) dog some commands in German. Chris commands the dog with "sprechen sie" (speak), "platz" (down), and "steh auf" (stand up). The dog complies with each command. This little exchange is not captioned verbatim but described with two captions, a two-speaker caption for the first command and a single caption spread over the second and third commands:

- [speaks German]
- [barks]

- [speaking German]

Even without the captions, we can see what's going on and guess that Chris is commanding the dog to speak, lie down, and stand up. He also uses hand gestures to help us see what we can't hear or understand, since the audience is presumed to speak English only. But caption readers deserve more than a vague reference to German. We need a transcription in German of what Chris is saying. Depending on the context and significance, we may also need some information about how the speaker is pronouncing his or her words. This information is crucial in *Inglourious Basterds* when the Americans try to speak Italian but aren't fooling anyone. It is less important in this example from *Parks and Recreation* to know the manner in which Chris pronounces the commands, but Andy's attempt to mimic the German commands, captioned appropriately as [speaking gibberish], might have benefited, space permitting, from the addition of a loose phonetic transcription of the nonsense sounds that weren't captioned and sound like "roxy," "roxit," and "chechindish."

This critique has proceeded from the ground up, but we also need to work from the top down, searching databases to identify foreign speech more systematically. For example, we can search large collections of caption files for nonspeech language identifiers, using search terms such as Spanish, French, Latin, Farsi, and "foreign language." In *Argo* (2012), for example, there are forty-four nonspeech references to "foreign language" in the captions. None of the parenthetical references to "foreign language" are paired with English subtitles. Such nondescript placeholders may be partly justified, as in the case of muffled or indecipherable background speech: [MAN SPEAKING IN FOREIGN LANGUAGE OVER PA] and [OFFICER SPEAKS IN FOREIGN LANGUAGE]. But we won't know for sure until we study each example of "foreign language" in context. At the very least, however, language identifiers must usually be more specific than a vague reference to "foreign language." People speak specific languages with few exceptions (gibberish, mumbling). When captioners label speech vaguely as "foreign" (rather than, for example, Farsi), their ideological commitments rise to the surface through the division of the world simplistically and ideologically into us and them—those who speak English and everyone else. In an American film such as *Argo*, in which a foreign and hostile country (Iran) is infiltrated by Americans to rescue American hostages, the repeated references to "foreign language" as a shortcut make sense on one level. The captions remind us (the presumed American audience) that the others are foreigners, plain and simple. That's all we presumably need to know. The captions suggest that neither a verbatim transcription of their speech nor an identification of the specific language is warranted. Subtitles are used sparingly throughout *Argo* as well (and never with nonspeech identifiers for "foreign language"), which heightens the viewer's identification with the Americans in the film who find this world and its language to be alien. (It's appropriate and ironic that the Americans' ruse in the film is that they are in Iran to scout locations for a fictitious film about aliens from another world.) We can only guess, along with many of the American characters in the film, what the guard is saying during the tense customs scene at the end of the film. But *Argo* also muddies this clear, ideological division of the world into us and them by captioning some foreign speech as Farsi. Indeed, there are twenty-three nonspeech identifiers of Farsi in *Argo*, such as [GUARD SPEAKS IN FARSI], some of which are paired with English subtitles. In one scene, a guard gestures to an American to remove his sunglasses. The unsubtitled Farsi command is captioned as [SPEAKING IN FARSI] and followed by a one-word reply, presumably also in Farsi, from the American: [SPEAKS IN FOREIGN LANGUAGE]. We might thus tentatively conclude that the Americans' efforts to pass as fluent Farsi speakers are being purposely marked in the captions as "foreign" to them, while the native speech of the Iranians is being identified as Farsi. This might have been a compelling approach were it used consistently. But there doesn't appear to be any pattern driving the language identifiers in *Argo*. The Iranian guards are identified as speaking both unsubtitled Farsi and an unsubtitled "foreign language" which is presumably also Farsi: see [GUARD SHOUTING IN FOREIGN LANGUAGE] and [GUARD SPEAKS IN FARSI]. The same is true of the Americans: [SPEAKS IN FOREIGN LANGUAGE] and [SPEAKING IN FARSI]. Why Farsi speech is captioned as "foreign" in some cases and "Farsi" in other cases is not clear and generates unnecessary confusion. Verbatim transcriptions could remove some of this confusion by presenting the speech to readers in the way that it's presented to listeners. In at least one case, *Argo* follows the same unfortunate pattern as "hola" in *Moneyball* when an exchange of "salâm" between two speakers (one of whom is played by Ben Affleck) is captioned only as [BOTH SPEAK IN FOREIGN LANGUAGE]. Meaning becomes unnecessarily obscured when caption readers are prevented from directly accessing the words being spoken, even if they are not in English. Caption readers should not have special access to foreign speech—note that I'm not asking for English subtitles where the film doesn't offer them—but access to the foreign words being spoken. In the case of Farsi, this would involve the transliteration of the Farsi alphabet, based on Arabic, into the Latin alphabet. **Media:** http://ReadingSounds .net/chapter8/#argo.

Since the English language makes use of few diacritical marks (and even those we have imported from other languages are becoming optional, such as naive for naïve), we might also search closed captioning databases for accent marks and diacritics in order to identify foreign speech—and then look at each example in context. This strategy is hit or miss and only works on closed captions, since subtitles are burned into video and can't be searched directly, at least not by outside scholars who, like me, have manually extracted plain text tracks from DVDs. But it would catch those instances in which a snippet of accented foreign speech in English closed captioning is conveyed in verbatim form. It is also worth noting again (see chapter 1) that such databases do not exist yet, at least not publicly. How captioners handle snippets of foreign speech in English-language programming will help us answer larger questions about language ideology in closed captioning. Sociolinguists define linguistic ideologies as "any sets of beliefs about language articulated by the users as a rationalization or justification of perceived language structure and use" (Silverstein 1979, 193). According to Kathryn Woolard and Bambi Schieffelin (1994, 72), a "wealth of public problems hinge on language ideology." Writing in the mid-1990s, they identified a number of issues in their comprehensive review of research on language ideology that continue to be relevant today:

Examples from the headlines of United States newspapers include bilingual policy and the official English movement; questions of free speech and harassment; the meaning of multiculturalism in schools and texts; the exclusion of jurors who might rely on their own native-speaker understanding of non-English testimony; and the question of journalists' responsibilities and the truthful representation of direct speech.

This last issue—the responsibility of journalists to represent speech truthfully—applies equally well to closed captioners working with short bursts of foreign speech in English-language programming. Too often, as we've seen, English speech gets the verbatim treatment while unsubtitled non-English speech is erased, reduced to a placeholder, marked as Other. When captioners take shortcuts on unsubtitled foreign speech— opting for (READING IN LATIN) or (SPEAKING SPANISH) over a verbatim transcription or transliteration—they subtly reinforce the notion that only

English really matters. Out of a multilingual text, a new text is created in the closed captions, one that may be narrowly monolingual, shaped by an English-language bias, or, more specifically, an American English bias. Amy Lueck (2013) sums up the problem this way:

Closed captioning is a striking example of a site of language use that is (mis)represented as a straightforward and unmediated transcription of language, when it is actually a complex and political translation with complicated ties to monolingual English-Only language politics (Lueck, 2012; Horner & Trimbur, 2002). Although closed captions are usually considered to be a transcription of the audio track, and transcription is assumed to be a matter of transparent rendering of language from one medium to another, captioners actually make significant rhetorical decisions about how voice, identity, language(s), and meaning are represented on the screen. Whose voice does the captioner mark as accented? When does one standardize speech to make it more readable, and what values and assumptions are operating in that choice? How are non-standard varieties of English and non-English languages represented when the captioner does not know the language in its original and is advised not to translate it to standard English?

This chapter has provided tentative answers to Lueck's (2013) important questions in terms of the pull towards standard English in pop-culture captioning, the reduction of voice and manner to simple parenthetical identifiers, and the privileging of English speech at the expense of foreign-language expressions. Any speech that crosses the presumably rigid borders of standard American English—including widely recognizable international expressions such as "hola" and "salâm"—runs the risk of being made invisible and marked, literally, as foreign.

Closed captions tend to reinforce the monolingual landscape of popular American cinema. More often than not, people from other countries and aliens from other planets miraculously and unexplainably speak English, usually with accents thick enough to mark them as foreign but not so foreign that subtitles are required. In the disaster film *2012* (2009), for example, everyone speaks English when it counts, regardless of where in the world a scene is located. The movie reaches a climax in Tibet as world leaders and wealthy passengers from around the world prepare to board giant "arks" that have been built to withstand an apocalyptic megatsunami headed their way. While the world leaders inside the arks speak English with accents and are persuaded by one American's impassioned plea (in English, of course) to open up one of the arks to the less fortunate masses clamoring outside, the general PA announcements outside the arks remind us that not everyone in the

world speaks English: [WOMAN 1 SPEAKING SPANISH OVER SPEAKERS], [MAN 2 SPEAK-
ING RUSSIAN OVER SPEAKERS], [WOMAN 2 SPEAKING ITALIAN OVER SPEAKERS]. But these
non-English language identifiers never become any more specific than a
vague nod to a multilingual world. Consider the verbatim English language announcement that appears after these three non-English PA captions: "WOMAN 3 [OVER SPEAKERS]: All green-card holders, please proceed to
Gallery D-4." Something similar happens in *Man of Steel* (2013) when an
alien voice begins speaking through all the televisions of the world to
demand that Kal-El/Superman turn himself in. The first caption of the
speech prematurely gives away Zod's identity in a Speaker ID: "ZOD [ON
TV]: You are not alone." (See chapter 5 on premature speaker identifiers).
While we might have guessed that the speaker is Zod, because we were
introduced to him at the beginning of the film as the murderer of Kal-El's
father, the disembodied voice is gravelly and distorted enough to suggest that the speaker's identity was intended to be obscured until he announces fifty-two seconds later: "My name is General Zod." Regardless,
the speaker repeats "You are not alone" four times in English before the
camera cuts to four different unnamed world locations, each one accompanied by the same nonspeech caption: [ZOD SPEAKING IN FOREIGN LANGUAGE
ON TV]. Zod presumably repeats the same line—"You are not alone"—in
each unsubtitled foreign language before the focus turns back to Kansas
for Zod's full speech in English. So we aren't missing much if we assume
that Zod's speech is being simultaneously translated for these foreigners around the world. But the point of this scene's quick multilingual
tour around the world is disguised when the captions fail to identify
the specific languages spoken or to distinguish the languages from each
other. The world becomes smaller in the closed captions, unnecessarily
reduced to a binary of English and Foreign, which defeats the scene's attempt to relay an image of Zod speaking to a multicultural, multilingual
planet. **Media:** http://ReadingSounds.net/chapter8/#2012-zod.

Placeholders for non-English speech seem to be least problematic in
movies that pit simple good against evil. In *True Lies* (1994), starring
Arnold Schwarzenegger and Jamie Lee Curtis, the American hero fights
the Middle Eastern threat in a simple case of Us vs. Them. When the
foreign leader of an extremist group is yelling and gesticulating in the
presence of the group's newly acquired nuclear weapon, his unsubtitled
words are captioned as [SPEAKING ARABIC], which dehumanizes them and
reinforces the clear and simple division between the familiar American
point of view and the untranslatable Other. While a verbatim transliteration would have provided caption readers with important access to the
sonic flavor of what he is saying, a language identifier by itself actually

reinforces the stark contrast between right and wrong. They are foreign. What else do we need to know? Later in this scene, a portion of the bad guy's speech is translated in real-time by government spy Harry Tasker (Arnold Schwarzenegger) to his English-only spouse (Jamie Lee Curtis). The audience identifies with Curtis in this scene, whose access to the meaning of the foreign speech must, like the audience's access, also be channeled through Schwarzenegger, the main star. In this way, [SPEAKING ARABIC] might be excused (even if I have suggested that more is needed to provide access to the non-English sounds): the bad guy's speech can't be known until the American hero translates it into English for us. **Media:** http://ReadingSounds.net/chapter8/#truelies.

These examples have implications for the study of comparative rhetoric, an area of study that has received increasing attention from rhetorical scholars over the last decade as a number of international issues take center stage, including globalization and localization, the global South, outsourcing, and war and terrorism. In comparative rhetorical studies, scholars "reflect on how one's cultural make-up influences the outcome of research and how networks of power asymmetry and interdependency shape and define discursive practices at all levels" (Mao 2013, 212). Scholars are encouraged to "recogniz[e] the influence of one's own cultural and ideological make-up on the study of the other, and valu[e] the importance of the historical and cultural contexts to rhetorical performances at all levels" (215). By "studying non-euroamerican rhetorics in their own terms" (215), scholars aim to avoid Eurocentric and ethnocentric biases and adopt a stance of "self-reflexivity" about the values and assumptions that shape their research. Mary Garrett (2013, 254), a self-described "Eurocentrically trained, non-Chinese academic studying Classical Chinese materials," offers "three ways to become more self-reflexive": practice mindfulness, "ask the natives" in order to learn their "differing perspectives," and engage in an "exercise of empathy" (252–53). The goal of these practices is to become more aware of the "unthinking projection[s]" (253) that shape one's implicit cultural attitudes and values. Comparative rhetoric intersects with work in disability studies in the areas of globalization and postcolonialism, especially disability rights in developing and non-European countries, including the global South (Barker and Murray 2013; Chouinard 2012; Connell 2011; Kim 2010; Meekosha and Soldatic 2011). Applied to caption studies, a comparative perspective doesn't ask captioners to stop serving as proxy agents for the content creators but to be mindful of their choices and how they impact readers' access to the material. Just because the captioner doesn't know any languages other than English doesn't excuse the captioner from "asking the natives" who will be able to provide, where needed, a translation more accessible than a vague placeholder like (SPEAKING SPANISH). Mindfulness, when applied to closed captioning, recognizes that, even when caption readers can't be assumed to know what the foreign words mean, they still deserve a full transcription of those words in the original language. What listeners hear, captioning users must be able to read.

Truthfully representing direct speech, whether English or not, is an ethical responsibility that most likely requires new or revised organizational arrangements for captioning companies. The captioners I have interviewed all stressed the importance of having excellent writing and grammar skills. Being a careful editor is assumed to go hand in hand with being a good offline captioner. But the preceding analysis has suggested that captioning also requires multilingual awareness. Hidden within discussions of good writing skills for captioners is the assumption that writing means English. This assumption may limit caption readers' access to verbatim foreign speech. To address this problem, captioners may need to seek out the assistance of professionals who can transcribe foreign speech when it is not transcribed in the script or already subtitled. At the least, addressing this problem may require tighter coordination between captioners and subtitlers.

NINE

The Future of Closed Captioning

Reading Sounds grew out of my fascination with closed captioning, both as a daily user and a scholar interested in reading and writing practices. Closed captioning remains an endless source of curiosity, delight, and, sometimes, frustration to me. I set out to break new ground in caption studies. Much work still remains. I hope this book can serve as a roadmap for future scholars and others interested in contributing to caption studies and, specifically, to our understanding of how closed captions create meaning, interact with the soundscape, and interface with readers. Our current definitions of caption quality, which have focused on accuracy, completeness, and formal issues such as typeface preferences and screen placement, only scratch the surface. While accuracy remains a key concern for captioning advocates (e.g., see YouTube's heavily criticized automatic speech recognition and transcription technology), we shouldn't stop short of trying to make sense of how captions transform the soundscape into a new text, just as subtitling theorists have explored the ways in which foreign language subtitling creates new texts through translation (cf. Nornes 2007, 15).

Closed captioning has too often served as an afterthought. The task of captioning a movie or television show is typically handed off to an independent company or team. This disconnect between producer and captioner does a disservice to everyone involved but especially to those readers who depend on quality closed captioning. If captions

9.1 **Blending form and content in closed captioning.**
In this frame from *The Three Musketeers*, a line of dialogue is manipulated to appear like it sounds—as an echo. The frame features Athos (Matthew Macfadyen), one of the Muske-teers, who has just been poisoned, sitting on the floor, looking down, while clutching the scrolled document he was bargaining for before being poisoned. Constantin Film, 2011. DVD. Echo effect produced by the author with Adobe After Effects. Caption: "Well, just so you don't leave empty-handed." http://Reading Sounds.net/chapter9/#figure1.

enable complex transformations of meaning, then both producers and captioners should work together to realize the producers' vision (see Udo and Fels 2010). In the college classroom, captioning is not well integrated into our multimodal video assignments (if captioning is dis-cussed at all). It is typically something to be performed at the end of a student's video project, after the "real" work has been completed. But captioning has the potential to change the meaning of the text in subtle and sometimes dramatic ways. Captioning demands rhetorical sensi-tivity and creativity. Classroom instructors need to talk with students at length about the challenges of making soundscapes accessible. I've found that some aspects of captioning, such as describing nonspeech sounds, including instrumental music, don't come naturally to hearing students. Treating captioning as an afterthought, a routine transcription exercise, or not at all denies our students a learning experience that's central to the composing practices we value, one that encourages them to think through the complexities of mode shifting and draw upon their powers of description in the process. When instructors discuss the ef-fective uses of onscreen text in video composition—headings, titles, bullets, credits—they should fold closed captioning and its affordances into these discussions. If closed captioning is transformative, it deserves more than passing mention. Indeed, caption studies has an important role to play in our understanding of how modes interact in digital video composition.

Reading Sounds complements existing books on closed captioning, including Gregory John Downey's (2008) historical study, Karen Peltz Strauss's (2006) legislative and economic history of telecommunications access, and Gary D. Robson's (2004) technical account of captioning equipment, standards, and specifications. It offers a humanistic rationale that moves us beyond captioning as a set of prescriptions or standards and toward a more flexible view that recognizes the creative work of captioners, the mediating influence of the captions, and the interpretative power of readers. Closed captioning is a rhetorical practice because it involves human choices about the best course of action to take under specific contexts and constraints of space and time. The most important decisions about meaning, context, and significance can't be reduced to a list of decontextualized prescriptions. Likewise, an ideology that presents captioning as simple transcription—so easy, even speech recognition software can do it!—dissuades us from exploring the complex nature of translating sounds into written texts for the purposes of timed reading. This complexity is reflected, as I have argued in these pages, in the degree to which captions contextualize, clarify, formalize, equalize, linearize, time-shift, and distill the soundscape.

My own perspective has been admittedly partial and biased. As a hearing advocate and parent of a deaf child, I have come to embody the perspective of universal design that, regardless of hearing status, closed captioning can potentially benefit all viewers. I wanted to focus this book on some of these benefits, because the dominant view (captioning as mere transcription) has left little room to discuss the differences between listening and reading. I've avoided discussing glaring errors or so-called "caption fails," even though the topic of accuracy is more important than ever in the era of autocaptioning. I haven't devoted much space to late, ill-timed captions either, even though delays of five to seven seconds are standard for live programming such as television news. Late captions are the flip side of what I have called "captioned irony" (chapter 6), but I wanted to focus on captioned irony over caption fails and slow captions because reading ahead and having advance knowledge are new contributions to our understanding of how captions make meaning.

Naturalizing and Celebrating Closed Captioning

Even if we don't watch closed captions, closed captions are watching us. The future of Internet video is being built on a foundation of robust closed captioning (or, more specifically, subtitling). Because closed captions on the Web are stored as separate plain text files, they can be fed to search engines and retrieved by keyword searches. Search engines are not very good at indexing the content of audio or video files. Indeed, Google's search engine has been metaphorically compared to both a blind and a deaf user (see Chisholm and May, 2009, 14). But search engines thrive on plain text: tags, keywords, text descriptions, text transcripts, and, of course, closed captions (Ballek 2010; Stelter 2010; Sizemore 2010). Google can process and index closed captions, allowing users to search for content inside YouTube videos that is difficult for search engines to process without the textual information that captions provide. No wonder search engine optimization (SEO) consultants promote closed captioning as a way for their clients to increase their search engine rankings and bring more visitors to their sites (e.g., Ballek 2010; Sizemore 2010). In education, closed captioning allows students to search for content inside recorded lectures. Without the added benefit of searchable text captions, students would have to manually scan lecture videos looking for that one example, anecdote, or solution that they vaguely remembered from class but cannot locate quickly or easily in the recorded video lectures.

Interactive transcripts raise the value of captions further by allowing users to click on a single word in a video transcript and be transported to that moment in the accompanying video where that word is spoken. I first became aware of, and then immediately recognized the immense game-changing power of, interactive transcripts on TED.com. Each word in an interactive transcript is time-stamped and clickable. The transcript is fully searchable and automatically scrolls in time with the video. Individual words are highlighted as they are spoken. Because captions on TED.com are crowdsourced out to regular users (TED 2014), many of the videos on the site are available in an impressive number of languages. One could, for example, listen to Aimee Mullins (2009) speaking in English, read the captions in a second language such as French, and browse the interactive transcript in a third language such as Japanese. (Or one could simply load captions and interactive transcript in English, which is what I do.) In the case of Mullins's (2009) TED talk, users can choose from thirty-two languages. YouTube also offers interactive transcripts for the captioned videos in its collection (Chitu 2010). Companies such as 3Play Media and ProTranscript also provide, as part of their regular video transcription service, a video player plugin that serves up interactive, clickable transcripts alongside closed captions. In addition, 3Play Media supports "archive searching" across a website's video collection, and has developed a "clipping plugin" that allows users to "[c]lip video segments simply by highlighting the text. Rearrange clips from multiple sources and create your own video montages" (3Play Media). The video clipping plugin will output a web link for sharing montages with other users. When these users are college students attending the same university, or enrolled in the same course, the video montage—fully accessible because it is built on closed captions—could be a powerful, accessible learning tool indeed. In these ways, then, captioning serves as a potential common ground upon which video indexing and retrieval are possible. Without captions, users are not able to search inside videos or make connections between videos based on keyword searches. Captioning solves the problem of video indexing by transforming video into something that search engines thrive on: plain text. Even if we don't watch web video with closed captioning, and even if some web videos are not yet captioned, captioning is already playing a vital role in creating the future of web video search. Captioning advocates must ensure that autocaptions generated by imperfect speech recognition technologies are not used to index the content of web videos. A fully searchable web must be built on accurate captions, not automated and less-than-perfect ones.

Interactive transcripts already provide users with an excellent way to search for and find information within a single video. As video becomes more popular and captioning technology provides a way to index large databases of video context, students will be able to search the video collection of an entire course, or even across all of the videos produced in all of the courses of a department, college, or university. In this near-future learning environment, captions will enable students to use keywords not only to find and review course content across multiple videos but also insert their own "margin" notes, which could take the form of time-stamped text comments or pop-up idea bubbles (as at BubblyPly .com), their own video responses or notes produced on the fly with their web cams, links to other related video moments in the course's video collection, links to external web resources, and comments from other students that have been made public. This added content may or may not be searchable/captioned, but it would at least be tagged and easier to find as visible nodes in the student's personalized video stream. The instructor's lecture videos would thus be transformed into the student's personalized study guide and an opportunity for collaborative learning. In addition, keyword searches would not return simply a list of matching video clips but also, perhaps, a single mash-up composed of all the clips that satisfied the search query, plus any accompanying student commentary. The inherent limitations of uncaptioned video would thus be addressed by a robust video captioning and search system that allows students to personalize and reconfigure the content of a course according to their needs. The promise of universal design could be achieved, in other words, by an accessible system that levels the playing field for all students—deaf, hard of hearing, and hearing. We need to continue to push for and applaud advances in captioning technology that will leverage the power of searchable text to provide a more inclusive, more accessible learning environment for our students. While it is naïve to think that a fully accessible library of university lecture videos is cheap or easy to achieve, it is nevertheless important for web accessibility advocates to continue to promote all of the reasons (ethical, legal, business, user-centered, etc.) that accessibility makes sense for our students and our pedagogies. As the number of distance learning, video-enriched courses grows on our college campuses, educators and students will require solutions that combine the richness of video with the data mining benefits of text-based captions and transcripts.

When closed captioning is seen as a widely desirable and vital component of the digital infrastructure—even if it sometimes operates below the surface, as in video search optimization—it becomes naturalized, marked not merely as a special accommodation but as a universal benefit and right. In this way, closed captioning can potentially "trouble the binary between normal and assistive technologies" (Palmeri 2006, 58) by showing us how captioning can serve all of us. To put it another way, when captioning supports universal design principles, it reminds us that "all technologies [are] assistive" (58). In addition to interactive transcripts and search engine optimization, closed captioning enters the mainstream through a variety of channels: enhanced episodes, easter eggs, caption fails, animated GIFs, parody videos, creative or humorous captioning, fictional captions and Internet memes, occasional English subtitles, direct "fourth wall" references to the captions, and animated captions.

Enhanced episodes and "pop-up" bubbles on television shows make use of onscreen text (similar in form to subtitles) to provide additional information to viewers about an episode or music video. VH1's "Pop Up

Video" program is perhaps the most well known use of onscreen textual enhancement, followed by enhanced episodes of the TV show *Lost*. An example of an enhanced caption on *Lost*: "And based on the iteration count in the tower the survivors have been on the island for 91 days" (Lostpedia n.d.). These enhancements are not closed captions because they are not intended to stand in for audio content but rather to supplement it with additional information. Nevertheless, enhanced episodes place similar cognitive demands on readers and may raise awareness of the needs of readers who depend on closed captioning. **Media:** http://ReadingSounds.net/chapter9/#enhanced.

Easter eggs, while rare in captioning, may nevertheless increase the level of interest in captioning among the general public. Dawn Jones, who runs the "I Heart Subtitles" blog, writes about a 2013 episode of BBC's *Sherlock* that included, in the upper left corner of the subtitle track, "letters that acted as clues to viewers and was part of the promotion to encourage repeated viewing and speculation about the new series" (Jones 2013). As the letters H-I-S appeared one at a time over the course of the episode, they spelled out a hint that only viewers who were watching with captions could see. **Media:** http://ReadingSounds.net/chapter9/#eastereggs.

Caption fails enter the mainstream through discussions of the limits of autocaptioning. For example, Rhett and Link's "caption fail" videos on YouTube average one million views each. Rhett and Link, two self-described "Internetainers," describe their comedic experiment with autocaptioning this way: ". . . we use YouTube's audio transcription tool, which doesn't always do the best job translating. We write a script. Act it out, then upload it. Let the tool translate it. Then make that into a script. Act it out. Upload it. Then let that be made into a script" (Rhett and Link 2011). The results are always hilarious and remind us that autocaptioning is still in its infancy and should never be a substitute for human-generated captions. People who wouldn't otherwise be aware of closed captioning, let alone autocaptioning, are introduced to it through an entertaining experiment that also contains an informative lesson about the imperfect state of speech recognition technology. **Media:** http://ReadingSounds.net/chapter9/#captionfails.

Animated GIFs, when they include text captions from TV shows and movies, require readers to read lips and read captions at the same time, thus mimicking in a small way how deaf and hard-of-hearing people process information on the screen. An animated GIF is an image format composed of a series of frames that simulate the movement of video when the frames play in sequence and loop automatically. The GIF file format does not support sound. When short clips from TV shows or movies are made into animated GIFs, the official closed captions may be included. Alternatively, verbatim text can be added by the GIF author in the style of meme text (Impact typeface, black stroke with white fill, sometimes set in all caps). For example, see the popular animated meme of Ron Burgundy (Will Ferrell) saying/mouthing "I don't believe you" in *Anchorman* (2004). As discussed in chapter 6, animated GIFs encourage a kind of *lip-reading culture* that positions closed captions as vital to the process of making sense of animated GIFs for all viewers. **Media:** http://ReadingSounds.net/chapter9/#animatedgifs.

In parody videos, subtitling may play an instrumental role. In "Bad Lip Reading" videos, for example, new audio is dubbed over the original audio and synched up roughly with characters' lip movements. One such video, "The Walking (and Talking) Dead—A Bad Lip Reading of *The Walking Dead*" (Bad Lip Reading 2013), dubs a new audio track in which characters appear to be saying such ridiculous things as "Hey, do you remember that costume party? / You went as a penguin / And I went as a pink shark." Subtitles assist in delivering the parody, which is why they are enabled or burned in by default. Other examples of parody that rely on subtitling include "literal music videos," in which new lyrics are written and performed to accompany the visuals in the official music video. The original literal music video was Dustin McLean's (2008) rewrite of A-ha's "Take on Me." Countless imitators and new literal music videos have followed, always with open subtitles. The wildly popular *Downfall* parody videos should be included here too. Each parody video presents new English subtitles for a scene from *Downfall* (2004), a movie about Hitler's final days in which Hitler (Bruno Ganz), speaking in German, goes on an animated rant. The *Downfall* parodies have become "so ubiquitous on YouTube that they have even spawned self-referential meta-parodies—jokes about Hitler learning about his internet fame" ("Hitler Downfall Parodies" 2009). Video parody and subtitles seem to fit together seamlessly. I've tried my hand at subtitle-delivered parody in a playful analysis of the limited vocabulary of one talk show host. **Media:** http://ReadingSounds.net/chapter9/#parody.

Creative or humorous captioning grabs the attention of viewers who may then be compelled to post screenshots on social media sites such as Reddit. I shared one such example in chapter 5, originally posted to Reddit, of a single closed caption from *The Tudors* that lingered on the screen during a commercial for IKEA (jeredhead 2013). The poster's screenshot captured the ironic tension between a lingering adult-oriented caption, [Sexual moaning], and the image of a young boy riding his trike in an IKEA kitchen. Adolescent humor may be perceived as more novel when it is encountered in closed captioning, at least for those viewers who are not accustomed to watching TV with captions. A Reddit user (secularflesh 2013) describes waking his girlfriend with his "adolescent giggling" after reading a captioned description of a bodily function in a documentary: (COLOBUS MONKEYS FART). Other popular captions posted to social media sites include [silence] and [silence continues] from *The Artist* (2011). Silence captions call into question our assumptions about the very nature of captioning itself and thus raise awareness of captioning's potential complexity and value. To this category I would also add experimental forms of textual representation. For example, Accurate Secretarial, a transcription and web captioning company, has been experimenting with new forms of textual representation on their YouTube channel, "The Closed Captioning Project," including new notational systems for music captioning. **Media:** http://ReadingSounds.net/chapter9/#creativecaptioning.

Fictional captions and internet memes call attention to the cognitive work that captions perform. The "[Intensifies]" meme can help us think through the problem of how to adequately describe modulating noise (see Know Your Meme, "[Intensifies]"). But the "[Intensifies]" meme, which often makes use a vibrating GIF image and static subtitle, describes actions that aren't clearly linked with sounds at all. For example, the caption [Shrecking Intensifies] (sic) accompanies a vibrating GIF image of Shrek and Donkey from the *Shrek* movies. The animation culminates in an increasingly distorted image of the two characters. Another example of this meme is [TARDIS INTENSIFIES], a subtitle that describes a vibrating image of the time machine police call box from *Doctor Who*. Fictional captions continue this theme of describing actions and playing on characters' personality quirks or traits: an image of a distraught Spock (Leonard Nimoy) from *Star Trek* [SOBBING MATHEMATICALLY]; an image of an emotional Eleventh Doctor (Matt Smith) from *Doctor Who* as he [ANGRILY FIXES BOW TIE]; and an animated image of a seething John Dorian (Zach Braff) from *Scrubs* as he [screams internally]. These examples are intended to be playful and even absurd. In some cases, they don't provide access to sounds to all. In others, they exploit characters' personality or sartorial traits. **Media:** http://ReadingSounds.net/chapter9/#memes.

Occasional English subtitles in English programming remind hearing viewers that onscreen transcriptions of what people are saying can benefit all viewers regardless of hearing status. Consider three examples: On the DVD for *Snatch* (2001), viewers can select a "pikey" track, which only displays subtitled translations of the hilariously thick English accent of Mickey O'Neil (Brad Pitt). On *Swamp People* (2013), an American reality TV series that follows a group of alligator hunters in Louisiana, the English but accented speech of these Cajun hunters is sometimes accompanied with hard-coded subtitles. On season 12 of *Project Runway* (2013), one of the contestants, Justin LeBlanc, wears a cochlear implant, identifies himself as deaf, and is accompanied on the show by a sign-language interpreter. His speech is also subtitled early in the season, much to the confusion of some viewers who didn't feel his clear English speech warranted English subtitles. One blog writer, in a recap of an early episode in this season, directed a question to the show's producers: "Why are you subtitling Justin when Sandro [another contestant] is much harder to understand?" (Toyouke 2013). One could ask the same question of the producers who unnecessarily subtitled the clear English speech in *Swamp People*. Later in the *Project Runway* season, after the producers presumably stopped subtitling LeBlanc's speech, LeBlanc himself tweets: "Yay for no subtitles under me! haha @ProjectRunway" (LeBlanc 2013). The examples from *Swamp People* and *Project Runway* suggest that decisions about subtitles may be driven, in part, by the mere perception of difference rather than any real need to provide access to accented English speech. These subtitles reinforce differences by marking certain speakers as not normal. **Media:** http://ReadingSounds.net /chapter9/#englishsubs.

Medium awareness applies to captioning when speakers make explicit references to the subtitle track, something that fictional characters aren't supposed to be aware of. For example, in a 2012 Dairy Queen commercial, the speaker literally hops on the subtitle track and rides it as it chugs off the screen. "These aren't just subtitles," he says. "These are subtitles I like to ride on." This example and others elevate subtitles to a topic of discussion. Subtitles become integral, meaningful elements of the program in their own right. They don't support or translate the primary meaning of the program, or try to sit unobtrusively at the bottom of the screen. Instead, they make their own meaning. We are asked to *look at* them, not merely *look through* them. These examples break through the so-called fourth wall. The imaginary fourth wall separates the audience from the action on the screen or stage. When the audience suspends its disbelief, the events are taken as real and believable. When fictional characters show an awareness of the medium (e.g., by talking directly into the camera, commenting on the soundtrack, bumping into or referring to the subtitles, etc.), they break through the fourth wall that enables the audience's suspension of disbelief. Put simply, fictional characters are not supposed to see subtitles. When they do, it's usually in the service of a joke. **Media:** http://ReadingSounds.net /chapter9/#mediumawareness.

Animated and enhanced captioning, similar to creative captioning, draws our attention to innovative and experimental forms, including kinetic typography and visual captions. For example, Raisa Rashid et al. (2008) have tested animated text captions with hearing and hard-of-hearing participants, comparing traditional captioning with "enhanced" and "extreme" forms. In the enhanced condition, kinetic type animates select words in the caption to signal emotional content. In the extreme condition, "some of the text was animated dynamically around the screen while static captions were displayed at the bottom of screen" (509). Participants found the enhanced captions to be preferable to both the traditional and extreme forms (516). Quoc Vy and Deborah Fels (2009) have explored the use of visual captions as a way to aid readers in identifying who is speaking (see also Vy 2012). Specifically, Vy and Fels (2009) tested the use of headshots (or avatars) decked out with color-coded borders. The avatars were placed alongside their respective captions, and traditional screen placement techniques were used as well. Results were mixed. Some participants felt "overwhelmed with the amount of information available on screen" (919). Nevertheless, these experiments need to continue if only because "closed captioning guidelines have remained virtually unchanged since the early days, and these guidelines actually discourage the use of colors, animation, and mixed case lettering in captions" (Rashid et al. 2008, 506). The traditional all-caps and center-aligned environment needs to be infused with alternatives grounded in healthy critique and a willingness to test out new forms. I've explored animated captioning in a revision of one scene from *The Three Musketeers* (2011) in which a character is drugged and experiences the speech of others as distorted and echoic. How might we visualize sonic distortion in the captions themselves? Using Adobe After Effects, I applied a small number of text treatments to the captions in this scene, while being mindful of the need to make the captions legible above all else. Finally, caption studies might address other representational challenges, such as how to embody sonic directionality and perspective when expressed, for example, in stereo or surround-sound. In one scene from *BloodRayne 2: Deliverance* (2007), someone or something is moving quickly outside a family's cabin. The mother follows the sound with her head, first turning her head from the viewer's right to left as the sound is captioned as (wind rushing), and then turning her head back again as the sound is captioned as (rushing continuing). The children also follow the sound with their heads, thus providing a number of visual clues to the sound's movement. The rushing wind turns out to be vampiric Billy the Kid. But these captions don't capture the directionality of the whooshing force as it moves from the listener's right ear to left ear and back again. The wind isn't rushing; it's moving with intentionality. Captioners must account for the ways in which film sound is dimensional and stereophonic and seek to counter the monophonic world suggested by the typical caption file, a world in which every sound is centered, static, fully present, and equally loud. **Media:** http://ReadingSounds.net/chapter9/#animated.

When captioning enters the mainstream, even if an author's intentions are satiric or absurd, captioning becomes more natural and less strange, more universal and less marginal, more central to our theories, pedagogies, and viewing habits and less likely to be overlooked or forgotten. In short, the more often we see or hear about captioning in the mainstream, the less often it becomes something we can write off as the purview of a seemingly narrow group. My hope is that closed captioning will be increasingly folded into and inform our scholarship on reading and writing practices, multimodality, and the future of Internet video. Accessibility is more than a transcript, afterthought, legal requirement, or set of prescriptions.

Acknowledgments

This book could not have been written without the support of a number of colleagues, friends, and family members. My colleagues in the Technical Communication and Rhetoric program at Texas Tech University have been incredibly supportive of my research and teaching interests in disability studies and web accessibility: Ken Baake, Craig Baehr, Kelli Cargile Cook, Joyce Carter, Sam Dragga, Angela Eaton, Michael Faris, Miles Kimball, Amy Koerber, Susan Lang, Kristen Moore, Rich Rice, Becky Rickly, Abigail Selzer King, Brian Still, and Greg Wilson. Teaching a graduate course on disability studies and web accessibility every summer since 2007 has allowed me to keep my teaching and research in close conversation. A number of PhD students developed dissertation topics in this course and subsequently asked me to chair their dissertation committees. I've learned so much from them about what it means to study disability from a rhetorical perspective: Diane Allen, Cris Broyles, Kim Elmore, Melissa Helquist, Maria Kingsbury, Ashlynn Reynolds, and Angela Shaffer. I've also had the pleasure of working with and learning from some other amazingly bright and dedicated PhD students, including Glenn Dayley, Sue Henson, Steve Morrison, Todd Rasberry, Danielle Saad, Emil Towner, and Charity Tran.

I owe a very special thanks to Brenda Brueggemann, who invited me to Ohio State University in February 2012 on behalf of OSU's Literacy Studies Program, The Digital Media Studies Program, The Digital Union, and The Disability Studies Program. Brenda's support of my research over the last few years has meant so much to me. It was during my preparations for this invited talk that I began to think seriously for the first time about writing a book on closed captioning. Within a month of my OSU visit I was working on the first sample chapter (which eventually became chapter 3). I don't think this book would exist today without Brenda's support. Other colleagues at OSU deserve special mention too: Ken Petri, Cindy Selfe, Amy Shuman, and H. Lewis Ulman.

Alec Hosterman and the Communication Studies faculty at Indiana University South Bend deserve my special thanks too. Alec invited me to IUSB in March 2014 to give the spring talk in their Arts Lecture Series. The first full draft of this book was nearly ready to send to the publisher at that point, and I used this talk to do something I hadn't done before: I offered a broad but complete overview of the book's argument, organized around the seven transformations of meaning I discuss in chapter 1. It was a valuable experience for me, and I got some great feedback on my presentation from the audience at IUSB.

A number of colleagues in disability studies, web accessibility, and rhetorical studies have motivated and inspired me. Some provided feedback along the way. Stephanie Kerschbaum graciously reached out to me over email to give thoughtful feedback on one of my blog posts and to offer additional examples from her experience. I credit Stephanie in chapter 8 for providing examples that inform my discussions of onomatopoeia and foreign speech. (Of course, I take full responsibility for all of the perceived strengths and faults in this book.) Other colleagues whose work has been influential to me or have helped me to situate this project in a larger teaching or research context include Lora Arduser, Kristen Betts, Amanda Booher, Jay Dolmage, Krista Kennedy, Lisa Meloncon, Sushil Oswal, Margaret Price, Whitney Quesenbery, Geoff Sauer, and Melanie Yergeau.

Outside of academia, captioning advocacy thrives—and for good reason. I want to thank Joe Clark, who gave occasional feedback on my blog and shared a number of industry contacts with me during a memorable phone conversation. But more importantly, Joe has studied captioning longer and been more influential than any other captioning advocate or critic. My own critique of the captioning style guides in chapter 2 is indebted to Joe's earlier critiques of the captioning industry. I also want to thank the members of the Collaborative for Communication Access via Captioning (CCAC) mailing list, who have taught me so much about the captioning needs, occasional frustrations, and preferences of deaf and hard-of-hearing viewers. The members of this list also helped me advertise my survey to viewers of closed captioning. A special thanks to the founder of the CCAC list, Lauren Storck. Other advocates and professionals I want to thank: Mike Lockrey (@mlockrey), Dawn Jones (@iheartsubtitles), Mirabai Knight (@stenoknight), and Karen Mardahl (@stcaccess). I also want to mention three captioning/transcription companies that have helped me at different stages of this project: Accurate Secretarial, Line 21 Media Services, and 3Play Media. The ten professional captioners I interviewed, and the 110-plus people who took my survey, deserve my thanks as well. There are a lot of people around the world who care passionately about quality captioning, and I've benefitted from productive interactions with a few of them.

The enthusiasm of David Morrow, my editor at the University of Chicago Press, kept me excited during those times when I wondered whether anyone in the humanities would be interested in an entire book devoted to closed captioning. David provided steady leadership and thoughtful feedback throughout the life of this project. Amy Koerber, my colleague at Texas Tech, gave me David's name when I was shopping the book proposal around. I'm in her debt for introducing me to David. I also want to thank the two anonymous peer reviewers who provided ample and critical feedback to the Press at two key stages along the way.

Three of my previously published articles provided early foundations and motivations for the current project. I have received permission, when necessary, to reprint these publications. Throughout this book, I draw selectively on examples and passages from "Which Sounds are Significant: Towards a Rhetoric of Closed Captioning" (*Disability Studies Quarterly* 31, no.3, 2011). In chapter 9, I draw heavily from "Personal Reflections on the Educational Potential and Future of Closed Captioning on the Web" (J. E. Aitken, J. Pedego Fairley, and J. K. Carlson, eds., *Communication Technology for Students in Special Education or Gifted Programs*, 221–29, Hershey, PA: IGI Global). Chapters 3, 4, and 6 include examples and passages originally published in "More Than Mere Transcription: Closed Captioning as an Artful Practice" (*User Experience Magazine* 14, no.1, 2014).

Finally, this book has been a family affair from the beginning. Captioning has been a daily part of our lives for the last fifteen years. My spouse, Denise, and our two boys, Liam and Pierce, played important roles in the book's development. Denise shared a number of examples with me over the years and provided regular feedback on my ideas. One example in particular stands out. Denise lead me to an important insight that is central to the argument in chapter 5. One August day in 2009, Denise called me in from the other room because she wanted to show me something in the closed captions of *Pirates of the Caribbean: Dead Man's Chest* (2006). A language identifier for the movie's native people— [speaks cannibals' language]—as well as other references to their identity such as [cannibals murmur] gave away the natives' cannibalism, an important plot point, a good eight minutes before the movie reveals it in dramatic fashion. Much later, I would group these cannibal captions together with other examples into what I eventually called "captioned irony."

My older son, Liam, has been programming computers since he was very young. When I needed a way to automatically extract nonspeech information (i.e., the captions in brackets or parentheses that describe sounds), isolate the speaker identifiers from the rest of the nonspeech captions, and then organize everything into sorted HTML tables, he wrote a handy little program in Perl. Separating speech from nonspeech captions by hand would have been prohibitively time-consuming for me. Liam also designed a web interface that allowed me to search my small but growing collection of DVD caption files and add new files easily. He also planted the seed for my analysis of the Hypnotoad character. Something he said one night while we were watching *Futurama* (with the captions turned on, of course) made me curious to find out whether the Hypnotoad's sound had been captioned in more than one way over ten years of episodes. As it turns out, I ended up finding so much diversity in the captions for the hypnosound that I devoted an entire chapter to it (see chapter 3).

My younger son, Pierce, is the reason I became intensely interested in closed captioning. This book exists because he does. He's a savvy caption reader and avid social media user who has a lot to teach us about the importance of inclusion, difference, and universal design. I've watched television programs with him at the maximum volume setting (volume level sixty-four on our TV), with the sound muted, and at every volume setting in between. I've sat beside him in the movie theater as he placed a CaptiView closed captioning display device into the drink holder and adjusted the snakelike device so the small LED display would intersect with his line of sight. Because he has always preferred to watch movies on a variety of devices, from iPhones to flat-screen televisions, I learned how to burn captions onto the surface of the movies themselves, thus forcing the display of captions on devices that, at the time, didn't yet support closed captioning. His current device of choice is a Kindle Fire, which provides good support for closed captioning and which he uses primarily to watch Netflix and Amazon Prime programming. Pierce has taught me that accessibility for deaf and hard-of-hearing viewers doesn't stop at DVD and TV captioning. But more importantly, it is through him that I am reminded every day of the importance of inclusion for all. I dedicate this book to my son, Pierce.

Sean Zdenek
Ransom Canyon, Texas
October 2014

Bibliography

3Play Media. 2014. "Interactive Transcript." Accessed January 28. http://www.3playmedia.com/services/interactive-transcript -plugins/.

Aberdeen Captioning. 2011. "The Stewie Effect." Accessed February 22, 2014. http://abercap.com/blog/2011/10/25/the-stewie -effect/.

Alexander, Jonathan. 2008. "Media Convergence: Creating Content, Questioning Relationships." *Computers and Composition* 25:1–8.

Allan, David. 2006. "Effects of Popular Music in Advertising on Attention and Memory." *Journal of Advertising Research* 46 (4): 434–44.

———. 2008. "A Content Analysis of Music Placement in Prime-Time Television Advertising." *Journal of Advertising Research* 48 (3): 404–17.

Allbritton, David. 2004. "Strategic Production of Predictive Inferences During Comprehension." *Discourse Processes* 38 (3): 309–22.

Amazon.co.uk. 2013. "DVD Regions." Accessed March 10. http:// amazon.co.uk/gp/help/customer/display.html?nodeId =502554.

Antonini, Rachele. 2007. "SAT, BLT, Spirit Biscuits, and the Third Amendment: What Italians make of cultural references in dubbed texts." In *Doubts and Directions in Translation Studies: Selected contributions from the EST Congress, Lisbon 2004*, edited by Yves Gambier, Miriam Shlesinger, and Radegundis Stolze, 153–67. Amsterdam/Philadelphia: John Benjamins.

———. 2008. "The Perception of Dubbese: An Italian Study." In *Between Text and Image: Updating Research in Screen Translation*, edited by Delia Chiaro, Christine Heiss, and Chiara Bucaria, 135–47. Amsterdam/Philadelphia: John Benjamins.

Archer, Arlene. 2006. "A Multimodal Approach to Academic 'Literacies': Problematising the Visual/Verbal Divide." *Language and Education* 20 (6): 449–62.

Arduser, Lora. 2011. "Warp and Weft: Weaving the Discussion Threads of an Online Community." *Journal of Technical Writing and Communication* 41 (1): 5–31.

Arrested Development Wikia. 2013. "Storming the Castle." Accessed May 1. http://arresteddevelopment.wikia.com/wiki/Storming_the_Castle.

Bad Lip Reading. 2013. "The Walking (and Talking) Dead—A Bad Lip Reading of *The Walking Dead*." YouTube video. Accessed January 30, 2014. http://www.youtube.com/watch?v=jR4lLJu_-wE.

Ballek, Matt . 2010. "FACT: YouTube Indexes Captions." *vidiSEO* (blog). Posted April 15. Accessed September 28, 2012. http://vidiseo.com/youtube-indexes-captions/.

Barker, Clare, and Stuart Murray. 2013. "Disabling Postcolonialism: Global Disability Studies and Democratic Criticism." In *The Disability Studies Reader*, edited by Lennard. J. Davis, 61–73. 4th edition. New York: Routledge.

Bayer, Nancy L., and Lisa Pappas. 2006. "Accessibility Testing: Case History of Blind Testers of Enterprise Software." *Technical Communication* 53 (1): 32–38.

Bazerman, Charles. 2004. "Intertextuality: How Texts Rely on Other Texts." In *What Writing Does and How It Does It: An Introduction to Analyzing Texts and Textual Practices*, edited by Charles Bazerman & Paul Prior, 83–96. Mahwah, NJ: Lawrence Erlbaum.

Bean, Rita M., and Robert M. Wilson. 1989. "Using Closed Captioned Television to Teach Reading to Adults." *Reading Research and Instruction* 28 (4): 27–37.

Berke, Jamie. 2009. "Deaf Culture—Jonah Syndrome." About.com Deafness. Posted April 10. Accessed August 21, 2013. http://deafness.about.com/cs/archivedarticles/a/jonahsyndrome.htm.

Bianchi, Diana. 2008. "Taming Teen-Language: The Adaptation of *Buffyspeak* into Italian." In *Between Text and Image: Updating Research in Screen Translation*, edited by Delia Chiaro, Christine Heiss, and Chiara Bucaria, 183–95. Amsterdam/Philadelphia: John Benjamins.

Big Bang Theory Transcripts. 2013. "Series 1 Episode 06—The Middle Earth Paradigm." Accessed March 22. http://bigbangtrans.wordpress.com/series-1-episode-6-the-middle-earth-paradigm/.

Bijker, Wiebe E. 1997. *Of Bicycles, Bakelites, and Bulbs: Towards a Theory of Sociotechnical Change*. Cambridge: MIT Press.

Bilton, Nick. 2010. "Mobile Uploads Spur Facebook Video Growth." *New York Times*. Posted June 11. Accessed January 29, 2014. http://bits.blogs.nytimes.com/2010/06/11/mobile-uploads-spurs-facebook-video-growth/.

Black, Edwin. 1978. *Rhetorical Criticism: A Study in Method*. Second edition. Madison: University of Wisconsin Press.

———. 1980. "A Note on Theory and Practice in Rhetorical Criticism." *Western Journal of Speech Communication* 44:331–36.

Booher, Amanda. 2011. "Defining Pistorius," in "Disability and Rhetoric," special edition of *Disability Studies Quarterly* 31 (3). Accessed September 27, 2012. http://dsq-sds.org/article/view/1673/1598.

Booth, Wayne. 2004. *The Rhetoric of Rhetoric: The Quest for Effective Communication.* Malden, MA: Blackwell.

Bowie, John Ross. 2012. "Answers to Kripke FAQ's." John Ross Bowie Dot Com. Posted February 2. Retrieved January 5, 2014. http://johnrossbowie.tumblr.com/post/16945123900.

Brueggemann, Brenda Jo. 1999. *Lend Me Your Ear: Rhetorical Constructions of Deafness.* Washington, DC: Gallaudet University Press.

———. 2009. *Deaf Subjects: Between Identities and Places.* New York: NYU Press.

Brummett, Barry. 1980. "Towards a Theory of Silence as a Political Strategy." *Quarterly Journal of Speech* 66:289–303.

Bruner, Gordon C., II. 1990. "Music, Mood, and Marketing." *Journal of Marketing* 54 (5): 94–104.

Bucholtz, Mary. 2007. "Variation in Transcription." *Discourse Studies* 9 (6): 784–808.

Bull, Michael, and Les Back. 2003. "Introduction: Into Sound." In *The Auditory Culture Reader*, edited by Michael Bull and Les Back, 1–18. New York: Berg.

Burnham, Denis, Greg Leigh, William Noble, Caroline Jones, Michael Tyler, Leonid Grebennikov, and Alex Varley. 2008. "Parameters in Television Captioning for Deaf and Hard-of-Hearing Adults: Effects of Caption Rate Versus Text Reduction on Comprehension." *Journal of Deaf Studies and Deaf Education* 13 (3): 391–404.

Calvo, Manuel G., and M. Dolores Castillo. 1998. "Predictive Inferences Take Time to Develop." *Psychological Research* 61:249–60.

Cambra, Cristina, Núria Silvestre, and Aurora Leal. 2008/9. "Comprehension of Television Messages by Deaf Students at Various Stages of Education." *American Annals of the Deaf* 153 (5):425–34.

Cameron, James. 2013. *Avatar* script. Undated draft. The Internet Movie Script Database (IMSDb). Accessed March 26. http://www.imsdb.com/scripts/Avatar.html.

Canadian Broadcasting Centre [CBC]. 2003. "The CBC captioning style guide." Accessed July 18, 2010. http://joeclark.org/access/captioning/CBC/images/CBC-captioning-manual-EN.pdf.

CaptionMax. 2013. "Why Do Captions Sometimes Stay on the Screen Through a Commercial Break?" *Viewer FAQs*. Accessed July 26. http://captionmax.com/faq-viewer/#commercial-break.

CaptionsOn. 2008. "Faces of Captioning—Closed Captions." YouTube video. Accessed September 28, 2012. http://www.youtube.com/watch?v=a_6n_aw1TqQ.

Casey, Michael A. 2001. "Sound Classification and Similarity Tools." In *Introduction to MPEG-7: Multimedia Content Description Language*, edited by B. S. Manjunath, P. Salembier, and T. Sikora, 309–23. New York: John Wiley & Sons.

Cavanaugh, Jillian R. 2005. "Accent Matters: Material Consequences of Sounding Local in Northern Italy." *Language & Communication* 25:127–48.

CBS News. 2012. "About Us: FAQs, Contact Info and Credits." *CBS News Sunday Morning*. Posted June 5. Accessed December 23, 2013. http://www.cbsnews.com/sunday-morning/about-us/.

CEA R4.3 Work Group 1. 2003/4. "DTV Closed Caption Test File Iteration 2A Version 1.0 (1080i)." Consumer Electronics Association and WGBH National Center for Accessible Media. Accessed June 5, 2014. http://ncamftp.wgbh.org/DTV/CEA%20test%20material/Iteration_2/It2A%20ReadMe%20V1.pdf.

Charlton, James I. 2000. *Nothing About Us Without Us: Disability Oppression and Empowerment*. Berkeley: University of California Press.

Chiaro, Delia, Christine Heiss, and Chiara Bucaria. 2008. *Between Text and Image: Updating Research in Screen Translation*. Amsterdam/Philedelphia: John Benjamins.

Chion, Michel. 2012. "The Three Listening Modes." In *The Sound Studies Reader*, edited by J. Sterne, 48–53. London: Routledge.

Chisholm, Wendy, and Matt May. 2009. *Universal Design for Web Applications*. Beijing: O'Reilly.

Chisnell, Dana E., Janice C. Redish, and Amy Lee. 2006. "New Heuristics for Understanding Older Users as Web Users." *Technical Communication* 53 (1):39–59.

Chitu, A. 2010. "YouTube's Interactive Transcripts." *Google Operating System Blog*. Posted June 4. Accessed January 28, 2014. http://googlesystem.blogspot.com/2010/06/youtubes-interactive-transcripts.html.

Chouinard, Vera. 2012. "Pushing the Boundaries of our Understanding of Disability and Violence: Voices from the Global South." *Disability & Society* 27 (6):777–92.

Christian, Jon. 2014. "How Netflix Alienated and Insulted Its Deaf Subscribers." *The Week*. Posted January 30. Accessed June 7, 2014. http://theweek.com/article/index/255618/how-netflix-alienated-and-insulted-its-deaf-subscribers.

Christopher, Tommy. 2011. "AP Reporter Responds To Chris Hayes Panel Debate On Racism Of Droppin' G's From Obama Speech." Mediaite. Posted September 25. Accessed January 4, 2014. http://www.mediaite.com/tv/ap-reporter-responds-to-panel-debate-on-racism-of-droppin-gs/.

Clark, Joe. 2002. "Reading the Tube: Typographic Atrocities Prevail in the World Of TV Captioning and Subtitling." *Print* March/April. Accessed July 30, 2013. http://joeclark.org/design/print/readingthetube.html.

———. 2003. Review of *Daddy Day Care*. *Joe Does the Movies: Accessible Movie Review in Toronto* (blog). Posted May 17. Accessed March 6, 2013. http://joeclark.org/access/cinema/reviews/daddydaycare.html.

———. 2004. "How standardization solves problems in captioning and beyond." *NAB Broadcast Engineering Conference Proceedings*, 295–301. Las Vegas, Nevada, April 17–22. Accessed July 18, 2010.

———. 2006a. "Comments on Josélia Neves's Captioning Thesis." Posted October 1. Accessed October 19, 2012. http://joeclark.org/access/captioning/Neves/.

———. 2006b. "What does the HDTV captioning spec say about typography?" Posted November 29. Accessed June 7, 2014. http://screenfont.ca/fonts/today/708/type-critique/.

———. 2008. "What do the CBC captioning manuals say?" Posted April 10. Accessed September 27, 2012. http://joeclark.org/access/captioning/CBC/manuals/.

Clark, Lorraine. 2009. "Genre and Communication: Why You Can't Leave the Knowledge Out of Knowledge Education." *Pedagogy: Critical Approaches to Teaching Literature, Language, Composition, and Culture* 9 (3): 509–19.

Colebrook, Claire. 2004. *Irony: The New Critical Idiom*. London: Routledge.

Colletta, Lisa. 2009. "Political Satire and Postmodern Irony in the Age of Stephen Colbert and Jon Stewart." *Journal of Popular Culture* 42 (5): 856–74.

Collins, Karen. 2008. *Game Sound: An Introduction to the History, Theory, and Practice of Video Game Music and Sound Design*. Cambridge: MIT Press.

Comcast. 2013. "About Closed Captioning." Comcast Official Site. Posted July 19. Accessed September 1, 2013. http://customer.comcast.com/help-and-support/cable-tv/what-is-closed-captioning/.

Condit, Celeste Michelle. 1993. "The Critic as Empath: Moving Away from Totalizing Theory." *Western Journal of Communication* 57:178–90.

———. 2013. "The Rhetorical Limits of Polysemy." In *The Routledge Reader in Rhetorical Criticism*, edited by Brian. L. Ott and Greg Dickinson, 644–59. New York: Routledge.

Connell, Raewyn. 2011. "Southern Bodies and Disability: Re-Thinking Concepts." *Third World Quarterly* 32 (8): 1369–81.

Cook, Paul G. 2009. "The Rhetoricity of Cultural Literacy." *Pedagogy: Critical Approaches to Teaching Literature, Language, Composition, and Culture* 9 (3): 487–500.

Cornell Lab of Ornithology. 2014. "Great Blue Heron." All About Birds. Accessed February 14. http://www.allaboutbirds.org/guide/great_blue_heron/sounds.

Cowling, Michael. 2004. "Non-Speech Environmental Sound Classification System for Autonomous Surveillance." PhD dissertation, Griffith University.

Cutts, Matt. 2009. "Show and Translate YouTube Captions." *Matt Cutts: Gadgets, Google, and SEO* (blog). Posted March 5. Accessed January 18, 2013. http://www.mattcutts.com/blog/youtube-subtitle-captions/.

Daboll, Peter. 2011. "Celebrities in Advertising are Almost Always a Big Waste of Money." Advertising Age. Posted January 12. Accessed January 11, 2014. http://adage.com/article/cmo-strategy/celebrities-ads-lead-greater-sales/148174/.

DaeOh. 2008. "Lost in Translation Ending Whisper." YouTube video. Accessed October 12, 2014. https://www.youtube.com/watch?v=xgkZHHKSYzQ.

Daily Show with Jon Stewart, The. 2011. "Trey Parker and Matt Stone." Published June 15. Accessed September 27, 2013. http:// thedailyshow.cc.com/videos /smsr0m/trey-parker---matt-stone.

Davies, Alexander. 2012. "World's Quietest Room Will Drive You Crazy in 30 Minutes." Discovery News. Posted April 9. Accessed September 1, 2013. http://news.discovery.com/human/life/worlds-quietest-room-will-drive -you-crazy-in-30-minutes.htm.

Davis, Lennard J. 2010a. "Constructing Normalcy." In *The Disability Studies Reader*, edited by Lennard. J. Davis, 3–19. Third edition. New York: Routledge.

———. 2010b. "The End of Identity Politics: On Disability as an Unstable Category." In *The Disability Studies Reader*, edited by Lennard. J. Davis, 301–315. Third edition. New York: Routledge.

———. 2013. *The End of Normal: Identity in a Biocultural Era.* Ann Arbor: University of Michigan Press.

Deal, Mark. 2007. "Aversive Disablism: Subtle Prejudice Toward Disabled People." *Disability & Society* 22 (1): 93–107.

Delo, Cotton. 2012. "Does Facebook Know You're Pregnant?" Advertising Age. Published September 10. Accessed July 30, 2013. http://adage.com/article /digital/facebook-pregnant/237073/.

Described and Captioned Media Program [DCMP]. 2009. *Captioning Key for Educational Media: Guidelines and Preferred Techniques.* Accessed July 18, 2010. http://www.captioningkey.org/captioning-key.pdf.

———. 2011a. "Presentation Rate." Accessed November 18, 2012. http://www .dcmp.org/captioningkey/presentation_rate.html.

———. 2011b. "Sound Effects." Accessed November 20, 2012. http://www.dcmp .org/captioningkey/sound_effects.html.

———. 2011c. "Speaker Identification." Accessed December 16, 2013. http:// www.dcmp.org/captioningkey/speaker_identification.html.

———. 2011d. "Text." Accessed November 18, 2012. http://www.dcmp.org /captioningkey/text.html.

———. 2011e. "Welcome to the DCMP *Captioning Key*." Accessed April 19, 2013. http://www.dcmp.org/captioningkey/.

Díaz Cintas, Jorge, ed. 2009. *New Trends in Audiovisual Translation.* Bristol: Multilingual Matters.

Díaz Cintas, Jorge, and Gunilla Anderman, eds. 2009. *Audiovisual Translation: Language Transfer on Screen.* New York: Palgrave Macmillan.

Dickinson, Greg, Brian L. Ott, and Eric Aoki. 2013. "Spaces of Remembering and Forgetting: The Reverent Eye/I at the Plains Indian Museum." In *The Routledge Reader in Rhetorical Criticism*, edited by Brian L. Ott and Greg Dickinson, 338–53. New York: Routledge.

Dolmage, Jay. 2005. "Disability Studies Pedagogy, Usability and Universal Design." *Disability Studies Quarterly* 25 (4). Accessed March 1, 2014. http://dsq -sds.org/article/view/627/804.

Dotsub. 2009a. "How to Transcribe a Video on dotSUB." Posted May 20. Accessed January 18, 2013. http://dotsub.com/view/e6562923-a6eb-4c6b-a6c7-ffed4a2848a5.

———. 2009b. "How to Translate a Video on dotSUB." Posted May 20. Accessed January 18, 2013. http://dotsub.com/view/30764f64-ffd1–4867–9151–35681 79037e7.

———. 2012. "30% Audience Increases Make Video Captioning a 'Must Have.'" *dotSUB Blog*. Posted June 30. Accessed January 18, 2013. http://blog.dotsub.com/2012/06/30/1335/.

———. 2013a. "Dotsub Tutorials." Accessed January 18. http://dotsub.com/tutorials.

———. 2013b. "U.S. Accessibility Regulations for Online Video Captions." Accessed January 18. http://dotsub.com/enterprise/laws.

Dove. 2013. "The Dove Campaign for Real Beauty." *Dove* (website). Accessed July 30. http://www.dove.us/social-mission/campaign-for-real-beauty.aspx.

Dow, Bonnie J. 2013. "Criticism and Authority in the Artistic Mode." In *The Routledge Reader in Rhetorical Criticism* edited by Brian L. Ott and Greg Dickinson, 141–49. New York: Routledge.

Downey, Gregory John. 2008. *Closed Captioning: Subtitling, Stenography, and the Digital Convergence of Text with Television*. Baltimore: Johns Hopkins University Press.

Dubois, Sylvie, and Barbara Horvath. 2002. "Sounding Cajun: The Rhetorical Use of Dialect in Speech and Writing." *American Speech* 77 (3):264–87.

Duerstock, Brad, and Rohit Ranchal. 2011. "Assistive Notetaking Using Speech Recognition Technology." Accessed February 21, 2014. https://iashub.org/resources/15.

Duffy, John, and Melanie Yergeau, eds. 2011. "Disability and Rhetoric," special edition of *Disability Studies Quarterly* 31 (3). Accessed September 27, 2012. http://dsq-sds.org/issue/view/84.

Dunn, Patricia A., and Kathleen Dunn De Mers. 2002. "Reversing Notions of Disability and Accommodation: Embracing Universal Design in Writing Pedagogy and Web Space." *Kairos: A Journal of Rhetoric, Technology, and Pedagogy* 7 (1). Accessed September 17, 2012. http://english.ttu.edu/kairos/7.1/binder2.html?coverweb/dunn_demers/index.html.

Dyer, Richard. 2012. *In the Space of a Song: The Uses of Song in Film*. London: Routledge.

Dyson, Frances. 2009. *Sounding New Media: Immersion and Embodiment in the Arts and Culture*. Berkeley: University of California Press.

Earley, Sharon. 1978. "Captioning at WGBH-TV: A Paper Presented at the 1978 Symposium on Research and Utilization of Educational Media for Teaching the Deaf." Accessed September 27, 2012. http://www.dcmp.org/caai/nadh147.pdf.

Ebert, Roger. 2003. Review of *Lost in Translation*. RogerEbert.com. Posted September 12. Accessed October 12, 2013. http://www.rogerebert.com/reviews/lost-in-translation-2003.

———. 2012. Review of *Hick*. RogerEbert.com. Posted May 23. Accessed August 3, 2014. http://www.rogerebert.com/reviews/hick-2012.

Ellis, Tim. 2007. "Amusing Ad Juxtaposition." *Seattle Bubble* (blog). Accessed July 26, 2013. http://seattlebubble.com/blog/2007/06/17/850/.

Ellwanger, Adam. 2009. "Bloom and his Detractors: The Academic Polemic and the Ethics of Education." *Pedagogy: Critical Approaches to Teaching Literature, Language, Composition, and Culture* 9 (3): 475–86.

Ellwanger, Adam, and Paul G. Cook. 2009. "Disciplinarity, Pedagogy, and the Future of Education: Introduction." *Pedagogy: Critical Approaches to Teaching Literature, Language, Composition, and Culture* 9 (3): 471–74.

Federal Communications Commission. 1999. Outland Sports, Inc. request for undue burden exemption. Published September 23. Accessed August 15, 2013. http://transition.fcc.gov/cgb/dro/outland.txt.

———. 2000. "FCC Adopts Technical Standards for Display of Closed Captioning on Digital Television Receivers." Accessed September 28, 2012. http://transition.fcc.gov/Bureaus/Mass_Media/News_Releases/2000/nrmm0031.html.

———. 2010. "Part 79—Closed captioning of video programming: Current through February 19, 2010." Accessed January 30, 2012. http://www.fcc.gov/cgb/dro/captioning_regs.html.

———. 2012. "Closed Captioning of Internet Protocol-Delivered Video Programming." *Federal Communications Commission*. Document # 11–154. Issued January 13. Accessed December 21, 2012. http://www.fcc.gov/document/closed-captioning-internet-protocol-delivered-video-programming-0.

———. 2013. "Order on Reconsideration and Further Notice of Proposed Rulemaking." MB Docket #11–154. Issued June 13. Accessed August 10, 2013. http://transition.fcc.gov/Daily_Releases/Daily_Business/2013/db0614/FCC-13-84A1.pdf.

———. 2014. "Closed Captioning on Television." Accessed February 16. http://www.fcc.gov/guides/closed-captioning.

Fidyk, Steve. 2012. "Focus on the Hi-Hat: The Classic Swing Sound." *Modern Drummer* (website). Posted January 31. Accessed February 26, 2013. http://www.moderndrummer.com/site/2012/01/jazz-drummers-workshop-swing-hi-hat/.

Flint, Joe. 2012. "Celebs Muffle the Voice of Experience." *Los Angeles Times*. May 7. Accessed January 11, 2014. http://articles.latimes.com/2012/may/07/business/la-fi-ct-voiceover-20120507.

Foy, George Michaelson. 2012. "Experience: I've Been to the Quietest Place on Earth." *Guardian*. Posted May 18. Accessed September 1, 2013. http://www.theguardian.com/lifeandstyle/2012/may/18/experience-quietest-place-on-earth.

Futurama Wiki. 2012a. "Bender Should Not Be Allowed On Television." Accessed December 20. http://futurama.wikia.com/wiki/Bender_Should_Not_Be _Allowed_on_Television.

———. 2012b. "Hypnotoad." Wikia Entertainment. Accessed December 20. http:// futurama.wikia.com/wiki/Hypnotoad.

Garland-Thomson, Rosemarie. 2010. "Integrating Disability, Transforming Feminist Theory." In *The Disability Studies Reader*, edited by Lennard. J. Davis, 353–73. Third edition. New York: Routledge.

Garman, Judith. 2011. "Autistic Spectrum, Captions and Audio Description." *Mindful Research* (blog). Posted August 29. Accessed September 11, 2012. http://mindfulresearch.co.uk/2011/08/29/autistic-spectrum-captions-and -audio-description/.

Garrett, Mary. 2013. "Tied to a Tree: Culture and Self-Reflexivity." *Rhetoric Society Quarterly* 43 (3): 243–55.

Giora, Rachel, Shani Federman, Arnon Kehat, Ofer Fein, and Hadas Sabah. 2005. "Irony Aptness." *Humor* 18 (1): 25–39.

Glenn, Cheryl. 2004. *Unspoken: A Rhetoric of Silence.* Carbondale: Southern Illinois University Press.

Glenn, Cheryl, and Krista Ratcliffe, eds. 2011. *Silence and Listening as Rhetorical Arts.* Carbondale: Southern Illinois University Press.

Goldberg, Larry. 2007. "When Good Captions Go Bad: HDTV Accessibility." *Hearing Loss Magazine* September/October: 20–24.

Goldman, M. and Goldman, S. 1988. "Reading with Closed Captioned TV." *Journal of Reading* 31 (5): 458–61.

Goldwasser, Dan, and Mike Brennan. 2005. Review of *Team America* soundtrack. Soundtrack.net. Posted January 24. Accessed March 1, 2013. http://www .soundtrack.net/albums/database/?id=3651.

Goodman, Steve. 2012. "The Ontology of Vibrational Force." In *The Sound Studies Reader*, edited by J. Sterne, 70–72. London: Routledge.

Google. 2009. "Automatic Captions in YouTube." *Google Official Blog.* Posted November 19. Accessed November 9, 2012. http://googleblog.blogspot.com /2009/11/automatic-captions-in-youtube.html.

———. 2013. "Ads in Google." Posted March 29. Accessed July 26, 2013. https:// support.google.com/mail/answer/6603?hl=en.

Goya Martinez, Mariana. 2012. "Juxtaposition and News: Analysis of Juxtaposition and its Evolution in U.S. Newspapers and Online News." Paper presented at the annual meeting of the International Communication Association, Sheraton Phoenix Downtown, Phoenix, Arizona, May 23–28.

Grabiner, Ellen. 2012. *I See You: The Shifting Paradigms of James Cameron's "Avatar".* Jefferson: McFarland.

Gray, Jean. 2008. "Close Encounters: Tones and Signs." The Straight Dope. Posted January 11. Accessed March 22, 2013. http://boards.straightdope.com/sdmb /showthread.php?t=451162.

Griffin, Jodie. 2011. "Copyright Does Not Trump Disability Rights Law." *Public Knowledge* (blog). Posted November 2. Accessed August 14, 2014. https://www.publicknowledge.org/news-blog/blogs/copyright-does-not-trump-disability-rights-law.

Grimes, Tom, and Robert Drechsel. 1996. "Word-Picture Juxtaposition, Schemata, and Defamation in Television News." *Journalism & Mass Communication Quarterly* 73 (1): 169–80.

Gross, Alan G. 1996. *The Rhetoric of Science*. Reprint edition. Cambridge: Harvard University Press.

Guo, Guodong, and Stan Z. Li. 2003. "Content-Based Audio Classification and Retrieval by Support Vector Machines." *IEEE Transactions on Neural Networks* 14 (1): 209–15.

Harkins, Judith E., Ellie Korres, Beth Singer, and Barbara M. Virvan. 1995. "Non-Speech Information in Captioned Video: A Consumer Opinion Study With Guidelines for the Captioning Industry." Technology Assessment Program. Gallaudet University. Accessed February 1, 2014. http://www.dcmp.org/caai/nadh126.pdf.

HathiTrust. 2014. "HathiTrust Statement on Authors Guild v. HathiTrust Appeal Ruling." HathiTrust Digital Library. Posted June 10. Accessed August 16, 2014. http://www.hathitrust.org/authors_guild_lawsuit_appeal_ruling.

Hazanavicius, Michel. 2013. *The Artist* script. Undated draft. The Internet Movie Script Database (IMSDb). Accessed October 6. http://www.imsdb.com/scripts/Artist,-The.html.

Hemstreet, Tyler. 2010. "Hearing Loss Number One Diagnosis for Military." Northwest Military. Posted April 6. Accessed September 28, 2012.

Hewett, Beth L., and Cheryl Ball, eds. 2002. In "Disability—Demonstrated by and Mediated Through Technology," special issue, *Kairos: A Journal of Rhetoric, Technology, and Pedagogy* 7, no. 1 (Spring). Accessed September 27, 2012. http://english.ttu.edu/kairos/7.1/coverweb.html.

Hirsch, E. D., Jr. 1988. *Cultural Literacy: What Every American Needs to Know*. 1st Vintage Books edition. New York: Vintage Books.

Hoberman, J. 1999. "All Droid Up." *The Village Voice*. Published May 18. Accessed January 5, 2014. http://www.villagevoice.com/1999–05–18/film/all-droid-up/.

Internet Movie Database. 1993. "*Groundhog Day* Trivia." Accessed March 26, 2015. http://www.imdb.com/title/tt0107048/trivia.

———. 2009a. "*Inglourious Basterds* Goofs." Accessed March 26, 2015. http://www.imdb.com/title/tt0361748/goofs.

———. 2009b. *Taken* movie poster. Posted January 8. Accessed July 5, 2013. http://www.imdb.com/media/rm3195046400/tt0936501.

———. 2010. "Full Cast and Crew for *The Other Guys*." Accessed January 31, 2013. http://www.imdb.com/title/tt1386588/fullcredits.

———. 2011. "Paul." Accessed August 19, 2013. http://www.imdb.com/title/tt1092026/.

———. 2012. *"The Cabin in the Woods* Trivia." Accessed November 28. http://www.imdb.com/title/tt1259521/trivia.

Ives, Nat. 2012. "One Path to Brand Buzz: Advertise Fast Food After 'Walking Dead' Zombie Feast." Advertising Age. Posted December 4. Accessed July 26, 2013. http://adage.com/article/trending-topics/kfc-lights-twitter-walking-dead-juxtaposition/238598/.

Jackson, Lisa J. 2011. "Grammar-Ease: Ellipsis Versus the Em-Dash." Posted September 13. Accessed July 8, 2013. http://nhwn.wordpress.com/2011/09/13/grammar-ease-ellipsis-versus-the-em-dash/.

Jamieson, Kathleen Hall, and Karlyn Kohrs Campbell. 1982. "Rhetorical Hybrids: Fusions of Generic Elements." *Quarterly Journal of Speech* 68:146–57.

Jarina, Roman, and Ján Olajec. 2007. "Discriminative Feature Selection for Applause Sounds Detection." In *Image Analysis for Multimedia Interactive Services*. Eighth International Workshop on Image Analysis for Multimedia Interactive Services (WIAMIS'07). 4 pages. doi: 10.1109/WIAMIS.2007.34.

Jensema, Carl. 1998. "Viewer Reaction to Different Television Captioning Speeds." *American Annals of the Deaf* 143 (4): 318–24.

Jensema, Carl, and Robb Burch. 1999. "Caption Speed and Viewer Comprehension of Television Programs." Final report submitted to the Office of Special Education Programs. Federal Award Number: H180G60013. Accessed July 15, 2013. http://www.dcmp.org/caai/nadh135.pdf.

Jensema, Carl J., Ramalinga Sarma Danturthi, and Robert Burch. 2000. "Time Spent Viewing Captions on Television Programs." *American Annals of the Deaf* 145 (5): 464–68.

Jensema, Carl, R. McCann, and S. Ramsey. 1996. "Closed-Captioned Television Presentation Speed and Vocabulary." *American Annals of the Deaf* 141 (4): 284–92.

jeredhead. 2013. "So the captions from my show stayed on screen during the commercial break . . ." Posted July 17. Accessed August 12, 2013. http://www.reddit.com/r/funny/comments/1ign2h/so_the_captions_from_my_show_stayed_on_screen/.

Jones, Dawn. 2013. "Sherlock—The Clue is in the Subtitles." *I Heart Subtitles* (blog). Posted July 15. Accessed January 25, 2014. http://iheartsubtitles.wordpress.com/2013/07/15/sherlock-the-clue-is-in-the-subtitles/.

Kain, Donna J. 2005. "Constructing Genre: A Threefold Typology." *Technical Communication Quarterly* 14 (4): 375–409.

Kiefaber, David. 2013. "Taco Bell Gives Notorious B. I. G.'s 'Big Poppa' the Spanish Treatment, Too: Another Job for Google Translate." Posted February 8. Accessed December 21, 2013. http://www.adweek.com/adfreak/taco-bell-gives-notorious-bigs-big-poppa-spanish-treatment-too-147182.

Kim, Eunjung. 2010. "Minority Politics in Korea: Disability, Interraciality, and Gender." In *The Disability Studies Reader*, edited by Lennard. J. Davis, 417–31. 3rd ed. New York: Routledge.

Kim, Hyoung-Gook, Nicolas Moreau, and Thomas Sikora. 2004. "Audio Classification Based on MPEG-7 Spectral Basis Representations." *IEEE Transactions on Circuits and Systems for Video Technology* 14 (5): 716–25.

Kjellmer, Göran. 2009. "Where Do We Backchannel?: On the Use of *mm, mhm, uh huh* and Such Like." *International Journal of Corpus Linguistics* 14 (1): 81–112.

Know Your Meme. 2012. "Hypnotoad." Accessed December 21. http://know yourmeme.com/memes/hypnotoad.

———. 2014. "[Intensifies]." Accessed January 31. http://knowyourmeme.com /memes/intensifies.

Koskinen, Patricia, Robert Wilson, Linda Gambrell, and Susan Neuman. 1993. "Captioned Video and Vocabulary Learning: An Innovative Practice in Literacy Instruction." *Reading Teacher* 47 (1): 36–43.

Kostelnick, Charles, and David Roberts. 1998. *Designing Visual Language: Strategies for Professional Communicators*. Needham Heights, MA: Allyn & Bacon.

Kravets, David. 2012. "Facebook Agrees to Pay $10 to Each 'Sponsored Stories' Victim." Published October 9. Accessed July 30, 2013. http://www.cnn.com /2012/10/09/tech/social-media/facebook-settlement-wired.

Kress, Gunther. 2010. *Multimodality: A Social Semiotic Approach to Contemporary Communication*. London: Routledge.

Leber, Rebecca. 2013. "UPDATED: 15 Companies Drop Facebook Advertising Over Domestic Violence Content." Posted May 28. Accessed July 30, 2013. http:// thinkprogress.org/health/2013/05/28/2064461/13-companies-drop -facebook-advertising-over-domestic-violence-content/.

LeBlanc, Justin. Twitter post. October 24, 2013. 5:32 pm. https://twitter.com /JLeBlancDesign/status/393535454928056320.

Leff, Michael. 1980. "Interpretation and the Art of the Rhetorical Critic." *Western Journal of Speech Communication* 44:337–49.

Leibs, Andrew. 2014. "What is Closed Captioning?" Accessed February 16. http:// assistivetechnology.about.com/od/ATCAT8/f/What-Is-Closed-Captioning.htm.

Leppihalme, Ritva. 1997. *Culture Bumps: An Empirical Approach to the Translation of Allusions*. Clevedon, UK: Multilingual Matters Ltd.

Lewiecki-Wilson, Cynthia, and Brueggemann, Brenda Jo. 2007. *Disability and the Teaching of Writing: A Critical Sourcebook*. Boston: Bedford/St. Martin's.

Liao, Wen-Hung, and Yu-Kai Lin. 2009. "Classification of Non-Speech Human Sounds: Feature Selection and Snoring Sound Analysis." In *Proceedings of the 2009 IEEE International Conference on Systems, Man, and Cybernetics*, 2695–700. San Antonio, TX. October 11–14. doi: 10.1109/ICSMC.2009. 5346556.

Lindemann, Stephanie. 2005. "Who Speaks 'broken English'? US Undergraduates' Perceptions of Non-Native English." *International Journal of Applied Linguistics* 15 (2): 187–212.

Lindsay, Erin. 2011. "President Obama at the Congressional Black Caucus Foundation Annual Dinner: 'March With Me and Press On.'" *The White-*

house Blog. Posted September 25. Accessed January 4, 2014. http://www
.whitehouse.gov/blog/2011/09/25/president-obama-congressional-black
-caucus-foundation-annual-dinner-march-me-and-pre.

Linebarger, Deborah L. 2001. "Learning to Read from Television: The Effects of
Using Captions and Narration." *Journal of Educational Psychology* 93 (2):
288–98.

Linebarger, Deborah L., Jessica Taylor Piotrowski, and Charles R. Greenwood.
2010. "On-Screen Print: The Role of Captions as a Supplemental Literacy
Tool." *Journal of Research in Reading* 33 (2): 148–67.

Linton, S. 1998. *Claiming Disability: Knowledge and Identity.* New York: NYU
Press.

Loomis, John. 2010. "Close Encounters of the Third Kind." Accessed March 22,
2013. http://www.johnloomis.org/ece303L/notes/music/Close_Encounters
.html.

Los Doggies. 2008. "Close Encounters of the Major Third Kind." *Los Doggies Blog.*
Accessed March 22, 2013. http://www.losdoggies.com/archives/17.

Lostpedia. 2013. "Enhanced Episodes." Accessed March 5, 2013 http://lostpedia
.wikia.com/wiki/Enhanced_episodes.

Lovett-Graff, Bennett. 1995. "Culture Wars II: A Review Essay." *Modern Language
Studies* 25 (3): 99–124.

Lueck, Amy. 2011. "Writing Without Sound: Language Politics in Closed Captioning," in "Writing with Sound," special issue, *Currents in Electronic Literacy.* Accessed September 26, 2012. http://currents.cwrl.utexas.edu/2011
/writingwithoutsound.

———. 2013. "Writing a Translingual Script: Closed Captions in the English
Multilingual Hearing Classroom." *Kairos: A Journal of Rhetoric, Technology,
and Pedagogy,* 17 (3). Accessed October 18, 2014. http://kairos.techno
rhetoric.net/17.3/praxis/lueck/index.html.

Macaulay, Ronald. 1991. " 'Coz it izny spelt when they say it': Displaying Dialect
in Writing." *American Speech* 66 (3): 280–91.

MacInnis, Deborah J., and C. Whan Park. 1991. "The Differential Role of Characteristics of Music on High- and Low-Involvement Consumers' Processing of
Ads." *Journal of Consumer Research* 18:161–73.

Mao, LuMing. 2013. "Beyond Bias, Binary, and Border: Mapping out the Future
of Comparative Rhetoric." *Rhetoric Society Quarterly* 43 (3): 209–25.

Markham, Paul. 1989. "The Effects of Captioned Television Videotapes on the
Listening Comprehension of Beginning, Intermediate & Advanced ESL
Students." *Journal of Educational Technology* 29 (10): 38–41.

Mbariket, Rich. 2011. "Jamie Berke Advocates for Closed Captioned Web Series."
Posted February 13. Accessed September 28, 2012. http://webseriesnetwork
.com/profiles/blogs/jamie-berke-advocates-for.

McGee, Michael Calvin. 2013. "Text, Context, and the Fragmentation of Contemporary Culture." In *The Routledge Reader in Rhetorical Criticism,* edited by
Brian L. Ott and Greg Dickinson, 227–38. New York: Routledge.

McKee, Heidi. 2006. "Sound Matters: Notes Toward the Analysis and Design of Sound in Multimodal Webtexts." *Computers & Composition* 23:335–54.

McLean, Dustin. 2008. "Take on Me: Literal Video Version." YouTube video. Accessed January 30, 2014. http://www.youtube.com/watch?v=8HE9OQ4FnkQ.

Meacham, Jon. 2007. "What You Need to Know Now." *Newsweek* 150 (2): 34–37.

Media Access Group. 2002a. "Captioning FAQ." Accessed September 1, 2013. http://main.wgbh.org/wgbh/pages/mag/services/captioning/faq/.

———. 2002b. "Suggested Styles and Conventions for Closed Captioning." Accessed August 11, 2013. http://main.wgbh.org/wgbh/pages/mag/services/captioning/faq/sugg-styles-conv-faq.html.

Meekosha, Helen, and Karen Soldatic. 2011. "Human Rights and the Global South: The Case of Disability." *Third World Quarterly* 32 (8): 1383–98.

Meloncon, Lisa, ed. 2013. *Rhetorical Accessability: At the Intersection of Technical Communication and Disability Studies*. Amityville: Baywood.

Miller, Carolyn R. 1979. "A Humanistic Rationale for Technical Writing." *College English* 40 (6): 610–17.

Morain, Matt, and Jason Swarts. 2012. "YouTutorial: A Framework for Assessing Instructional Online Video." *Technical Communication Quarterly* 21:6–24.

Moran, Charles. 1999. "Access: The 'A' Word in Technology Studies." In *Passions, Pedagogies, and 21st Century Technologies*, edited by G. Hawisher and C. Selfe, 205–20. Logan: Utah State University Press.

Mullin, Joe. 2012. "Netflix Settles with Deaf-Rights Group, Agrees to Caption All Videos by 2014." Ars Technica. Posted October 10. Accessed February 21, 2014. http://arstechnica.com/tech-policy/2012/10/netflix-settles-with-deaf-rights-group-agrees-to-caption-all-videos-by-2014/.

Mullins, Amy. 2009. "Aimee Mullins and her 12 pairs of legs." Accessed January 28, 2014. http://www.ted.com/talks/lang/eng/aimee_mullins_prosthetic_aesthetics.html.

National Association of the Deaf. 2009. "NAD Calls out Netflix on Captions." Posted October 5. Accessed September 28, 2012. http://www.nad.org/news/2009/10/nad-calls-out-netflix-captions.

———. 2012. "NAD, NFL, and NBC Team for Super Bowl Captioning Experience." Posted February 3. Accessed September 29, 2012. http://www.nad.org/news/2012/2/nad-nfl-and-nbc-team-super-bowl-captioning-experience.

National Institute on Deafness and Other Communicative Disorders. 2010. "Quick Statistics." Posted June 16. Accessed September 28, 2012. http://www.nidcd.nih.gov/health/statistics/quick.html.

Netflix. 2009. "Closed Captions and Subtitles." *US & Canada Blog*. Posted June 12. Accessed September 28, 2012. http://blog.netflix.com/2009/06/closed-captions-and-subtitles.html.

———. 2010. "Subtitles Now Available for Some Titles for PC/Mac Viewing." *US & Canada Blog*. Posted April 15. Accessed September 28, 2012. http://blog.netflix.com/2010/04/subtitles-now-available-for-some-titles.html.

———. 2011. "30% of Netflix Streaming Content has Subtitles; 80% by End of 2011." *US & Canada Blog.* Posted February 24. September 28, 2012. http://blog.netflix.com/2011/02/30-of-netflix-streaming-content-has.html.

Neves, Josélia. 2005. "Audiovisual Translation: Subtitling for the Deaf and Hard-of-Hearing." PhD dissertation, Roehampton University, University of Surrey. Accessed October 19, 2012. http://roehampton.openrepository.com/roehampton/bitstream/10142/12580/1/neves%20audiovisual.pdf.

———. 2008. "10 Fallacies about Subtitling for the d/Deaf and the Hard of Hearing." *Journal of Specialised Translation* 10:128–43.

Nornes, Abé Mark. 1999. "For an Abusive Subtitling." *Film Quarterly* 52 (3): 17–34.

———. 2007. *Cinema Babel: Translating Global Cinema.* Minneapolis: University of Minnesota Press.

Office for Civil Rights, US Department of Education. 1999. "Impact of the Civil Rights Laws.". Accessed September 28, 2012. http://www2.ed.gov/about/offices/list/ocr/impact.html.

O'Hara, Karen. 2004. "Curb Cuts on the Information Highway: Older Adults and the Internet." *Technical Communication Quarterly* 13:423–45.

Oswal, Sushil. 2013. "Exploring Accessibility as a Potential Area of Research for Technical Communication: A Modest Proposal." *Communication Design Quarterly* 1 (4): 50–60.

Palmeri, Jason. 2006. "Disability Studies, Cultural Analysis, and the Critical Practice of Technical Communication Pedagogy." *Technical Communication Quarterly* 15 (1): 49–65.

Palmer-Mehta, Valerie, and Alina Haliliuc. 2011. "The Performance of Silence in Cristian Mungui's *4 Months, 3 Weeks, and 2 Days.*" *Text and Performance Quarterly* 31 (2): 111–29.

Park, C. Whan, and Mark Young. 1986. "Consumer Response to Television Commercials: The Impact of Involvement and Background Music on Brand Attitude Formation." *Journal of Marketing Research* 23:11–24.

Peltz Strauss, Karen. 2006. *A New Civil Right: Telecommunications Equality for Deaf and Hard of Hearing Americans.* Washington, DC: Gallaudet University Press.

Perez, Thomas E. 2010. "Assistant Attorney General Thomas E. Perez at the National Disability Rights Network Annual Conference." The United States Department of Justice. Posted June 10. Accessed September 28, 2012. http://www.justice.gov/crt/opa/pr/speeches/2010/crt-speech-100610.html.

Perlman, Merill. 2011. "Gonna Wanna: When Dialects Collide." Posted July 5. Accessed January 4, 2014. http://www.cjr.org/language_corner/gonna_wanna.php.

Pexman, Penny M., Melanie Glenwright, Andrea Krol, and Tammy James. 2005. "An Acquired Taste: Children's Perceptions of Humor and Teasing in Verbal Irony." *Discourse Processes* 40 (3): 259–88.

Pickett, G. D. 1986. "Reading Speed and Literature Teaching." In *Literature and Language Teaching*, edited by Christopher J. Brumfit and Ronald A. Carter, 262–82. Oxford: Oxford University Press.

Pond, Steve. 2010. "Hollywood Outsources DVD Captions to . . . India?" Posted April 27. Accessed March 22, 2013. http://www.thewrap.com/movies/ind -column/dvd-captions-made-india-16727.

Porter, James E. 2009. "Recovering Delivery for Digital Rhetoric." *Computers and Composition* 26:207–24.

Pražák, Aleš, Zdeněk Loose, Jan Trmal, Josef V. Psutka, and Josef Psutka. 2012. "Captioning of Live TV Programs through Speech Recognition and Re-speaking." *TSD 2012, LNCS 7499*, edited by P. Sojka et al., 513–19. Berlin: Springer-Verlag.

Price, Margaret. 2011. *Mad at School: Rhetorics of Mental Disability and Academic Life*. Ann Arbor: University of Michigan Press.

Rashid, Raisa, Quoc Vy, Richard Hunt, and Deborah I. Fels. 2008. "Dancing with Words: Using Animated Text for Captioning." *International Journal of Human-Computer Interaction* 24 (5): 505–19.

Ratcliffe, Krista. 2006. *Rhetorical Listening: Identification, Gender, Whiteness*. Carbondale: Southern Illinois University Press.

Reed, Jo. 2011. Interview with John Williams. Podcast. Published March 3. Accessed March 22, 2013. http://www.nea.gov/av/avCMS/Williams-podcast -transcript.html.

Reid, Alex. 2008. "Disposable Video Comments." *Digital Digs* (blog). Posted May 15. January 29, 2014. http://alexreid.typepad.com/digital_digs/2008/05 /disposable-vide.html.

Reid, Blake E. 2014. "Third Party Captioning and Copyright." The Global Initiative for Inclusive Information and Communication Technologies (G3ict). Policy white paper. Published March 17. Accessed August 16, 2014. http:// papers.ssrn.com/sol3/papers.cfm?abstract_id=2410661.

Rhett and Link. 2011. "CAPTION FAIL: Jamaican Vacation Hoax." YouTube video. Accessed January 30, 2014. http://www.youtube.com/watch?v=23H8IdaS3tk.

Rickert, Thomas. 2013. *Ambient Rhetoric: The Attunements of Rhetorical Being*. Pittsburgh: University of Pittsburgh Press.

Robo Panda. 2010. "Fifteen Funny Advertisement Juxtapositions." UPROXX. Posted August 16. Accessed July 26, 2013. http://www.cmo.com/content/cmo-com /home/articles/2010/8/19/fifteen-funny-advertisement-juxtapositions .frame.html.

Robson, Gary D. 2004. *The Closed Captioning Handbook*. Amsterdam: Focal Press.

Romero-Fresco, Pablo. 2009. "More Haste Than Speed: Edited Versus Verbatim Respoken Subtitles." *Vigo International Journal of Applied Linguistics* 6:109–33.

Rosen, Stuart, and Paul Iverson. 2007. "Constructing Adequate Non-Speech Analogues: What *is* Special about Speech Anyway?" *Developmental Science* 10 (2): 165–68.

Rotten Tomatoes. 2012. "Whiteout (2009)." Accessed December 23. http://www
.rottentomatoes.com/m/10008655-whiteout/.

Salvo, Michael. 2001. "Ethics of Engagement: User-Centered Design and Rhetorical Methodology." *Technical Communication Quarterly* 10 (3): 273–90.

Schacter Lintz, Janice. 2013. "Help Needed on Closed Captioning." *Multichannel
News* 34 (29): 43.

Schafer, R. Murray. 1977. *The Soundscape: Our Sonic Environment and the Tuning of
the World*. Rochester: Destiny Books.

Schilperoord, Joost, Vanja de Groot, and Nic van Son. 2005. "Nonverbatim Captioning in Dutch Television Programs: A Text Linguistic Approach." *Journal
of Deaf Studies and Deaf Education* 10 (4): 402–16.

Scholes, Robert. 1988. "Review: Three Views of Education: Nostalgia, History,
and Voodoo." *College English* 50 (3): 323–32.

Scott, Linda M. 1990. "Understanding Jingles and Needledrop: A Rhetorical
Approach to Music in Advertising." *Journal of Consumer Research* 17 (2):
223–36.

Section 508. 1998. "Amendment to the Rehabilitation Act of 1973." Accessed September 28, 2012. http://section508.gov/index.cfm?fuseAction=1998Amend.

secularflesh. 2013. "GF was sleeping so I watched a documentary with subtitles
turned on. She woke up to adolescent giggling." Posted November 2. Accessed January 30, 2014. http://www.reddit.com/r/funny/comments/1prhix
/gf_was_sleeping_so_i_watched_a_documentary_with/.

Seidel, Matt. 2014. "Tale of the Transcription Tape." Published February 6. Accessed February 12, 2014. http://www.themorningnews.org/article/tale-of
-the-transcription-tape.

Shipka, Jody. 2006. "Sound Engineering: Toward a Theory of Multimodal Soundness." *Computers and Composition* 23:355–73.

Silverstein, Michael. 1979. "Language Structure and Linguistic Ideology." In *The
Elements: A Parasession on Linguistic Units and Levels*, edited by Paul R. Clyne,
William F. Hanks, and Carol L. Hofbauer, 193–247. Chicago: Chicago Lin-
guistic Society.

Simkins, Michael. 2013. "Actors Should Pay More Than Lip Service to Clear
Speaking." *Telegraph*. Published July 16. Accessed August 31, 2013. http://
www.telegraph.co.uk/culture/tvandradio/10182999/Actors-should-pay
-more-than-lip-service-to-clear-speaking.html.

Sizemore, Everett. 2010. "Google Video Test Results: Captions vs. Description
vs. Speech Recognition." Accessed September 28, 2012. http://www.esize
more.com/google-video-test-results-captions-vs-description-vs-speech
-recognition/.

Slattery, Laura. 2013. "Facebook Hate Speech Should Have Advertisers Running."
Irish Times. Published May 23. Accessed July 30, 2013. http://www.irish
times.com/business/sectors/media-and-marketing/facebook-hate-speech
-should-have-advertisers-running-1.1404164.

Smith, Bruce R. 2004. "Tuning into London c. 1600." In *The Auditory Culture Reader*, edited by Michael Bull and Les Back, 127–35. New York: Berg.

Smith, Mark S. 2011. "Obama Tells Blacks to 'Stop Complainin' and Fight." Associated Press. Published September 25. Assessed January 4, 2014. http://news.yahoo.com/obama-tells-blacks-stop-complainin-fight-015928905.html.

Soundtrack Info Project, The. 2013. "Team America World Police (2004)." Accessed March 26. http://www.soundtrackinfo.com/OST/teamamerica/.

SpecialK1417. 2012. "Walking Dead KFC Ad Placement." YouTube video. Accessed July 26, 2013. http://www.youtube.com/watch?v=gRXNce4unN8.

Stableford, Dylan. 2011. "Was the Associated Press Transcription of Obama's CBC Speech 'Racist'?" *The Cutline* (blog). Posted September 26. Accessed March 26, 2015. http://news.yahoo.com/blogs/cutline/associated-press-transcription-obama-cbc-speech-racist-173438340.html.

Stelter, Brian. 2010. "On Web Video, Captions are Coming Slowly." *New York Times*. Published June 21. Accessed September 28, 2012. http://www.nytimes.com/2010/06/21/business/media/21captions.html.

Summers, Nick. 2014. "Vimeo's New Player Is Twice As Fast, Supports Captions, Subtitles, On Demand Payments And More." Posted January 7. Accessed January 29, 2014. http://thenextweb.com/insider/2014/01/07/vimeos-new-player-twice-fast-supports-demand-payments-closed-captioning/.

Sydik, Jeremy. 2007. *Design Accessible Websites: 36 Keys to Creating Content for All Audiences and Platforms*. Raleigh: Pragmatic Bookshelf.

Szarkowska, Agnieska, Izabela Krejtz, Zuzanna Klyszejko, and Anna Wieczorek. 2011. "Verbatim, Standard, or Edited? Reading Patterns of Different Captioning Styles Among Deaf, Hard of Hearing, and Hearing Viewers." *American Annals of the Deaf* 156 (4): 363–78.

Taivalkoski-Shilov, Kristiina. 2008. "Subtitling *8 Mile* in Three Languages: Translation Problems and Translator license." *Target* 20 (2): 249–74.

Tarantino, Quentin. 2014. *Django Unchained* script. Undated draft. The Internet Movie Script Database (IMSDb). Accessed January 5. http://www.imsdb.com/scripts/Django-Unchained.html.

Technology Access Program. 1994. "NSI Executive Summary." Accessed April 13, 2012. http://tap.gallaudet.edu/Captions/nsi_sum.asp.

TED. 2014. "TED Open Translation Project." Accessed January 28. http://www.ted.com/translate/about.

Telegraph. 2009. "Hitler Downfall Parodies: 25 Worth Watching." Posted October 6. Accessed January 31, 2014. http://www.telegraph.co.uk/technology/news/6262709/Hitler-Downfall-parodies-25-worth-watching.html.

Texas Tech University. 2013. "Americans with Disabilities Act (ADA) and Section 504 of the Rehabilitation Act (Section 504)." Operating Policies and Procedures 10.08. Published March 26. Accessed February 23, 2014. http://www.depts.ttu.edu/opmanual/OP10.08.pdf.

Texas Tech University Student Disability Services. 2014. "Captions & Subtitles for the Classroom." Accessed February 23. http://www.depts.ttu.edu/students/sds/Captions.asp.

theamishaugur. 2008. "English Downfall." *Kairos: A Journal of Rhetoric, Technology, and Pedagogy* 13 (2). Accessed September 28, 2012. http://www.technorhetoric.net/13.2/disputatio/theamishaugur/index.html.

University of Chicago Press Staff, ed. 2010. *The Chicago Manual of Style.* 16th edition. Chicago: University of Chicago Press.

Theofanos, Mary F., and Janice Redish. 2003. "Guidelines for Accessible and Usable Web Sites: Observing Users Who Work with Screen Readers." *Interactions* 10 (6): 36–51.

———. 2005. "Helping Low-vision and Other Users with Web Sites That Meet Their Needs: Is One Site for All Feasible?" *Technical Communication* 52 (1): 9–20.

Tip Top Music. 2008. "John Williams—Cantina Band." Posted July 4. Accessed March 26, 2013. http://tiptopmusic.com/pinkblog/john-williams-cantina-band/.

Titchkosky, Tanya. 2011. *The Question of Access: Disability, Space, Meaning.* Toronto: University of Toronto Press.

Tittler, Jonathan. 1985. "Approximately Irony." *Modern Language Studies* 15.2: 32–46.

Toyouke. 2013. "Project Runway 8/1/13—'An Unconventional Coney Island' Summary." *My Monkey Could Do That* (blog). Posted August 1. Accessed January 31, 2014. http://mymonkeycoulddothat.blogspot.com/2013/08/project-runway-8113-unconventional.html.

Travers, Peter. 2012. Review of *Cabin in the Woods. Rolling Stone.* Published April 12. Accessed November 28, 2013. http://www.rollingstone.com/movies/reviews/the-cabin-in-the-woods-20120412.

Tulley, Christine. 2011. "IText Reconfigured: The Rise of the Podcast." *Journal of Business and Technical Communication* 25 (3): 256–75.

Turner, David M. 2010. "Why VTA Suggests Captioning Commercials." *VTA* (blog). Posted March 18. Accessed July 26, 2013. http://vtainc.wordpress.com/2010/03/18/why-vta-suggests-captioning-commercials/.

TV Tropes. 2013. "Film: Close Encounters of the Third Kind." Accessed March 22. http://tvtropes.org/pmwiki/pmwiki.php/Headscratchers/CloseEncountersOfTheThirdKind.

Udo, J.P., and D. I. Fels. 2010. "The Rogue Poster-Children of Universal Design: Closed Captioning and Audio Description." *Journal of Engineering Design* 21 (2–3): 207–21.

US Census Bureau. 2008. "National Population Projections." Accessed September 28, 2012. http://www.census.gov/population/www/projections/summarytables.html.

Van Der Geest, Thea. 2006. "Conducting Usability Studies with Users Who Are Elderly or Have Disabilities." *Technical Communication* 53 (1): 23–31.

Van Goor, Roel, and Frieda Heyting. 2006. "The Fruits of Irony: Gaining Insight Into How We Make Meaning of the World." *Studies in Philosophy and Education* 25:479–96.

Van Horen, F. M., C. Jansen, A. Maes, and L. G. M. Noordman. 2001. "Manuals for the Elderly: Which Information Cannot Be Missed?" *Journal of Technical Writing & Communication* 31 (4): 415–31.

Vimeo. 2012. "Closed Captioning." Forums/Feature Requests. Accessed December 17. http://vimeo.com/forums/topic:4072.

VITAC. 2013. "Resources." Accessed September 1. http://www.vitac.com/resources/index.asp.

Vouloumanos, Athena, and Janet F. Werker. 2004. "Tuned to the Signal: The Privileged Status of Speech for Young Infants." *Developmental Science* 7 (3): 270–76.

Vy, Quoc V. 2012. "Enhanced Captioning: Speaker Identification Using Graphical And Text-Based Identifiers." Master's thesis, Ryerson University. Paper 1702. http://digitalcommons.ryerson.ca/dissertations/1702/.

Vy, Quoc V., and Deborah I. Fels. 2009. "Using Avatars for Improving Speaker Identification in Captioning." In *INTERACT 2009, Part II, LNCS 5727*, edited by T. Gross et al., 916–19. Berlin: International Federation for Information Processing.

Wald, Mike and Keith Bain. 2007. "Enhancing the Usability of Real-Time Speech Recognition Captioning through Personalised Displays and Real-Time Multiple Speaker Editing and Annotation." *Universal Access in Human-Computer Interaction: Applications and Services Lecture Notes in Computer Science* 4556: 446–52.

Walters, Shannon. 2010. "Toward an Accessible Pedagogy: Dis/ability, Multimodality, and Universal Design in the Technical Communication Classroom." *Technical Communication Quarterly* 19 (4): 427–54.

Ward, Phillip, Ye Wang, Peter Paul, and Mardi Loeterman. 2007. "Near-Verbatim Captioning Versus Edited Captioning for Students who are Deaf or Hard of Hearing: A Preliminary Investigation of Effects on Comprehension." *American Annals of the Deaf* 152 (1): 20–28.

Warner Bros. 2012. "Synopsis." Official website for *Cloud Atlas*. Accessed November 27, 2013. http://cloudatlas.warnerbros.com/about.php.

Wei, Eugene. 2009. "Search Captions on Hulu." *Hulu Blog*. Posted December 21. Accessed January 29, 2014. http://blog.hulu.com/2009/12/21/search-captions-on-hulu/.

WhatIs.com. 2005. "Closed Captions." Last modified September 2005. http://whatis.techtarget.com/definition/closed-captions.

White, Cindy. 2007. "Director David Slade shines a light on Josh Hartnett and Melissa George to bring a new vision of vampires to the screen in *30 Days of Night*." *SciFi Weekly*. Posted October 15. Accessed November 7, 2013. http://web.archive.org/web/20080420150746/http://www.scifi.com/sfw/interviews/sfw17143.html.

Wikipedia. 2013a. "File: Paul poster.jpg" Accessed March 22. http://en.wikipedia
.org/wiki/File:Paul_poster.jpg.

————. 2013b. "List of British words not widely used in the United States." Accessed March 22. http://en.wikipedia.org/wiki/List_of_British_words_not
_widely_used_in_the_United_States.

————. 2013c. "List of words having different meanings in American and British English: A–L." Accessed March 22. http://en.wikipedia.org/wiki/List_of
_words_having_different_meanings_in_British_and_American_English:
_A%E2%80%93L.

————. 2014. "Closed Captioning." Last modified February 7. http://en.wikipedia
.org/wiki/Closed_captioning.

————. 2015. "*Soap* (TV series)." Last modified March 14. http://en.wikipedia
.org/wiki/Soap_%28TV_series%29.

Wilson, James C. 2000. "Making Disability Visible: How Disability Studies Might Transform the Medical and Science Writing Classroom" *Technical Communication Quarterly* 9 (2): 149–61.

————. 2010. "Disability and the Human Genome." In *The Disability Studies Reader*, edited by Lennard J. Davis, 52–62. 3rd ed. New York: Routledge.

Wilson, James C., and Cynthia Lewiecki-Wilson, eds. 2001. *Embodied Rhetorics: Disability in Language and Culture*. Carbondale: Southern Illinois University Press.

Wingfield, Arthur, Sandra L. McCoy, Jonathan E. Peelle, Patricia A. Tun, and L. Clarke Cox. 2006. "Effects of Adult Aging and Hearing Loss on Comprehension of Rapid Speech Varying in Syntactic Complexity." *Journal of the American Academy of Audiology* 17 (7): 487–97.

Winke, Paula, Susan Gass, and Tetyana Sydorenko. 2010. "The Effects of Captioning Videos Used for Foreign Language Listening Activities." *Language Learning & Technology* 14 (1): 65–86.

WiseGeek. 2014. "What is Closed Captioning?" Accessed February 16. http://
www.wisegeek.com/what-is-closed-captioning.htm.

Woolard, Kathryn A., and Bambi Schieffelin. 1994. "Language Ideology." *Annual Review of Anthropology* 23:55–82.

Yergeau, Melanie. 2009. "aut(hored)ism." Computers and Composition Online. Accessed September 27, 2012. http://www.bgsu.edu/departments/english
/cconline/dmac/index.html.

Zaillian, Steven, and Aaron Sorkin. 2013. *Moneyball* script. Undated draft. The Internet Movie Script Database (IMSDb). Accessed November 26. http://
www.imsdb.com/scripts/Moneyball.html.

Zdenek, Sean. 2009. "Accessible Podcasting: College Students on the Margins in the New Media Classroom." Computers and Composition Online. Accessed September 27, 2012. http://accessiblerhetoric.com/article-accessible
-podcasting/.

————. 2011a. "Personal Reflections on the Educational Potential and Future of Closed Captioning on the Web." In *Communication Technology for Students* in *Special Education or Gifted Program*, edited by J. Aitken, J. Pedego Fairley, and J. K. Carlson, 221–29. Hershey, PA: IGI Global. Accessed September 27, 2012. http://tinyurl.com/future-of-cc.

————. 2011b. "Which Sounds Are Significant? Towards a Rhetoric of Closed Captioning," in "Disability and Rhetoric," special edition of *Disability Studies Quarterly* 31 (3). Accessed September 27, 2012. http://www.dsq-sds .org/article/view/1667/1604.

————. 2013. "The main factor that drives captioning quality is what clients are willing to pay for it." *Accessible Rhetoric* (blog). Posted January 19. Accessed March 5. http://seanzdenek.com/2012/05/09/the-main-factor-that-drives -captioning-quality-is-what-clients-are-willing-to-pay-for-it/.

————. 2014. "More Than Mere Transcription: Closed Captioning as an Artful Practice." *User Experience Magazine* 14 (1). http://www.usabilityprofessionals .org/uxmagazine/more-than-mere-transcription/.

Index